Contested Energy Futures

Stuart Rosewarne

Contested Energy Futures

Capturing the Renewable Energy Surge in Australia

palgrave
macmillan

Stuart Rosewarne
Department of Political Economy
University of Sydney
Newtown, NSW, Australia

This study was funded under the Australian Research Grant Discovery Program.

ISBN 978-981-19-0223-9 ISBN 978-981-19-0224-6 (eBook)
https://doi.org/10.1007/978-981-19-0224-6

Cover illustration: Marina Lohrbach_shutterstock.com

This Palgrave Macmillan imprint is published by the registered company Springer Nature Singapore Pte Ltd.
The registered company address is: 152 Beach Road, #21-01/04 Gateway East, Singapore 189721, Singapore

To James O'Connor and Barbara Laurence and the Capitalism Nature Socialism *project*

Preface

US President Joe Biden's Climate Summit on Earth, 22 April 2021, highlighted just how embedded the Australian Coalition government is in an increasingly redundant energy and emissions reduction policy. The intransigence had already cost the government, with the Prime Minister being 'disinvited' to speak at the United Nations Climate Action Summit held in December 2020. When given the opportunity to advertise Australia's commitment to reducing emissions, the Prime Minister reiterated the long-held position that Australia is more than on track to meet a target of net-zero emissions by 2050, which had been adopted by most other nations:

> Australia is on the pathway to net zero. Our goal is to get there as soon as we possibly can, through technology that enables and transforms our industries, not taxes that enables and transforms the jobs and livelihoods they support and create, especially in our regions. For Australia, it is not a question of if or even by when for net zero, but importantly how.
>
> That is why we are investing in priority new technology solutions, through our Technology Investment Roadmap initiative ….
>
> We are investing around $20 billion to achieve ambitious goals that will bring the cost of clean hydrogen, green steel, energy storage and carbon capture to commercial parity. We expect to leverage more than $80 billion in investment in the decade ahead. Prime Minister Scott Morrison, 22 April 2021.

Two days later, Morrison announced another half a billion dollars would be allocated to support the regeneration of coastal fora as a means of sequestering CO_2. Mangroves would be regenerated, marshes resuscitated and sea grass beds rejuvenated in the embrace of blue carbon. Set alongside the commitments shared with other national leaders at the Biden summit, a little more than half a billion dedicated to support further research and development of carbon capture and storage and to establish hydrogen development hubs; the most striking feature of the developing climate change policy agenda is that its success hinges on burying carbon and not on reducing the generation of greenhouse gases.

Australian governments have been flirting with burying carbon as the core of energy and emissions reduction policy for at least two decades and it has been an unequivocal failure. There is nothing much to show for the billions thrown at carbon capture and storage (CC&S), and certainly no commercially viable instances of carbon being captured and stored, and most of the funds that were set aside to reward agricultural industry for soil sequestration remain unallocated with Treasury. Establishing the magnitude of carbon sequestered and secured in the soil has proven such an expensive exercise that the cost of sampling can exceed the failure of carbon permits that could be issued.

Peak business groups have been quick to question the substance of the Prime Minister's claims about the progress that Australia has been making on emissions mitigation, The Investor Group on Climate Change, which represents large institutional investors, has pointed out that Australia now has the third highest level of per capita emissions behind Russia and Saudi Arabia. Australia's emissions reduction target has been set at a level that falls behind most other nations' set targets. The Business Council of Australia, the Australia Industry Group and the National Farmers Federation have all called for the federal government to adopt the 2050 Net-Zero Emissions Target, as have many major corporations.

Still, there is some progress in evidence. Almost to spite the Coalition government's intransigence, households are increasingly taking charge of their energy destiny, leading the world in the rate of photovoltaic solar panel installations. Business and especially international energy corporations are following suit, encouraged by energy users negotiating Purchasing Power Agreements with renewable energy generators in order to secure carbon-free power and providing the market certainty for prospective renewable energies that justifies investment. Energy generation in Australia is being transformed and this, in turn, is effecting a radical transformation in the organisation of the energy industry. From one in which the energy market liberalisation project was supposed to engender a market milieu that would result in power being freely traded and at competitive prices, there have emerged more social forms of contractual relations, of which households as electricity generators and PPAs are but two. These aim to eliminate carbon emissions as the *raison d'être* of the energy system.

Newtown, NSW, Australia

Stuart Rosewarne

References

Foley, M. (2021) "Blue Carbon' A Drop in the Ocean' *The Sydney Morning Herald* 26 April, 11.

Lee, M. (2021) 'From Clean Growth to Climate Justice' in William K. Carroll (ed) *Regime of Obstruction: How Corporate Power Blocks Energy Democracy* Edmonton: Athabasca University Press.

Shaw, K. (2021) 'Flashpoints of Possibility: What Resistance Reveals About Pathway Toward Energy Transition' in William K. Carroll (ed) *Regime of Obstruction: How Corporate Power Blocks Energy Democracy* Edmonton: AU Press, Athabasca University.

Belliveau, E., J.K. Rowe, and J. Dempsey (2021) 'Fossil Fuel Divestment, Non-reformist Reforms, and Anti-capitalist Strategy' in William K. Carroll (ed) *Regime of Obstruction: How Corporate Power Blocks Energy Democracy* Edmonton: AU Press, Athabasca University.

Contents

Abbreviations

ABS	Australian Bureau of Statistics
ACCC	Australian Competition and Consumer Commission
ACOSS	Australian Council of Social Services
AEMC	Australian Energy Market Commission
AEMO	Australian Energy Market Operator
AER	Australian Energy Regulator
ALGA	Australian Local Government Association
APRA	Australian Prudential Regulation Authority
ARENA	Australian Renewable Energy Agency
ASIC	Australian Securities and Investments Commission
BCA	Business Council of Australia
CEFC	Clean Energy Finance Corporation
CC&S	Carbon Capture and Storage
CEFP	Clean Energy Future Package
COAG	Council of Australian Governments
CPRS	Carbon Pollution Reduction Scheme
ESB	Energy Security Board
ESG	Environmental, Societal and Governance
ETS	Emissions Trading Scheme
GERSAs	Green Electricity Retail Sale Agreements
IEA	International Energy Agency
IRENA	International Renewable Energy Agency
ISP	Integrated System Plan
MCA	Minerals Council of Australia
MREP	Melbourne Renewable Energy Project
MRET	Mandatory Renewable Energy Target
NEM	National Electricity Market

NEO	National Electricity Objective
NRECA	The National Rural Electric Cooperative Association
OECD	Organisation for Economic Co-operation and Development
PPAs	Power Purchase Agreements
RECs	Resource Energy Credits
SEMREP	The South East Melbourne Project
SMUD	Sacramento Municipal Utility District
TRY	Totally Renewable Yackandandah
UNGI	Underwriting New Generation Investments

List of Figures

CHAPTER 1

Introduction

There has been a quiet energy revolution unfolding in Australia driven by a surge of investment in renewable energy. This is all the more remarkable because the surge has unfolded in spite of the determined resistance of the Coalition government and the resource sector to adopting any measures that might lessen the reliance on coal as the paramount fuel source for generating electricity or coal as the second most important commodity export. Yet, in the contest over the future of energy, renewable energy is beginning to prevail at an ever-accelerating pace. This reflects the force of a climate politics that is challenging the hegemony of fossil fuels that have set the focus and pace of capital accumulation in Australia. It is what Timothy Mitchell (2011) has captured in his analysis of the transition from the reliance of coal, the spark that ignited the industrial revolution, to the more fluid age of oil, and the politics that began to redefine the political economy of class relations, national affairs, diplomatic and imperial ambitions. This marked the political power of oil and in the transition from coal the pre-eminence of 'carbon democracy'.

The Australian story is propelled by a momentum of quite contrasting proportions, one that is framed by the prospective threat of an existential crisis. This has its origins in the combustion of coal as the principal input in the generation of electricity and the manufacture of steel and the conservative push throughout many parts of the world to free energy markets from those formed on the foundations of state-owned and controlled forces. Liberalisation would, it was contended, spur an injection of capital

S. Rosewarne, *Contested Energy Futures*,
https://doi.org/10.1007/978-981-19-0224-6_1

that would modernise the industry, ferment competition that would drive down the cost of power and reignite the momentum of capital accumulation. Beginning in the 1980s, the liberalisation project was unleashed across the world with state power utilities being disassembled and corporatised and these unbundled entities subsequently privatised. It was frequently the case that this process of corporatisation and privatisation was linked to the lifting of state or nation borders to expand the reach of the competitive market, and in Australia this heralded the establishment of the National Electricity Market (NEM) that serviced the States of Queensland, New South Wales, Victoria, Tasmania and South Australia and the Australian Capital Territory. By creating an environment that intensified the competitive impulse, the cost of electricity as an essential input in production would be driven down and this would enhance the international competitive advantage of industry more generally.

In Australia, as was the case in many other liberalised energy markets, the hubris with which this rhetoric was sold did not survive for long. Power prices increased steadily over the first two decades of the National Electricity Market. Fossil fuel capital and elements within the state coalesced around an organisational and lobbying force, the *coal-industrial complex*, that successfully secured the enduring backing of successive Coalition governments' backing of coal as the pivot of the economy and the capital accumulation dynamic.

There has been a second string in the bow of the liberal energy stalwarts that sat most uncomfortably at the centre of the energy and emissions policy debate. The conservatives advocating the liberalisation of the energy market were quite clear-sighted in setting out the terms that would define the operation of the market. To distinguish the liberalised energy market from the state monopoly counterpart, clear rules had to be formulated to govern the market and the governance architecture established to ensure its smooth functioning. The *National Electricity Law 1996*, which was designed at the same time that debate on management of global greenhouse gas emissions was being aired in international forums in the lead up to the negotiation of the Kyoto Protocol, made no reference to emissions. Completely disregarding the critical contribution of conventional environmental economic theory, that a mechanism be adopted to provide a proxy estimate of the cost of environmental externalities, the economists and legal eagles engaged in the National Electricity Market (NEM) governance architecture interpreted the National Electricity Law's silence on this by the *National Electricity Law* as not requiring any reckoning of the greenhouse gas emissions produced by the coal-fired power plants. Because the liberalised energy market did not explicitly countenance such a charge, there was no justification for introducing a carbon tax or a price that would arise in an emissions trading scheme. This principle became the cornerstone of the conservative resolve in opposing any attempt to introduce an emission trading scheme or a carbon tax. This refusal to countenance some restrictions was the source of some of the most bitter and spiteful exchanges in Australian political history and set the tone of *climate politics* and what would emerge in the struggle to forge a *climate democracy*.

The hollowness of the competition rhetoric became increasingly apparent with the escalation in power prices. The conventional wisdom was that the issue of greenhouse gas emissions was a small cost when the combustion of coal as a cheap source of power underwrote Australia's international competitive advantage, and that it delivered substantial additional benefits with Australia assuming the mantle of the world's largest exporter of coal. However, the link between emissions and climate change according to the considered wisdom of the ever-growing number of physicists, meteorologists and other scientists was unassailable. This new reality was translating into popular discourse. The realisation that the idealist *laissez faire* model on which the NEM is based affords no meaningful mechanism, nor reflects the will, to contain the price inflation of coal-fired power that had become the subject of calls for change. The narrative of the benefits of competition that could be derived from liberalising the power industry and establishing the National Electricity Market was being questioned. This was not simply a war of words; it was translating into a radical shift in how we do energy. The second decade of the second Millennium became witness to the NEM becoming a new reality; there was a radical transformation in energy generation and distribution and in patterns of use. There was a remarkable revival of investment, overwhelmingly in renewable energy technologies, and this unfolded across the political economy landscape and became the defining feature of the NEM.

Capturing the renewable energy surge with a view to arresting its consolidation or with the aim of enabling renewable energy to flourish introduced a new dynamic into the energy industry, but the transition faced extraordinary resistance. We commence the analysis of the ascendancy of renewable energy as an instrument for enhancing greater access to electricity by first considering the institutional and organisational forces that have set the terms of energy governance and sought to defend coal, and fossil fuels more generally, as the pivot around which energy practices are constituted and developed. The role of the federal state, under successive Coalition conservative governments, is interrogated and the walls of resistance to an energy transition consciously constructed. The formation of this fossil fuel story and privileging of coal in the defence of the Australian energy regime are explored in Chap. 2. The Labor governments' endeavours to incorporate climate change management policies are examined in Chap. 3, while Chap. 4 focuses on the substance of the National Electricity Market with a view to developing a more critical understanding of how the varied forces seeking to define the energy trajectory reinforce one another or collide.

The questioning of the confidence in the competitive impulse is the subject of the analysis that follows. The profound shortcomings of the NEM construct to deliver on price competition or the liberalised electricity system's capability to reduce the magnitude of greenhouse gases it is generating ignited a range of counter forces that have dramatically transformed the energy landscape. The political campaigns to physically halt the extraction and combustion of coal in the generation of electricity is one aspect of the counter-movement that peaked around 2009 (Rosewarne et al. 2014). More sustained and socially and physically transformational developments began to emerge and consolidate in the wake of the peaking of climate politics. This marked a reorientation in what we can encapsulate in the notion of *energy democracy*.

The transformation in residential Australia's engagement with energy is one illuminating manifestation of the mobilisation of the progressive forces that emerged to challenge the hegemony of the industrial-scale, emissions-intensive energy model. Oligopolistic pricing that established Australian power prices climbing to be among the highest in the OECD and the resistance of fossil fuel generators to invest in renewable energy technologies spawned an extraordinary movement across the residential landscape in a uniquely Australian phenomenon. Households became active agents for change investing in rooftop solar voltaic panels at unprecedented levels. Investment in this sector has established Australia as having the highest per capita installation of solar panels in the world, and the high rate of the installations has endured with over three million systems now installed with the capacity to meet 40 per cent of the demand for electricity. This challenge to fossil capital's dominance is examined in Chaps. 5, 6 and 7. The dominant form of households' traditional engagement with the energy has been through the market, based on transactional relations, whereas the investment in solar panels portends two possible developments: household entrepreneurialism becomes part of the everyday lived experience and a more social form of engaging with the energy system as residents become listed on the energy value supply chain. While the challenge can be regarded as principally an assault on the oligopolistic pricing practices of the energy corporations, observing residential investment in decarbonising makes this more than simply a consumer-led revolution. It can be regarded as much as the fermenting of a politics of climate change, a cultural transformation to enable residential Australia taking control of its energy destiny. This can be interpreted as giving expression to a progressive framing of energy democracy. The politicisation of climate change manifests as a lived experience that directly challenges capital's dominance of power generation and the fetishised representation of electricity as more than a mere commodity. The uniqueness of this development in households' embrace of rooftop solar panel installations demands critical scrutiny as much because it gives voice to energy democracy and because there is also a story to be told about capital's endeavours to reassert its place in the energy generation and distribution chain to claim the benefits of the investments.

A second climate politics movement that emerged has been linked to the engagement of municipal councils in taking the lead on decarbonising projects. The genesis of this lies with what can be described as the purchase of climate politics invoked through the exercise of citizenship. The municipality has proved to be an important site in which climate agency is projected to influence the form in which energy is generated, distributed and utilised. Chapters 8 and 9 explore this third tier of the state as an institutional site that is responsive to the will of the local community, to the organised voicing of energy democracy, and which draws upon the organisational capacities and resources of local government to mark another moment in the democratisation of energy. We first focus on some European and American examples of municipal councils as providing models and initiatives that have been drawn on in Australia, especially through the forging of international networks. These include municipal councils adopting measures to mitigate emissions and support clean energy through simple projects, such as organising schemes for households replacing incandescent light globes with LED lighting, organising community-wide solar panel installations or subsidising installations for low-income households. Re-municipalisation campaigns have also endeavoured to effect more substantive developments, such as council resuming the role of electricity generation displacing fossil fuel-focused energy corporations to replace fossil fuel-fired power plants with renewable energy technologies that establish the bases for municipality energy self-reliance. The traction in driving decarbonisation has also been reinforced through forging links with the local, national and global municipal communities in a classic display of the environmentalists' catchcry 'act locally think globally'. This takes the socialisation of the climate project to a more organised level compared with the household rooftop solar panel installations and which builds on the formation of quite comprehensive and far-reaching partnerships.

A third climate movement that has been established and which presents a potentially more radical challenge to the future of the NEM has been the formation of energy cooperatives. The cooperatives are formed with the objective of establishing communities that are energy self-reliant, generating their power requirements by means of renewable energy technologies. In becoming energy self-reliant they aim to defect from the grid. Defections can erode the integrity and cohesion of the larger energy system, although most energy cooperatives are concentrated in regional and rural locations and tend to be on the periphery of the conventional energy network. Energy democracy and climate justice figure prominently in the cooperatives' *raison d'être*. There are comparatively few cooperatives in Australia, but there is evidence of mounting interest, many drawing on the experience in Europe and North America where community-based energy cooperatives have been an important feature of the energy landscape. Chapter 10 considers the role of, and challenges faced by, European and North American cooperatives that have been provided models for how local energy cooperatives can mount the resources necessary for building community interest in the establishment of cooperatives in Australia in Chap. 11. There has been some discussion in the wake of the summer wildfires of late 2019 and early 2020 that the localisation of energy generation in establishing more secure and reliable energy supply networks provides a logic for forming local energy cooperatives and that experience provides evidence of these becoming a foundation for developing stronger communities. Communities along the eastern seaboard devastated by bushfires have been exploring the potential of rebuilding their communities around the establishment of off-the-grid power systems and forming decentralised, self-reliant and self-sufficient economies which inscribe energy democracy in the decarbonising equation.

The launch of the political project to privatise and liberalise Australia's energy system in the early 1990s involved a range of forces that were as much focused on preserving some vestiges of the past as they were preoccupied with building anew an order that would inject new momentum into the economy and reinvigorate the pace of capital accumulation. Whereas residential Australia's engagement in energy generation has mostly been brought back into the centralised governed system, and municipal councils' investments generally complement this engagement, the cooperative movements tend to weaken the integrity of the system albeit at the periphery of the market. Still, to the extent that residential Australia, municipal councils and electricity cooperatives have been carving out spaces to enhance their voice in the energy system, there is potential for energy system participants to defect from the grid, and this has introduced some disorder and uncertainty. However, while the Australian Energy Market Operator (AEMO) and the Energy Security Board (ESB) were tasked with the job of mapping a pathway for the future development of the energy system, the Minister for Energy and Emissions Reduction was hovering in the background developing his *Roadmap*. While each proposal claimed to be embedded in a free market, the competing scenarios gave cause for some considerable tension over the interventionist role of the state vis-à-vis the importance of the 'invisible hand' of the free market as the means of restoring some order into the market, and how this tension has been unfolding is explored in Chap. 12.

Another critical element driving the transformation and an additional source of instability has been the investments in the renewable energy sector by international capital. Global capital—not wedded to the old industrial-scale, fossil fuels, emissions-intensive power generators—has been a dominant force in investing in renewable energy technologies and is occurring on an ever-larger scale, both expediting the transformation and pulling it in another direction. Not contingent on a historical-cultural heritage that is tied to coal, international energy corporations have become a crucial source of not only the capital that is financing the transformation in energy generation but also the innovatory dimensions of the investment. The criticisms of the South Australian government for supposedly overreaching the reliance on renewable energy technologies, which was held to be the cause of the 2017 blackout, appeared to be without substance following Elon Musk's offer to build what was then a world first in the Australian energy market, the world's largest grid storage back-up battery. One of the intriguing aspects of this story of international capital investing in the National Electricity Market is just how much the momentum of the transformation has been linked to the diverse sources of global interest in the Australian transition experiment. This is in spite of the antipathy of the Coalition government towards renewable energy.

Underwriting international competitive advantage was held up as the arsenal to herald this new age, but this is a story that is more complex than was envisaged. Two key market regulators, the Australian Energy Market Operator in partnership with the Energy Security Board, were tasked with mapping a policy to restore some order by laying the foundations for an integrated system. They acknowledged the transition towards renewable energy and proposed an energy model that would build upon and incorporate the diverse ways of engaging with the energy market. At the same time, the Minister for the Energy and Emissions Reduction set about formulating an alternative model, the *Technology Investment* Roadmap, designed to secure coal, and increasingly gas rhetorically recast as a transition fuel, as being at the core of the liberalised energy market.

The competing positions, while comparatively insignificant, have resulted in some heated exchanges as the Minister tries to sell his *Roadmap* while still not having produced a policy to take before parliament, but they are emblematic of more fundamental struggles redefining the energy system, and these form the focus of Chaps. 12 and 13. Climate activist divestment campaigns to contain coal capital have had considerable traction in transforming corporate culture and, increasingly, business practices. Climate politics took a different turn. These campaigns have contributed to fresh thinking on the potential costs of not decarbonising which have emphasised the financial risks associated with investments in fossil fuel. Shareholder activism has reinforced demands for reporting on risk, and these reporting requirements are being mandated by various state agencies. The analysis draws out the tensions and contradictions in the efforts to resurrect the institutional force of the *laissez faire* liberalised energy market and considers the dramatic transformations in economic and governance environment that have prompted energy corporations and high-end industrial levels to explore their options to decarbonise. This is not a tale that can readily be resolved within the sphere of exchange, on terms framed by the economic iconography of transactional relations. The pressure to decarbonise is fostering more social forms of emissions governance.

In concluding with Chaps. 14 and 15, we re-examine the ascendancy of renewable energy over fossil fuels to reflect on the competing visions of the energy future canvassed by the established climate governance architecture. We explore how this has impelled a climate politics that contrasts with the competing trajectories envisaged by the Coalition government and the determination of residential Australia, and other forms of consumer activism seek to make their mark on the climate agenda. In observing the significance of the various investment strategies, and especially investment by international energy corporations investing to capture a place in the Australian sun, the focus returns to the role of State and Territory governments in asserting their constitutional responsibility for energy to help breaking with the Coalition's cultural conviction in coal and more recently gas. In mapping the success of the transition, the chapter reflects on the tension between the decentralisation of the generating system and the contradictions that have emerged with the successful transitioning and how this is prompting a renewed conviction in the need for an industrial-scale transmission system built around an expanded number of interconnectors closing off the possibility of alternative and more democratic, progressive energy regimes being forged.

References

Mitchell, T. (2011). *Carbon democracy: political power in the age of oil.* London: Verso.

Rosewarne, S., J. Goodman and R. Pearse (2014) *Climate Action Upsurge: The Ethnography of Climate Movement Politics.* London: Routledge.

CHAPTER 2

The Hegemony of Coal in the Australian Political Economy: Digging Deeper

Setting the Debate

Coal has held a historically privileged place in the Australian political economy. It is and has been the main source of generating electricity. Until relatively recently electricity generation and transmission were operated by State government owned and controlled the coal-fired power stations and the transmission and distribution networks, and the combustion of coal at these plants was the single most significant producer of greenhouse gases. The generators, now mostly privatised and owned and controlled by international energy corporations, still account for over two-thirds of the electricity generated in Australia. This makes coal the single most significant source of greenhouse gas emissions, responsible for one-third of Australia's emissions. The consequence is that Australia has the highest per capita level of emissions among OECD nations at a point in time when the international community is desperately trying to secure an agreement on cutting emissions.

The significance of coal in the Australian economy is also underwritten by its value as the second most valuable export earner, after iron ore, although coal's position on the export ladder is, from time to time, being usurped by exports of liquefied natural gas (LNG). This twofold significance has influenced successive Australian governments' opposition to measures proposed in global climate change forums that would restrict the combustion or extraction of coal. The Coalition governments have

S. Rosewarne, *Contested Energy Futures*,
https://doi.org/10.1007/978-981-19-0224-6_2

consistently maintained that nothing should stand in the way of Australia realising the value residing in the nation's coal deposits. The conventional wisdom endorsed the role of the free market to ensure unfettered extraction and combustion and exports should prevail.

Paradoxically, the conviction in the free market and the ideology of the *laissez faire* model has dominated the debate on energy in Australia and secured the continued significance of coal as the principal energy source in spite of it being by far the foremost producer of greenhouse gas emissions. The embrace of this ideological conviction also influenced decisions to privatise the state-owned utilities based on assumptions that competition would drive increased efficiencies in the generation and distribution of electricity. At the same time, there was mounting concern that the increasing atmospheric concentration of greenhouse gases was leading to global warming, and there was a developing consensus among scientists, endorsed by a growing number of governments and international organisations, that action to abate the growth of emissions was necessary to avoid climate change. Coincidentally, among the more popular policy-design proposals being canvassed for climate change looked to market-based mechanisms that could be introduced to remedy the challenge of climate change and which would likely be more politically palatable than unilateral state-determined interventions.

The respective convictions in the promise of the free market would set the parameters that defined debate on the liberalisation and development of the electricity generating system and climate change governance for the next 25-plus years. This was most obviously demonstrated with the establishment of the National Electricity Market following the privatisation of much of the State energy system. It has been evident in the deliberations over climate change policy, in the negotiation of the Kyoto Protocol. Free market principle informed several federal government policy initiatives: voluntary participation to meet emissions reduction commitments and renewably sourced energy targets was regarded as a more effective means of engaging enterprises in reducing their emissions than compulsion. In unpacking the principles that informed the conviction in the market, the chapter analyses the extent to which the advocates of coal sought to secure their place as the bedrock of the Australian political economy. Before doing so, it makes sense to make clear the enduring commitment of the two political parties that are in the position of being elected to office to govern the nation. Australian conservative Coalition governments in particular stand out as the champions of the *laissez faire* free market and as the laggards in not taking seriously the challenge of climate change. To this day, Coalition governments have failed to develop policies designed to address this challenge. In large measure, this can be explained in terms of the Coalition's representation as the party of business and the product of the broad coalition being of coal champions: the resource sector including the mining companies directly involved in extraction and exporting coal; the energy-intensive industries that have been determined to maintain the supply of low-cost energy; State governments whose budgets have been bolstered by the steady flow of royalties and who are overseeing economies built on the relatively cheap source of energy; sections of the workforce employed in mining and the regional communities that benefit from well-paid employment; as well as bureaucrats and politicians from across the ideological spectrum that have been won over by the wealth-generating benefits of coal combustion and coal exports. This broad coalition of interests, of industrial and political forces, has contributed to what can be best described as the coal-industrial complex, and this coalition of fossil fuel supporters has proved to be an enduring and dominant force in setting the political agenda over the past 25 years, successfully thwarting the national and global endeavours to place some checks on the polluting effects of coal.

Labor governments have proved to be more proactive in adopting measures that would mitigate emissions. The first act of the Labor government led by Kevin Rudd upon being elected in 2007 was to ratify the Kyoto Protocol. This government moved to introduce the *Carbon Pollution Reduction* Scheme. Unsuccessful, the successor Gillard Labor government over 2010–2013, with backing from the Greens, introduced the *Clean Energy Package* which put a price on emissions. Much of the legislation was abandoned when the Coalition returned to govern in 2014. Today, the Labor Party in opposition has no clear policy on the future of coal in the Australian political economy. It supports the continued development of the export industry but is prevaricating on initiatives that would reduce reliance on coal (and gas) in order to mitigate emissions.

In the two States with large coal deposits of coal, the New South Wales State Coalition government and the Queensland Labor government, both continue to approve the expansion of coal mining, and this is notwithstanding their efforts to promote investment in renewable energy sources. They oversee economies built on coal, a cheap fuel driving the nation's power stations, and they have had a vested economic interest in expanding extraction because from the moment licences are approved, the capitalisation of coal proves to be a revenue bonanza for them in the form of royalties. More generally, coal has been a crucial ingredient in the buoyancy of the Australian economy in the post-Global Financial Crisis period. Alongside iron ore, coal has more or less defined the reorientation of Australia's economic relations in the Asia-Pacific region. The consequence has been that coal has had a hold on the nation's psyche for most of the post-World War Two era.

The March of Economic Liberalisation: The National Electricity Market

Governments across many parts of the world had by the 1990s succumbed to the promised immediate budgetary benefits of privatising state-owned utilities. One of the first sectors to feel the full weight of this paradigmatic shift in the role of public enterprise and governance was the energy industry. It was envisaged that unbundling the utilities—the coal-fired power plants, the transmission networks, the distribution grid and the retail outlets—and corporatising these enterprises before selling them would provide a one-off boost to government budgets and reduce fiscal pressures. In fact, Australian governments were among the first to buy into the economic liberalisation project. Advocates of energy market liberalisation argued that unleashing market forces would liberate the spirit of competition by attracting new investment in cutting-edge technologies, foster innovation, increase supply reliability and drive down energy prices. This would prove of immeasurable benefit to businesses and households, reducing their energy costs and enlivening the economy more generally. Crucially, this went hand in hand with breaking down State borders that had largely restricted competition to within the spatial bounds of State-based energy markets. The proposal to establish the National Electricity Market (NEM) by integrating six discrete markets—the eastern and southern States of Queensland, New South Wales, Victoria, Tasmania and South Australia and the Australian Capital Territory—was motivated by the objectives of reinforcing the competitive impulse and enhancing the opportunities for scaling up generating capacity, improving supply reliability and reducing electricity prices.

From the early 1990s, concerned that Australian produced electricity prices were among the most expensive in the developed world, State and federal governments became convinced by the liberal notion that State governments should abandon their involvement in energy by unbundling and privatising the different components of the electricity system. Conservative discourse questioned the economic wisdom in maintaining the state's continued responsibility for the provision of electricity and pointed to deleterious consequences of energy managed by state bureaucrats: wholesale electricity prices that were among the highest in the OECD (Industry Commission 1991). Energy market reform became a focus of a special Premiers Conference held in October 1990, and State-owned and operated utilities were regarded as leading candidates for reform. The Conference recommended that serious consideration be given to reforming the organisation of state-owned power utilities and it established the Electricity Working Group to explore possible options. In May 1991, the national economic agency, the Industry Commission, published a substantial report *Energy Generation and Distribution* questioning the efficiency of State utilities and the implications this had for the international competitiveness of Australian industry. A precursor to the 1993 Hilmer *National Competition Policy Review*, this set in train a movement to reform the organisation of public monopolies. As federal and State governments became exercised with National Competition Policy, State electricity utilities became the testing ground for the launch of an economic liberalisation project based on unbundling, corporatising and privatising the utilities. The electricity sector was to become the model for economic reform.

The rationale that informed this structural and organisational transformation drew on the tradition of an economic paradigm that considers the 'free market', the *laissez faire* construct, as the superior and most efficient form for organising the allocation and distribution of finite resources. The conviction in this model translated into signing up State-based energy systems to privatise (although not all have) and to breaking down the borders to provide expanded opportunity for competition in the form of a National Electricity Market. This had ramifications that went beyond the issues of ownership and organisation because, at a time when governments were exploring the adoption of measures that would enable the state to intervene in the energy market and considering an impost to account for emissions, states were also signing onto the liberalisation project and vacating State energy enterprises' responsibilities to deliver of power. Meanwhile, the federal government's economic agency, the Industry Commission, reviewed the sector to recommend that the industry be placed on a quite different organisational footing. The Commission's report, a substantial three-volume report on *Electricity Generation and Distribution*, produced a template for reorganising the sector based on breaking up the state utility monopolies through a process of unbundling the generation, transmission and distribution of power and the retail arms, and establishing these as separate legal enterprises governed by established corporate norms. Interestingly, the Commission did not think that state involvement in the sector should be brought to an end with privatisation.

The conservative government of Victoria led the way in transforming the energy industry, unbundling the State-owned utility, the State Electricity Commission of Victoria, in 1993 and corporatising the separate entities to establish a cohort of enterprises competing with one another for market share and privatising these entities. The project of liberalising the energy market was pursued within the Council of Australian Governments, a forum that the then Labor government's Treasurer, Paul Keating, had established in 1992. The Council of Australian Governments (COAG) provided a platform to bring federal government leaders together with their State government counterparts, as well as a representative from the Local Government Association and a New Zealand government representative.

Treasurer Keating was an enthusiastic supporter of the liberalisation and competition agenda. In 1993 Keating had negotiated the Hilmer review *of National Competition Policy.* The review became the *sine qua non* standard for evaluating the preferred organisational form of economic enterprise, and although not specifically focused on energy, the sector identified as one that would benefit from reform. Seeking radical reform to give expression to the full force of the market, Keating turned to reform of the spatial constraints placed on the market because the scope of markets was coincident with state-owned utilities. Substituting State-based markets with an enlarged market that would encompass Queensland, New South Wales, the Australian Capital Territory, Victoria, Tasmania and South Australia would extend the reach of the market and make for more robust competition in the form of a 'national electricity market'.

The COAG endorsed the initiative to establish a national energy market, firstly, looking towards enabling cross-border competition in the delivery of power, a task that was to be undertaken by the newly formed National Grid Management Council, which would lead the construction of infrastructure with interconnectors planned for New South Wales, South Australia and Victoria, to establish some semblance of a national market. A National Electricity Code was registered to set the terms for managing access to a national market, and this was agreed before a Heads of Government meeting towards the end of 1994. A secretariat was formed to progress the formation of the market, and the *National Electricity (South Australia) Act* 1996 was passed to establish the governance and enforcement framework and key obligations surrounding the national electricity market and the regulation of access to electricity networks. The National Electricity Market Management Company Limited (NEMMCO) was formed in 1996, registered as a limited liability company in South Australia, to oversee trade in the wholesale electricity market. The governments of the Australian Capital Territory, New South Wales, Queensland, South Australia, Tasmania and Victoria were represented on the NEMMCO board, which was tasked with implementing, administering and operating the wholesale NEM. A National Grid Management Committee was formed as a prelude to investing in interstate interconnectors that would make it possible for electricity to be transmitted across State borders and recast the physical organisation of electricity generation and transmission. With the NEMMCO in place and a division of the Australian Competition and Consumer Commission (ACCC) charged with oversight for enforcing the regulations, the National Electricity Market was formally launched at the end of 1998 (Gallagher 2004).

The labour movement fought against the electricity privatisation project, State-by-State. Victoria became the first State to unbundle, corporatise and privatise under the leadership of the Premier, Jeff Kennett. Kennett had led the argument in support of energy market liberalisation and privatisation of State utilities at the Premiers' conferences, and the Liberal government defied community opposition and the anti-privatisation campaign organised by the labour movement riding roughshod over appeals to maintain State ownership. A strident and enduring anti-privatisation campaign involving the public sector unions and community ensured that reform would not be an open and shut case. These forces had tried to block the unbundling of the Electricity Trust of South Australia believing this would be a precursor to privatisation, and one assurance from both major political parties that, in government, they would not move to privatise the Trust. Indeed, in the lead up to the 1997 election, Liberal Premier John Olsen reassured voters that if returned to office, a Liberal government would not modify the status quo, which, with the support of an independent and two Labor parliamentarians who crossed the floor to vote with the government, is exactly what he did not do. In reneging on a pre-election promise, Olsen paved the way for the privatisation of the

State-owned utility n 1999. A determined opposition to privatising Queensland's State utility has held firm, whereas in New South Wales a strong show of the strength of the labour movement resulted in the staggered privatisation of different stages of the energy value supply chain of the State-owned utility.

With campaigns to block the privatisation of State-owned power utilities becoming a spent force, the communities and labour movement were shut out of the process of determining the form of the energy market or the responsibilities of private for-profit operators and the definition of the rules and roles of the governance agencies. This had consequences because one of the more distinguishing features of these agencies has been that their operating brief has been quite narrowly defined. The brief is infused with the ideational principles of conventional economic theory, the epistemological foundation of the *laissez faire* construct. This called for minimal intervention in markets by state agencies, and this philosophical position appears to be reflected in the extent to which the agencies were populated by a cadre that is imbued with the belief in the *laissez faire* market construct (Blume 2018; Parkinson 2018). This is also reflected in many of the agencies' evaluation of practices and policies being outsourced, undertaken by like-minded consultancies.

In December 2003, the Ministerial Council on Energy released a document entitled *Energy Reform* that recommended consolidating oversight and governance of the National Energy Market. the National Electricity Law explored options for consolidating the architecture governing the national market and for elevating the significance of the national market and governance arrangements over State-based electricity and gas markets. In 2005, the federal government increased funding to establish two new federal agencies, the Australian Energy Market Commission (AEMC), which was charged with setting the rules and regulations, and the Australian Energy Regulator (AER), which was tasked with enforcing compliance with the rules. This was a relatively economical addition to the state system, the cost of which was mostly paid for by an impost on energy operators. The AER was established as a department of the Australian Competition and Consumer Commission. The Australian Energy Market Operator replaced the NEMMCO and took over the role of the State agencies in 2009 which resulted in the incorporation of gas into the national energy governance architecture. The AEMC was mandated to set rules for trading electricity and gas, and the National Electricity Market morphed into the Australian Energy Market Commission. In what was a unique organisational arrangement, representatives of the energy operators were invited to join government representatives on the Australian Energy Market Operator (AEMO) board. This gave industry a direct voice in planning although not the ability to dictate decisions because the government representatives held a majority of 60 per cent voting rights. One of the distinguishing features of this reorganisation was that the *laissez faire* construct presumes a minimalist role for the state, one that draws a narrow oeuvre for the state. It did not extend the purview of the state to accommodate a reckoning of emissions as an integral factor shaping the dynamics of the energy market, at least not until very recently.

Charting a Sustainable Future for an Energy-/Emissions-Intensive Economy

Intriguingly, the liberalisation of the electricity market coincided with the growing intensity of debate on ecologically sustainable development, including how best to mitigate greenhouse gas emissions. The catalyst that incited interest by some Australian governments to begin formulating an emissions mitigation strategy was initially international in origin although domestic environmental concerns also played a part. Concerns about the imminent danger of global warming, and what was characterised as the likely effect of the increasing concentration of greenhouse gases in the global atmosphere, had been raised in international forums as one of a broad suite of environmental issues that had been identified in the World Commission on Environment and Development's *Our Common Future*. This was subject to a more measured research focus with the 1988 Toronto conference on the atmosphere, which brought policy makers and scientists together and recommended cutting emissions. It led to the establishment of the Intergovernmental Panel on Climate Change in 1989 and foreshadowed the United Nations Environment Program report on climate change, the 1992 Rio Earth Summit and the setting up of the United Nations Framework Convention on Climate Change to formulate international agreement on mitigating emissions. These developments had a special resonance in Australia because Australian scientists, meteorologists and climatologists were at the forefront of international research on the link between emissions and climate change. Australian governments were quickly swept up in the mounting concern. An Australian government could not simply ignore the evidence, especially when the State governments of Western Australia, New South Wales and Victoria had adopted the Toronto target of reducing emissions to 80 per cent of 1988 levels by 2005.

A more substantial movement was evident nationally, with the Hawke Labor government seeing political advantage in embracing the ecologically sustainable development cause. Labor's election to office in 1983 was very much the result of its decision to support the first truly national environmental movement campaign to block the damming of the Franklin River, in an area in Tasmania that was to be listed under the World Heritage List. Seeking to demonstrate its environmental credentials as a responsible global citizen, the Hawke government established a multi-party forum that included nine industry-focused working groups to lay down the foundations of ecological sustainable development policy (Rosewarne 2003). For the first time, an Australian government announced its intention to develop policies to support sustainable development and, in particular policy measures that would explore how best to address the challenge of climate change. In 1990, it adopted an interim target to stabilise emissions by 2000 and to reduce emissions by 20 per cent by 2005 based on 1988 levels. The main strategic focus for reducing emissions emphasised energy efficiency rather than, for instance, substituting emissions-intensive industries such as the energy from coal-fired power stations with energy produced from renewable sources. However, a note of caution accompanied these initiatives and made clear that policy proposals that could have a net negative impact on the international and national competitiveness of industry would not be countenanced. Business could be assured that decisions to mitigate emissions would be based on economic benefits and the principle of 'no regrets'. These caveats were to define all subsequent attempts to establish a climate change policy.

The energy sector figured prominently in the Ecologically Sustainable Development (ESD) working groups, with two groups appointed to focus specifically on energy production and energy use. An overarching *National Renewable Energy Strategy* that incorporated recommendations from several industry reports promised to deliver policies to limit emissions, increase efficient use of energy and support the use of renewable energy; the *Strategy* did not question the continuing significance of coal in the generation of electricity (Council of Australian Governments 1992a). As much as these initiatives set in motion the foundations for developing a climate policy, a 1995 review concluded that there was little evidence the *Strategy* had any material effect (Hamilton 2007: 50). By then, the conviction in the need for taking action to mitigate Australia's greenhouse gas emissions was being overshadowed by a radical realignment of the ideological and political landscapes.

The ESD Working Group process provided a salutary lesson in how political fiat does not provide a ready pathway for transitioning. As much as the ESD process was a quite radical project, its failure to generate any concrete outcomes was emblematic of just how entrenched the energy-intensive pattern of accumulation was. The *Greenhouse Report* reiterated the Australian government's 'no regrets' caveat and accepted that coal would remain an important 'business-as-usual' source of energy generation (*Greenhouse Report*, 1991: 58–59). The *Report* also cautioned that Australia should not get ahead of other countries in mandating emissions reductions because this would likely be to the detriment of energy-intensive industries. In fact, it recommended that, given the vast reserves of coal, any attempt to meet the Toronto goal would compromise Australia's comparative advantage in energy-intensive industries that had been consolidated with the recent State investments in large-scale coal-fired power stations. It recommended extending the emissions reduction goal of 2005 through to 2030 (*Greenhouse Report*, 1991: 59–60). The *Greenhouse Report* anticipated greater reliance on renewable energy technologies as the means of reducing emissions but concluded that the cost of wind and solar sourced energy precluded the transition in the short-to-medium term because these sources could not compete with coal.

The absence of any developed sustainable energy policy belied the rhetoric of a *National Strategy for Ecologically Sustainable Development* (1992b). The *National Strategy*'s energy production and greenhouse gas mitigation goals advocated 'more efficient energy practices [that] include adoption of appropriate market and regulatory measures' and improving 'the availability, efficiency and affordability of alternative energy sources', but the mantle for progressing these priorities rested almost entirely with 'the public energy agencies' (National Strategy for Ecologically Sustainable Development 1992b: 44). There was no reference to the undertakings made by the Council of Australian Governments to proceed with the privatisation agenda and the dissolution of these public energy agencies.

In conceding the comparative economic advantage of coal as the key source of energy, the Labor government affirmed the importance of coal as the energetic foundation and, more generally, the pivotal place of resources in the momentum of accumulation. Labor had already committed to this nexus, having established in 1989 the Resources Assessment Commission. This was a body that was tasked with responsibility for assessing contentious development applications. The Commission was established to undertake more rigorous cost-benefit analyses of prospective mining projects, and this meant that evaluations became more of a technical exercise reducing the opportunity for community submissions (Galligan and Lynch 1992). This limited the scope for opposition to a particular development from being politicised. Furthermore, because projects were evaluated in terms of their individual worth, the likelihood that any emissions mitigation benchmarks would form a condition of development approval was reduced.[1]

Debating an Emissions Mitigation Policy

The question of an effective strategy for underwriting emissions mitigation loomed large in the deliberations of the ESD Working Groups. The court of opinion, proselytised by conventional economists, was that in light of the need to contain the increasing concentration of greenhouse gases in the atmosphere, the most effective and efficient emissions mitigation measures gravitated around the purported advantages of adopting free market-based instruments in the form of emissions trading systems.[2] This was represented as a relatively straightforward technique. Depending on community or government preferences, governments could issue permits, either by auction or for free by grandfathering the right to emit. The quantum of permits allocated in the first instance would generally be based on existing emissions levels, in line with the 'no regrets' principle, and

[1] In fact, it was not until 2020 that development planning authorities recognised that the evaluation of the costs and benefits of coal mine development should take into account the costs associated with the greenhouse gas emissions generated in the combustion of the coal, 'Scope3' emissions. In 2019, the NSW Independent Planning Commission and the Land and Environment Court imposed restrictions on the United Wambo coal project near Singleton, granting approval on the condition that coal be exported only to those countries that had committed to the Paris Accord. The government had brought forward legislation that will prevent the Commission and the Court from imposing such a condition.

[2] The 1992 Ecologically Sustainable Working Group Chair's *Greenhouse Report* endorsed the conventional approach with respect to pricing but qualified its application in Australia because of the uncertainty regarding other countries' commitment to enacting emissions governance policies as well as the technical challenges. The Australian Greenhouse Office researched and published a series of working papers in 2001–2002 on emissions trading systems. The 1996 Stern Review and the Garnaut Reports were critical in cementing this focus (Garnaut 2008, 2011).

The preferred market-like instruments that have been canvassed include introducing a tax on carbon; rewarding an individual agent for reducing emissions; subsidising alternative renewable or clean energy technologies or setting a cap on emissions and issuing property rights in emissions that can be traded.

then reduced over time to align with a preferred emissions reduction scenario. Framed as a technical process, such mechanisms have been argued to take the politics out of climate politics and to depoliticise climate governance (Swyngedouw 2010).[3]

However, the introduction of an emissions trading scheme (ETS) has been strenuously and successfully opposed, apart from a two-year period following the Gillard Labor government introducing a price on carbon emissions as an element in its *Clean Energy Package*. The election of the conservative Coalition government led by Tony Abbott brought that trial to a sudden end. The public claim in support of the move to abolish the ETS was that, in providing a reckoning of the cost of emissions, establishing a price on carbon would increase energy prices. Pricing carbon would be to the competitive disadvantage of coal-reliant industries, and this informed the enduring opposition to an ETS and this, in turn, blocked any developments in climate governance.

Notwithstanding the justification for the energy market liberalisation project and the emissions mitigation management proposals sharing common epistemological roots in neoclassical economic theory, the objective of deepening the liberalisation of the energy market was posited as the main means for breaking the barriers that seemed to stand in the way of more competitively priced electricity. While arguments about the 'market failure' associated with the failure to account for the cost of emissions could be readily set aside, there were other arguments proffered to explain the failure of the *laissez faire* market to deliver more competitive prices. Chief among these were the anti-competitive, oligopolistic practices that had become a structural feature of the market. The 'big three' energy corporations—AGL, EnergyAustralia and Origin Energy—had strengthened their market presence through vertical integration, a factor that is argued to be responsible for much of the sustained increase in energy prices. The Australian Competition and Consumer Commission took aim at those States that had completely privatised their utilities arguing in the instance of Queensland that the State's continued ownership and control of electricity utilities explained the escalation of power prices in that State, while high prices in New South Wales could be attributed to its retention of its interest in transmission infrastructure and the failure to call for

[3] Jessup and Mercer (2001: 20) observe, the Treasury had set the terms of the 'depoliticisation' by demanding that any policy considerations had to be expressed in the language of economics.

competitive tenders in the sale of State-owned generators. There were obvious contradictions in the selective application of conventional economic thinking to overlook one manifestation of market failure while justifying its critique of another instance of market failure.

Of greater significance in understanding the contradictory approaches to drawing on the conventional economic theory to explain the merits in liberalising the energy market is just how selective this was. Silences became embedded in the *raison d'être* of the energy market governance agencies. The setting aside of a key tenet of economic theory, that account should be taken of the cost of emissions wherever the market does not incorporate some reckoning of these externalities in its cost-benefit summations, is a fundamental flaw that was written into the *National Electricity Law* and governance practices. Market liberalisation and emissions governance have generally been regarded as distinct and discrete concerns, and this failing has been repeatedly endorsed. In the deliberations on energy market liberalisation, the question of emissions as an economic cost was at best an incidental consideration (Jessup and Mercer 2001). In submissions to the Finkel *Independent Review into the Future Security of the National Energy Market*, the Australian Energy Regulator, the Australian Energy Operator and the Australian Energy Market Commission were each quite forthright in arguing 'against the inclusion of any reference to environmental considerations in the NEM [National Electricity Market]' (Finkel et al. 2017: 176). This is despite the fact the electricity generating sector accounts for a substantial proportion of emissions in Australia—about one-third of aggregate land-based emissions—and that any consideration of the structure and organisation of the industry cannot, or should not, be undertaken without reference to emissions governance and how this would be incorporated into the design of electricity policy governance.

One of the ways in which the state sought to reconcile this tension was to encourage voluntary participation in emission reduction partnerships. The *Greenhouse Challenge Program* adopted by the Keating Labor government in 1995 and endorsed by the successor Coalition government led by John Howard held out the opportunity for emitters to voluntarily commit to cutting emissions, and this approach was consistent with Coase's theorem. Voluntarism in fact became the hallmark of Australia's emissions policy.

The refusal to accede to the case for adopting an economic reckoning of the real economic costs associated with emissions can be explained in terms of the energy market liberalisation project being premised on the state having a minimalist role, one that is more instrumental and functional in purpose, an impartial market administrator promoting the integrity and vitality of the 'free market' in energy. A decidedly anti-statist sentiment has framed the policy debate. This presents the state as an enforcer of regulations to ensure that the energy market drives the competitive impulse and meets the aspiration of increased efficiency to deliver power at affordable prices and to do so reliably. The canvassing of policies to incorporate an economic reckoning of the real costs associated with emissions is also based on a minimalist role for the state. It draws on the seminal work of Ronald Coase, 'The Problem of Social Cost' (1960), which advocates addressing market failure—that impositions on the environment emissions have an economic effect that is not the subject of an economic reckoning captured in price—by establishing property rights over the atmosphere. These rights can then be traded and priced, but the role of the state is limited to establishing and sanctioning the property rights.

However, unlike the acceptance of a role for the state as an institutional force that helps drive the liberalisation project, albeit grudgingly and sometimes not at all, there has been strong resistance to the idea of the state taking action to address market failure associated with emissions, and this has been justified by a selective drawing on conventional economic theory. Coase' positing of a limit on impositions on the environment being set in the declaration of private property rights opens up greater choice that is also economical. The role of the state is minimal once permits are issued. There are enhanced opportunities for those constituencies calling for emissions mitigation to buy and sell these permits and acquit them and thereby play a direct role in reducing emissions. Alternatively, it pays those enterprises that produce emissions to buy permits up to that point where the benefits of producing a product or service exceed the cost of buying emission permits. However, as much as energy-intensive industries and conservatives are advocates of the *laissez faire* market and accept the conventional economic theory as the model form for this market institution, they have generally adopted a defensive approach when responding to arguments that the integrity of the model requires stricter adherence to principles of the iconic perfectly competitive market. In defence of the model, they either defend the status quo as being the practical form of the model market, lobby for minimal changes, or in the event that some mitigation targets are imposed appeal for existing rights to be grandfathered or that they receive some compensation in the interests of equity. This points to the adoption of an economic fundamentalism that has defined the Coalition's opposition to meaningful emissions mitigation, the clearest example of this being the Australian Bureau of Agricultural and Resource Economics (ABARE) case that was presented in the negotiations for the Kyoto Protocol. Australia successfully negotiated endorsement of the 'Australia clause', handing Australia the right to increase emissions. Notwithstanding the fact that the right to extract coal is contingent on the imprimatur of the state, the fossil fuel lobby has waged a determined battle to block the introduction of a system that would extend the role of the state in the energy market to, for instance, set up a market that would trade in emissions or generate a pollution tax. The consequence has been a fundamental policy impasse. Policy prevarication and uncertainty have been the defining feature of successive Coalition governments.

In Defence of the Status Quo: The Coal-Industrial Complex and Policy Uncertainty

The dubious record of coal as the world's most significant source of greenhouse gases belies the challenges that have presented for the energy industry since the early 1990s. The energy industry has come under ever-mounting pressure to respond to the concerns that the burning of coal was driving the increase in the concentration of greenhouse gas emissions and having a global warming effect. Governments have been pressured to explore measures to mitigate emissions and underwrite a transition to renewable energy sources. Responses have ranged from those that could effect quite measured reductions through to little more than window dressing exercises. The determination of conservative political forces in opposing any measures that might impact on the scale of the industry underpins the much-derided consequence: continuing policy 'uncertainty'.

Australian Coalition governments in particular stand out as the laggards in not taking seriously the challenge of climate change and in failing to develop policies designed to address this challenge. In large measure, this can be explained as the product of the broad coalition of coal champions: the industry supporters, and especially energy-intensive industries, the resource sector, governments whose budgets have been bolstered by the steady flow of royalties and who are overseeing economies built on the relatively cheap source of energy, the mining companies directly involved in extraction and export of coal, sections of the workforce employed in mining and the regional communities that benefit from well-paid employment, as well as bureaucrats and politicians from across the ideological spectrum that have been won over by the wealth-generating benefits of coal combustion and coal exports. This broad coalition of interests, of industrial and political forces, has contributed to what can be best described as a cohering of common interests in the coal-industrial complex. This coalition of fossil fuel supporters has proved to be an enduring and dominant force in setting the political agenda over the past 30-plus years, successfully thwarting the national and global endeavours to place some checks on the polluting effects of coal.

To this day, the coal-industrial complex has proven quite effective in countering the influence of environmentalists, climate change activists and many others. This is despite the diversity and scale of citizens mobilised in support of concrete actions to reduce reliance on coal. This other story of the endeavours to break the enduring reliance on fossil fuels would not be complete without recognising the organisational and structural transformations that have occurred in the energy industry and energy market over this period. And this has underwritten the political influence of the coal lobby.

The role of key state agencies steeped in the free market tradition has been crucial in securing the interests of the coal-industrial complex, but even this has not been without its contradictions. The agencies that have been voicing concerns about the market failures arising from the exercise of oligopoly power and calling for more robust state intervention to secure the integrity of the liberalised energy market have sowed the seeds of uncertainty and helped to promote disorder. More recently, the lack of unanimity within the state has become more evident in the instances of federal agencies appealing for effective emissions governance to minimise the costs or risks associated with climate change. The divisions over climate governance within the state have contributed to the policy impasse.

This emphasis on voluntary engagement also has to be seen as a response to the increasingly assertive and organised opposition to any government initiatives that would restrict the continued development of the resources and energy-intensive sectors of the economy. The government's *Greenhouse Challenge Program* was publicised as a success in engaging some of the largest corporate emitters in mitigation programmes, but as several critics have observed, the negotiation of government-business *Challenge* partnerships also revealed a shift in government-business relations (Pearse 2007; Hamilton 2007; Wilkinson 2020). Capital had begun to dedicate more resources in advocacy to influence government policy design and policy decisions. Energy-intensive enterprises also committed funds to strengthen the lobbying capacities of industry associations, and among the more prominent were the Minerals Council of Australia, the Australian Coal Association, the Australian Aluminium Council and the Electricity Supply Association of Australia. The Australian Industry Greenhouse Network was formed as a peak umbrella organisation to protect the interests of resource and mining companies, industrial enterprises and the energy industry associations, and included as members of the peak business organisations, the Business Council of Australia and the Australian Industry Group. Adding to this coalition of the forces opposing substantive policy measures to reduce emissions were a number of conservative think tanks, including the Institute for Public Affairs, the H.R. Nicholls Society and some conservative economic consultants.

In the period preceding the Kyoto negotiations, the political environment became quite animated as opposition to the possible imposition of restrictions on emissions and support for drawing more electricity from renewable technologies became more focused and organised. The individual and broad coalition of organisations emerged as a veritable phalanx determined to instil in the consciousness of the nation a sceptical attitude towards climate change science that aimed to block or weaken any policy initiatives that could restrict the development of the resource sector and energy-intensive industries. As Hutton observed in dissecting the links between mining companies and senior bureaucrats in the planning and environment departments and agencies in the Queensland government, the process of approving and overseeing coal mine developments established a commonality of interests and sense of purpose across the business-government divide that was reinforced by government and ministerial pro-development agendas (Hutton 2013). A not dissimilar story can be told about mining and energy in New South Wales, the planned expansion of hydro-electricity schemes in Tasmania (which was compromising the integrity of pristine forests heritage areas) and brown coal-fired power stations in Victoria's La Trobe Valley.

The unfolding international negotiations on climate change injected a new level of intensity and purpose into the industry to defend coal as an essential element in the energy mix. This prompted those opposed to the adoption of any initiatives that would impede the development of resource sector and energy-intensive industries to launch more concerted and coordinated efforts to influence public opinion and capture state policy making. As Pearse and others have observed, the resource boom and the privatisation of the energy market resulted in comparable although more politicised patterns of business-government relations to those documented by Hutton. This could be observed across the spectrum of State and federal governance architecture. The links were both structural and personal (Pearse 2007; Wilkinson 2020). Development plans and approvals could be facilitated or lubricated by drawing on personal connections. Government ministers welcomed approaches by the representatives of individual corporations and industry groups. Energy and resource companies enjoyed open-door policies that provided direct access to ministers, their advisors and heads of departments, and established personal relations with politicians and senior officials within government that wrought a fundamental change in state-business relations.

Characterised as the 'greenhouse mafia', or more recently the 'carbon club', this has been a two-way street. The working relationship became a revolving door as industry invested more resources in lobbying by engaging former Ministerial staffers to lobby and providing employment sinecures for departmental and advisory staff as well as several Ministers when they left office. Governments similarly recruited staff from industry to advise on policy rather than drawing on public servants who could provide advice 'without fear or favour'. These practices were mostly beyond reproach even when these involved conflicts of interest, such as when lobbyists, some former staffers, drafted position papers or even contributed to drafting actual legislation. The multi-layered nature of the links and the material benefits that are being produced have been critical to the forging of the coal-industrial complex and helped to lock in the defence of a future for coal and the opposition to supporting a transition to renewable sources of energy.

Australia's distinctive negotiating position at Kyoto bears witness to the force of the coal-industrial complex. The government was determined to defend the resource sector from any restrictions that might impair its future, and economists came to its rescue. The government was represented by the Minister for the Environment, but the real work undertaken in support of Australia's interests was led by a team of economists from the conservative government agency ABARE, the Australian Bureau of Agricultural and Resource Economics. ABARE was, and continues to be, staffed by economists trained in neoclassical economics who looked upon the 'free market' as a springboard for growing Australia's resource and agricultural commodity exports. Behind the scenes ABARE economists were being financially backed by the resource industry which was paying for the privilege of being represented on the ABARE board and having a say in the modelling, and which was subsequently criticised for exaggerating the cost to Australia of the Kyoto Protocol (Bulkeley and Newell 2015: 116–117). The Australian delegation was supportive of the Protocol being based on the introduction of market-based mechanisms to address the problem of market failure. This included mandating a common emissions reduction target for developed (Annex 1) countries and, were this to be agreed, it would have had a profound effect in forcing changes in the focus of energy market governance. However, the Australian delegation made it clear that Australia could not agree to a common target because of the distinguishing feature of wealth creation in Australia and the heavy reliance on emissions-intensive industries and mineral and agricultural commodity exports. A common target would have a disproportionately negative impact on the economy and amount to an unfair burden. It successfully appealed to the principle of common but differentiated responsibilities and, under the terms of the Protocol, it was resolved that the emissions target for Australia would not require cutting emissions but permit an increase to 108 per cent of the 1990 baseline figure in 2000, the so-called Australia clause.

The traction achieved in the projection of the coal-industrial complex's political might is captured in Fig. 2.1. The Kyoto Protocol had a negligent impact on the significance of coal in the Australian political economy. The amount of coal used in the domestic generation of electricity reliance levelled off from 1997 and there was a gradual growth in the combustion of coal by generators in the new Millennium which continued for the next decade.

The really significant development was the expansion of coal exports from 1997, and this increase was strongly maintained from 1997. As Fig. 2.1 demonstrates, there was a doubling in the quantum of exports in the wake of the Kyoto Protocol.

The consequences of the 'Australia clause' for forestalling meaningful policy action on emissions mitigation are clear in emissions produced in the course of generating electricity as mapped in Fig. 2.2. The magnitude of electricity emissions increased by over 50 per cent over the time frame of the 'Australia clause'.

The industry has been among the most forthright in proselytising and endorsing the liberalisation project and supporting this as one aspect of a broader engagement with the 'free market' and the removal of barriers to

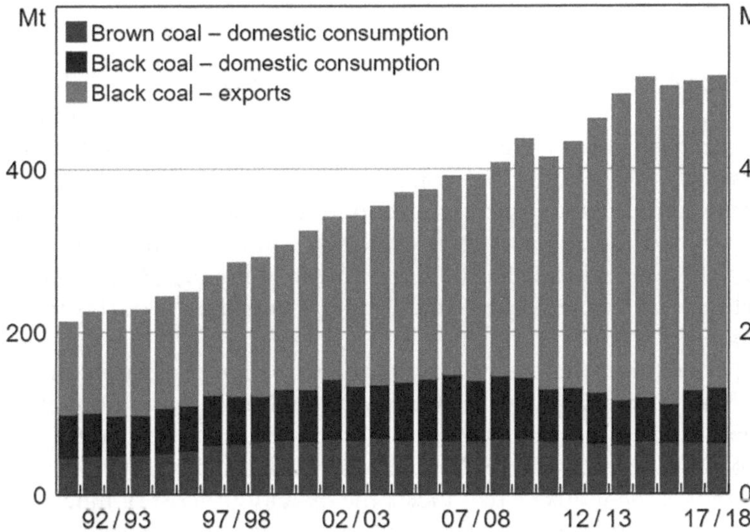

Fig. 2.1 Australian coal production and exports, 1990–2018. (Source: Graph 1—M. Cunningham, L. van Luffelen and M. Chambers, 'The changing global market for coal' *Reserve Bank Bulletin* September)

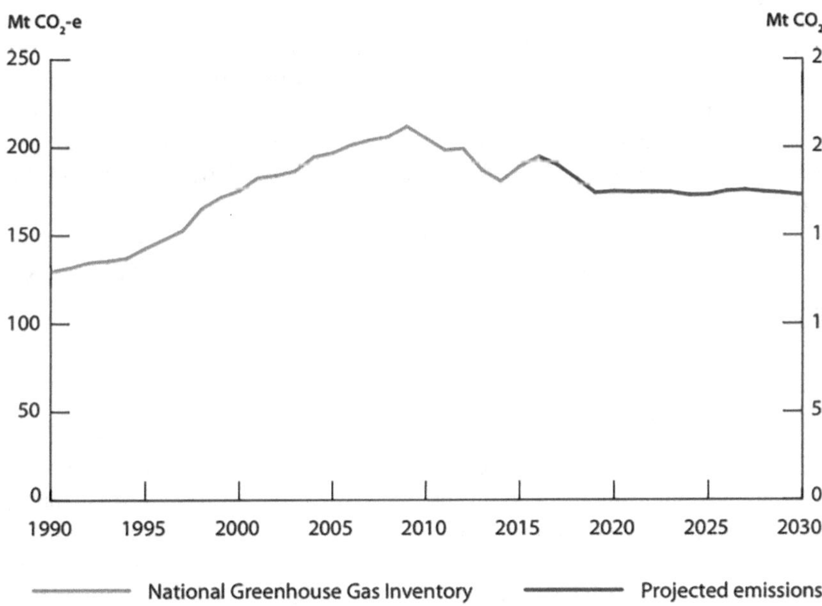

Fig. 2.2 Electricity emissions, Australia, 1990–2030. (Source: Department of Environment and Energy 'Australia's emission projections 2017', December)

international trade liberalisation. However, the recalibrating of the nexus between capital and the state has not been without its contradictions. The argument in support of a minimalist state has not resulted in a diminution in the role of the state in the economy. On the contrary, trade liberalisation has required government concentrating resources on consolidating relations with the leading export destination countries. The incorporation of the trade portfolio within the Department of Foreign Affairs, now the Department of Foreign Affairs and Trade, is emblematic of this transformation in the role of the state.

The contention that energy market liberalisation project would result in the market, supplanting the state and would make for a more efficient economy that would deliver power at a lower cost, has proved illusory. The governance architecture that has been established is quite substantial, although it is the industry that meets the operational cost of much of this governance. If efficiency is measured in accord with conventional economic theory and includes emissions as an environmental externality, then, as is captured in Fig. 2.2, supplanting the state has coincided with the electricity generation sector's emissions increasing by some 60 per cent over the first 20 years that emissions were recorded. As for the promise that competition would become a feature of the sector and that this would drive down energy prices Apart from the first couple of years of the NEM's establishment when Queensland and South Australia invested in interconnectors that enabled access to Victorian and New South Wales power, power prices increased in the first three years of the NEM, and in the five years that followed residential electricity prices increased by 91 per cent (AER 2007, 2012). Electricity price inflation has focused the energies of pricing and competition agencies to monitor the market with a view to identifying the causes of the inflationary trend and how this should be best managed.

In the National Interest and *Safeguarding the Future*: Digging Further in?

It was evident, however, that there was a waning of enthusiasm for confronting the challenge of greenhouse gas emissions, especially in the latter years of Labor government and particularly under Keating's Prime Ministership when the preoccupation with economic reform prevailed. There were few concrete initiatives being adopted by industries according to the National Greenhouse Advisory Panel's Report on the national greenhouse response strategy (1996), and this smoothed the transition from the Labor government in 1996 to a Coalition government wedded to climate change denial. At the same time as Australia was preparing the case it would present at the Kyoto Protocol negotiations the Coalition government, elected to office in 1996, was outlining its national greenhouse management policy. In August 1997, the Coalition government released the Energy White Paper *In the National Interest* which mapped the nation's international relations priorities. Topping the list was how best to position Australia to take advantage of the economic benefits of globalisation and the rise of East Asian economies. Expanding coal exports by promoting industry international competitiveness was foremost among the strategic goals for maintaining and extending energy markets in East Asia. 'Promoting open energy markets' was considered an important objective (Australia 1997: 46). The White Paper did not systematically regard as a priority its involvement in navigating international negotiations on how to confront the threat of climate change. Rather, it noted the potential dangers of international climate agreements. In one brief reference, it anticipated the need for early identification of 'the risks and possibilities inherent in international agreements', such as the United Nations Framework Convention on Climate Change and expressed its concerns with what might be resolved to flag the government's preparedness to 'stand aside' should an agreement 'not adequately protect Australia's national interests' (97). The Coalition signalled its intention to abandon the international order if it did not get its way.

In the National Interest presented a cursory hint of the newly elected Coalition government's opposition to any climate change agreement that impacted negatively on Australian business enterprise. A more emphatic declaration soon followed in the Prime Minister's *Ministerial Statement* to the Australian parliament. The *Statement* was a declaration of defiance against criticism of Australia's refusal to entertain setting an emissions reduction target that aligned with most OECD countries in the Kyoto Protocol. In defending 'the wealth creating efficient industries', and particularly the resources sector, the Prime Minister observed that Australia's continued economic development would result in continued increases in emissions and that this justified the demand for the right to increase emissions under the terms of the Protocol. The alternative would result in Australia being subject to an 'unfair burden'.

In the lead up to the climate negotiations in Kyoto in 1997, the Coalition government announced the $180 million 5-year package, *Safeguarding the Future: Australia's response to climate change*. The Prime Minister also sought to assuage criticism that Australia would be regarded as shirking its responsibilities in not confronting the climate change challenge. In November 1997, at the same time as the Kyoto Protocol was being negotiated, the Prime Minister unveiled what he represented as a major policy shift that challenged accusations of inaction on climate change. In outlining an expanded commitment to addressing climate change, *Safeguarding our future* announced a substantial increase in the allocation of funds to support the take-up of renewable energy technologies (Howard 1997). This, it was claimed, would result in the growth of emissions generated by the renewable energy sector from an estimated 40 per cent over 1990–2010 being reduced by as much as a third. Sixty-five million dollars was allocated to support the development of renewable forms of energy and would help bring the proportion of renewable sources in Australia's energy mix in line with Europe. An additional $21 million would establish a specialist renewable energy innovation investment fund, a $30 million loans and grants programme would bankroll 'the development and commercialisation of the renewable energy industry' and $10 million would be allocated for some leading-edge renewable energy 'showcase' projects. Most of this funding was designed to finance an extension of the voluntary *Greenhouse Challenge Program*.

The *Safeguarding our future* package was argued to provide a balanced approach to energy policy. According to the Prime Minister, the package marked an unparalleled commitment to supporting renewable energy which, in budget terms, far exceeded the $48 million spent by the predecessor Labor government. The package included an important institutional development with the establishment of a unique agency, the Australian Greenhouse Office. The first such agency in the world dedicated to monitoring greenhouse gases and engaging business and government in initiatives to increase efficiencies in energy use and to promoting the development of renewable energy, the Greenhouse Office proved ineffectual. The Prime Minister treated the Office as a distant artefact and did not once seek a briefing from the Office (Baker 2005), and as reported by one of the leading industry associations' environment portfolio officers in an interview conducted in 1999, the Greenhouse Office's director displayed an unnecessary timidity in developing initiatives and a reluctance to test the limits of what could be achieved. Industry associations expected more to come out of the Office and concluded that this was a missed opportunity.

Safeguarding the future was also viewed as a means of uniting the nation in common cause: 'the success of our domestic measures depends upon the goodwill, commitment and action by ordinary Australians, consumers, farmers, families, firms and industry groups, as well as governments federal, state, and local'. However, the end goal was decidedly unambitious. The most obvious reflection of this was the proposal to set a mandatory target for electricity retailers to source an additional 2 per cent of their electricity from renewable energy sources by 2010. This was presented as a concrete initiative ostensibly marking a shift in policy focus, albeit a largely symbolic one and a shift mostly aimed at garnering sentiment from a public perturbed by the government's intransigence in refusing to ratify the Kyoto Protocol. The 2 per cent target was a huge retreat from Australia's Toronto undertaking.[4]

Safeguarding the future hardly spoke to the issue of energy market reform, the other burning issue at the time. Beyond the call for maintaining energy market reform, and a recommendation to secure State agreements on setting efficiency standards for fossil fuel power generation plants, there was no evidence of a strategic vision as to how market reform would support policies designed to deliver on emissions mitigation. For all the promise of what was presented as a major policy shift, the reality was that there was no demonstrable effect on emissions generation. The generation of coal-fired power remained the largest single source of emissions, accounting for a third of the nation's emissions. The industry held its own as the major source of electricity generation through until towards the end of the 2000s. There was a steady increase in electricity generation that was almost wholly attributable to coal-fired power until towards the end of the decade. Data mapping the pattern of sectoral emissions indicates that the increase in the magnitude of emissions associated with electricity generation was greater than any other sector (Department of Industry, Science, Energy and Resources (2020) Data charts: Table 5: Percentage change in emissions of the economy).

[4] The *Renewable Energy (Electricity) Act* was legislated in December 2000, establishing the Mandatory Emissions Reduction Target. The MRET set a target of generating 9500 Gwh from renewable energy technologies by 2020, or some 2 per cent of national generating capacity. While a decidedly unambitious project, it laid a foundation for a more ambitious target that would later be legislated by a Labor government. It was a significant move, and not an entirely gratuitous one, because it was linked to another industry partnership project, the Renewable Energy Action Agenda, to support the ongoing development of Australia's renewable energy industry.

Challenging the Coal Hegemony?

Needless to say, those interests calling for serious action on climate change were quite critical of the position the Coalition government had pursued in the negotiation of the Protocol. The 'Australia clause', which conceded Australia's disproportionate reliance on emissions-intensive, trade-exposed industries, aroused a scepticism as to whether the Coalition was taking the climate change challenge seriously.

Among the first to seek a critical review of the Coalition's policy was the Greens which called for a Senate inquiry into the adequacy of the progress being made under the *National Greenhouse* Strategy. The resultant report, *The heat is on: Australia's Greenhouse Future*, which was published in 2000, reveals both the ad hoc and ineffectual nature of much government policy. Policy was essentially kowtowing to business-as-usual sentiments. The report also highlighted the limited purview of those demanding more serious initiatives to rein in emissions. For the Greens, the critical issue that had to be confronted was the inadequacy of measures that had been adopted to give tangible effect to Australia's Kyoto commitments.

The heat is on: Australia's Greenhouse Future examined the stationary energy sector to consider the scope for transitioning to lower emissions sources and technologies. It questioned the lack of transparency in the government's close relationship with business through the *Greenhouse Challenge Program* and criticised the lack of action to question the benefits of the centrepiece of government policy, the market-based and voluntarist approach of the *Greenhouse Challenge Program.* It also pointed to the plethora of other unambitious and poorly funded measures that had been adopted irrespective of whether they were complementary and called for more systematic action. The report also considered the form an emissions trading scheme could take.

The Coalition government responded to criticisms by reaffirming its opinion that engaging major corporate emitters on a voluntary basis was proving an effective means of progressing abatement. It argued that business was best placed to develop measures to mitigate emissions. This would be underwritten by continuing with the energy market reform programme. The standard principles were reiterated: 'Competition principles must remain the primary driver for continuing energy market reform if the full economic benefits are to be realised' (Government response 2001). The Coalition maintained that it had launched 'a whole of government approach' and that it was on track to meet its Kyoto emissions reduction target although this claim was hotly contested (Government response 2001). Wanting to allay criticism of its inaction, the Coalition response also endorsed the recommendation of *The heat is on* to introduce an emissions trading scheme, though this endorsement was on the condition that this be part of an international trading scheme and that the competitive advantage of trade-exposed and emissions-intensive industries not be compromised.

A Revitalised Energy Industry: Digging into Coal's Future

At the same time as the debate on climate change was providing a backdrop to political discourse on the energy-intensive and emissions-intensive character of the Australian economy, the national energy market was taking concrete form. This had been a relatively long gestation, and it was one distinguished by an architecture of agencies totally committed to honouring their mandate, a brief that was quite narrowly defined. The brief is infused with the ideational principles of conventional economic theory, the epistemological foundation of the laissez faire construct. This called for minimal intervention in markets by state agencies, and this philosophical position appears to be reflected in the extent to which the agencies were populated by a cadre imbued with the belief in the *laissez faire* market construct (Blume 2018; Parkinson 2018). This is also reflected in many of the agencies' evaluation of practices and policies being outsourced, undertaken by like-minded consultancies.

The *laissez faire* construct of the perfectly competitive free market continues to be regarded as providing the most economically efficient form of organisation, one instilled with the competitive impulse and thus the most likely to deliver electricity at a competitive price. However, the form of the NEM was hardly the mirror image of the *laissez faire* construct because there was only one State, Victoria, that had completed the process of withdrawing from direct involvement in energy generation, although South Australia soon followed with its privatisation programme. Other States have joined the NEM despite not having fully privatised their utilities. Queensland joined the NEM in 2001 following the completion of an interconnector linking Queensland and New South Wales, and Tasmania still later in 2006. Moreover, the market dominance of the 'big three' has meant that energy governance has had to manage around these structural constraints as well as meet the policy outcome preoccupations of governments, and the most important outcome that has occupied government has been the issue of price. This has resulted in what appears to have become an irresolvable contradiction. The objective of limiting the role of the state has been paramount but this has had to be reconciled with how to manage the extent to which the market was delivering on lowering prices. The evidence was clear: it was not. At least until very recently, when the surge in renewable energy being generated has severely reduced wholesale prices and raised the prospect of the ever-increasing vulnerability of power plants becoming stranded assets. The initial gains in lower electricity prices have been overshadowed by the price inflation that has occurred with the maturation of the market. The initial gains were partly a demonstration of the corporatised utilities' ability to deliver electricity at a lower price while still returning dividends to State governments. Prices could be cut because power generated in Victoria was being produced, and some exported to other States, through the combustion of brown coal, by far the cheapest energy source, and because there was next to no investment in the refurbishment of privatised power plants.

Power prices trended upwards and industry energy users began to express concerns about the cost implications of this. Responding to the mounting criticism, the Coalition conducted an inquiry into the development of the liberalised energy market. It was chaired by Senator Warwick Parer who had a longstanding business involvement in the resource sector. The resultant report, *Towards a truly national and efficient energy market*, was released in 2002. It acknowledged that unleashing the forces of competition in the belief that this would drive down prices to reinforce Australia's energy competitive advantage had placed too much confidence in the 'invisible hand' of the free market. The establishment of the market was a novel experiment and a risky one, given the material significance of energy as well as its value in economic terms. The reversal of the trend in energy prices over the latter period of the 1990s was not only the result of an initial competitive spurt linked to increased cheaper brown coal-fired power generation in Victoria. It also reflected the emergent industry structure because in opening a new arena for private capital accumulation, global energy corporations had entered the industry and they sought to profit from their position of market strength. The Parer-chaired report questioned the governance arrangements that enabled the exercise of such market power. In the Review's opinion this was a problem of the state's making. A trusting market regulator had endorsed price settings, especially for transmission companies, that aimed to incentivise private investment, but which was argued to have resulted in investment in excess capacity and above market-norm returns on those investments. Profits had become inflated, or 'gold plated', according to a growing chorus of criticism. The States that had not privatised their utilities were also considered to have contributed to the price inflation because the utilities were pricing electricity to generate surpluses that would be delivered in the form of dividends to State treasuries. The report recommended tightening energy market governance by strengthening rule-making and regulation by the NEM governing agencies and called for remaining State-controlled utilities, or the components still retained by State governments, to be fully privatised.

The Parer report proved to be an exercise in trying to come to terms with the inherent contradictions in the energy market liberalisation project: to underwrite orderly development but not over-regulate and thereby discourage investment. State oversight was a necessary institutional factor in the formation and governance of the NEM, yet the dominant theme of the report called into question the overreach of regulation. What the report considered over-regulation had introduced confusion into NEM governance, and this had contributed to price inflation because it created regulatory and sovereign risk. The practice of administered pricing for those enterprises that had invested in the transmission grid and distribution networks, the 'poles and the wires' of the industry, may well have created the opportunity for unfettered rent-seeking. Determining the price that could be charged to deliver a return sufficient to provide an incentive for further investment was always going to be a risky balancing exercise, especially once the private operator had secured ownership of the asset and was in the position to use the reliability of electricity supply as a bargaining chip. Reliability could be assured but at a premium. The challenge that presented was one of striking a balance in governance and regulatory arrangements at the same time as the liberalisation project was the pivot for an energy-led economic recovery (Outhred 2004). Given that Parer was personally wedded to the *laissez faire* 'free market' as the foundation for the flourishing of free enterprise, however much his corporate model of free enterprise bore little relation to the ideal construct, the dilemma that had to be confronted was how much authority should the state be able to exercise in energy market governance. This remained unresolved.

The Parer inquiry was novel in another related respect. For the first time, an inquiry into the energy sector addressed the issue of emissions governance and the liberalisation of the energy market. The debate on energy market liberalisation was for the most part completely silent on the question of the industry's status as the preeminent producer of greenhouse gas emissions. Throughout the debate on establishing the National Energy Market, the issue of electricity generation and emissions mitigation was completely sidelined. One chapter in the Parer report outlines the 'Options to Reduce Greenhouse Gas Emissions'. This could be considered to be something of a watershed were it not for the ad hoc manner in which the subject was approached. The report noted the plethora of schemes initiated by State and federal governments adopted to deal with emissions produced by stationary energy plants but questioned the usefulness of these because there was a lack of coherence and consistency in their design and because they were motivated by conflicting objectives based on excessive state interventions. The report recommended that these be abandoned and appealed to an orthodox economics explanation to recommend the adoption of a nationwide emissions trading scheme. The recommendation envisaged a highly qualified version of a trading scheme, one that would not have universal application, excluding the usual industries and its establishment contingent on the adoption of a global scheme. With these qualifications, it could be expected to give rise to the very economic inefficiencies that it was designed to address. The report was very much a rhetorical exercise, justifying the energy liberalisation project. With no strategic plan as to how to follow through on the adoption of effective emissions mitigation measures, the report was clearly lacking in political acumen and the recommendations that flowed from it attracted no further consideration (Outhred 2004; Pears 2004).

Missing from the Parer report was a critical appreciation of the structure of the energy industry. It bore little resemblance to the perfectly competitive model of *laissez faire* capitalism, and Parer would have been well acquainted with this given his history in the industry. The different stages of the energy value supply chain, of generation, transmission and distribution, had been taken over mainly by large energy corporations, some controlled by global investment funds, others state-owned and controlled utilities. Enterprises engaged in generation and transmission could exercise disproportionate bargaining power in the market because their position in the supply chain was akin to a natural monopoly. Energy corporations also sought to strengthen their position in the market by establishing vertically integrated links across the energy supply chain that were oligopolistic in nature and they were able to exercise their market power in supply when setting prices. The state of the regulatory apparatuses did not have the means to force changes in the structure of enterprises that could have forced limits on monopolistic and oligopolistic practices. The regulation of the market and the development of governance arrangements were still works-in-progress, and the limited resources at the disposal of the NEM agencies restricted the capacity to do much more than sanction this market power as a condition for ensuring reliability (Harper et al. 2015; Vertigan Review 2015). Moreover, markets continued to reflect their historical organisation constituted principally within the bounds of States rather than in a yet-to-be fully cohered national market. Industry energy users tended to be locked in. Consumers' reticence to switch suppliers highlighted the attenuated nature of the liberalisation project. Price inflation was not an unexpected result, and this presumably contributed to the opposition to the introduction of measures to provide some reckoning of environmental externalities by putting a price on emissions as recommended by the Parer Report. The idea of a carbon price was completely at odds with the Report's preoccupation with cutting power costs and made little political sense when there would be no immediate tangible benefit.

The two Senate reports—*The heat is on* (*Report of the Senate Environment, Communications, Information Technology and the Arts References Committee* 2000) and *Towards a truly national and efficient energy market*—which originated from opposite ends of the climate change political spectrum were not greatly at odds when it came to the role of the state in the energy market. *The heat is on* was more consciously aware that the developing industry bore little resemblance to the perfectly competitive model of *laissez faire* capitalism but did not regard the structure of the industry, dominated at each stage of the energy supply chain by large energy corporations and corporatised profit-maximising state utilities, as the main impediment to progressing emissions mitigation initiatives. The orthodox economic prescription of an emissions trading scheme could provide an acceptable policy solution. Those forces demanding the prioritisation of climate change action sought more robust measures and this would entail the state bringing these measures forward. The counter position was an ETS that was well-defined, would provide one mechanism that replaced a raft of policy instruments and narrowed the role of the state in emissions governance subject to the caveat that emissions-intensive industries should not be disadvantaged.

Consolidating the NEM and the Policy Breakthrough: Reinstating an Activist State

The construction of interconnectors increased interstate energy transmission capacity to cement the development of the NEM and this prompted more careful consideration of the nature and role of the state in energy market governance. In December 2003, the Ministerial Council on Energy released a document entitled Energy Reform that explored options for consolidating the architecture governing the national market and for elevating the significance of the national market and governance arrangements over State-based electricity and gas markets. In 2005, the federal government increased funding to establish two new federal agencies, the Australian Energy Market Commission (AEMC), which was charged with setting the rules and regulations, and the Australian Energy Regulator (AER), which was tasked with enforcing compliance with the rules. This was a relatively economical addition to the energy governance architecture, the cost of which was mostly paid for by an impost on energy operators. The AER was established as a department of the Australian Competition and Consumer Commission. The Australian Energy Market Operator replaced the NEMMCO and took over the role of the State agencies which resulted in the incorporation of gas under national energy governance. The AEMC was mandated to set rules for trading electricity and gas, and the National Electricity Market morphed into the National Energy Market. In what was a unique organisational arrangement, representatives of the energy operators were invited to join government representatives on the AEMO board. This gave industry a direct voice in planning although not the ability to dictate decisions because the government representatives held a majority 60 per cent voting rights. One of the distinguishing features of this reorganisation was that the agencies' operating brief of the NEM agencies has and continues to be quite narrowly defined. It is infused with the ideational principles of conventional economic theory that are embodied in the *laissez faire* construct. The issue of the role of the state has remained pivotal in charting the success of the liberalisation project. The Coalition government was concerned that with prices rising it would face increasing criticism from industry and households. The concerns were discussed in the Council of Australian Governments Energy Council, and in 2006 it resolved to launch an inquiry, the objective of which was made clear from the team commissioned to conduct the inquiry: four longstanding champions of the free market, who had spearheaded the case for market transformation from within the bureaucracy. Chaired by the Productivity Commission's Bill Scales, and including David Carmody, Treasury economist, Geoff Swift, electricity supply association, business development advisor, and Alan Rattray, industry analyst, the Energy Reform Implementation Group (ERIG) crystallised a decade and a half of *laissez faire* market advocacy.

When completed, the Scales' ERIG report proved to be the most detailed analysis of the energy industry to have been undertaken (Scales et al. 2007). Whether the report was a watershed in thinking is highly questionable. The dominant theme was that the energy reform project had not met expectations, and the report largely reiterated conventional reflections on the changes required to enhance efficiency, push down prices and ensure supply and secure the international competitiveness of Australian industry. The report called for greater market contestability, the removal of barriers to entry, questioned the anti-competitive practices and poor governance, particularly those associated with the State-owned utilities that had not been privatised, and queried the adequacy of the established regulatory order's ability to challenge anti-competitive practices. States should be pressured to privatise the remaining state-owned utilities or at least make some concessions by agreeing to partial privatisations.

The ERIG observed that the transmission grid was proving to be the critical weakness in the market infrastructure. Given its characterisation as a natural monopoly, the industry regulator set the amount it could charge for its services and thus returns on investments, and this was deemed to have been a key factor in the price inflation as well as resulted in investment in overcapacity. The ERIG report recommended a more strategic approach be adopted in managing the transmission grid given the crucial role the infrastructure played in laying the foundations of the NEM. There had already been some movement in this direction. In mid-2006, COAG amended the Australian Energy Market Agreement to align distribution and retail regulation more closely and to progressively remove some state incursions into an increasingly centralised system of regulation. Consolidating the NEM was regarded as critical to the liberalisation project if the benefits of competition were to be fully realised. The ERIG recommended a National Transmission Network Development Plan that would include encouraging private investment in the transmission network and establishing a more robust and cohered NEM.

The ERIG stressed the importance of restricting the ability of State governments to exercise any influence in the definition of markets because such interventions tended to frustrate the liberalisation of the energy market. A proposal for a National Transmission Network Development Plan would require an increased funding allocation and the strengthening of the NEM governance bodies' regulatory powers. The recommendation to bolster the authority of the National Energy Market Management Company while circumscribing the authority of State governments seemed somewhat paradoxical given that constitutional responsibility for energy resides with State governments and given that dedicating more resources to finance the NEMMCO would seemingly be at odds with the liberalisation project. There was, however, another agenda that was put in play, and this linked to augmenting the role of the energy industry's governance arrangements, whilst weakening that of State governments. Private energy operators were invited to assume a more active place in the market as well as participate in energy governance deliberations. Previous arguments against such potentially anti-competitive practices as vertical integration were not regarded by the ERIG as necessarily anathema to the vision of a perfectly competitive energy market. This defence of private capital vis-à-vis the state was underlined by the recommendation to include private operators in a redesigned governance architecture that resulted in the National Electricity Market Management Company being replaced by the Australian Energy Market Operator in 2009 and private operators being provided with a representation of 40 per cent of AEMO board positions, the balance being held by state officials. This made for a more emphatic capital-friendly energy market liberalisation process.

The enduring nature of the commitment to a *laissez faire* energy market has been embodied in the leadership of the AEMO and the more recently established Energy Security Board, which was formed to spearhead an energy reform agenda that was abandoned. AEMO and the ESB had been led by women who have the courage of their conviction in a *laissez faire* energy market and state minimalism to question the judgement of the Minister for Energy and Emissions Reduction who has had no qualms about riding roughshod over free market principles to push his personal agenda. The fact that they have stood up to the Minister to defend the free market model highlights just how much has and hasn't changed in the energy industry. They remain wedded to a narrow oeuvre that does not extend the purview to a reckoning of emissions as an integral factor shaping the dynamics of the energy market.

References

Australia (1997) Energy White Paper *In the National Interest* Canberra.

AER [Australian Energy Regulator] (2007) 'State of the Market' 2007.

AER [Australian Energy Regulator] (2012) 'Submission to Senate Select Committee on Energy Prices' 17 September.

Baker, R. (2005) 'How big energy won the climate battle'. *The Age* 30 July.

Bulkeley, H., and P. Newell (2015) *Governing Climate Change* London: Routledge.

Ronald Coase, R., (1960) 'The Problem of Social Cost' *The Journal of Law and Economics* 3: 1–44.

Blume, S. (2018) 'If any institution needs to go, should be AMEC, not AMEO' *Renew Economy* 14 February.

Council of Australian Governments (1992a) *National Renewable Energy Strategy* Canberra

Council of Australian Governments (1992b) *National Strategy for Ecologically Sustainable Development* Canberra.

Department of Industry, Science, Energy and Resources (2020) 'Support for Australian households' 6 July.

Ecologically Sustainable Working Group (1992) *The Chair's Greenhouse Report.*

ESD Working Group. (1991). *Greenhouse Report.*

Finkel, A. (2017) *Independent Review into the Future Security of the National Electricity Market, the Australian Energy.*

Galligan, B., and G. Lynch (1992) 'Integrating conservation and development: Australia's Resource Assessment Commission and the testing case of Coronation Hill' *Federalism Research Centre Discussion Papers*, A.N.U. 14, March.

Garnaut, R. (2011) *The Garnaut review 2011: Australia in the global response to climate change* Cambridge University Press.

Garnaut, R. (2008) *The Garnaut climate change review: final report* Cambridge University Press.

Hamilton, C. (2007) *Scorcher: The dirty politics of climate change* Melbourne: Black Inc.

Harper, I., P. Anderson, S. McCluskey, M. O'Bryan (2015) *The Australian Government competition policy review: final report*, Treasury, Canberra.

Hilmer, F., G. Tapperall and M. Rayner (1993) *National Competition Policy Review* Canberra.

Howard, J. (1997) *Safeguarding the Future: Australia's response to climate change—Ministerial Statement* Hansard, 20 November.

Hutton, D. (2013) *Mining: The Queensland Way* Brisbane: At a Glance.

Industry Commission (1991) report *Energy Generation and Distribution* Canberra.

Jessup, B., and D. Mercer (2001) 'Energy Policy in Australia: A comparison of environmental considerations in New South Wales and Victoria' *Australian Geographer* 32(1): 7–28.

National Greenhouse Advisory Panel (NGAP), (1996) *Report on the national greenhouse response strategy*, Environment Australia, Environment Protection Group, for the NGAP, AGPS, Canberra,

Outhred, H. (2004) 'The Evolving Australian National Electricity Market: An Assessment', in G. Hodge, V. Sands, D. Hayward and David Scott (eds) *Power Progress: An Audit of Australia' Electricity Reform Experiment* Melbourne: Australian Scholarly Publications, 154–170.

Parer, W., P. Breslin, R. Sims and D. Agostini (2002) *Towards a truly national and efficient energy market*, Canberra,

Parkinson, G. (2018) 'All we want for Christmas: Four "Disruptors" share their energy policy wishes' *Renew Economy* 21.

Pears, A. (2004) 'Energy Market Reform and the Environment', in G. Hodge, V. Sands, D. Hayward and David Scott (eds) *Power Progress: An Audit of Australia' Electricity Reform Experiment* Melbourne: Australian Scholarly Publications, 171–182.

Pearse, G. (2007) *High and Dry: John Howard, climate change and the settling of Australia's future* Camberwell: Viking Press.

Report of the Senate Environment, Communications, Information Technology and the Arts References Committee (2000) *The Heat Is On: Australia's Greenhouse Future*.

Rosewarne, S. (2003) "The Kyoto Protocol and the Australian State's Commitment to Capital Accumulation", *Capitalism Nature Socialism* 14 (1): 1–36.

Scales, B., D. Carmody, G. Swift and A. Rattray (Energy Reform Implementation Group) Energy Reform—The way forward for Australia—Final report (2007) Council of Australian Governments Energy Council.

Swyngedouw, E. (2010) 'Apocalypse forever? Post-political populism and the spectre of climate change' *Theory, Culture and Society* 27(2–3), 213–232.

Vertigan, M., G. Yarrow and D. Morton (2015) Review of Governance Arrangements for Australian Energy Markets -Final Report of Council of Australian Governments Energy Council Commonwealth of Australia

Wilkinson, M. (2020) *The Carbon Club Sydney* Allen and Unwin.

CHAPTER 3

Holding Back the Renewable Energy Surge: But Change Is in the Air

While the organisational foundations of the NEM were being consolidated, there was a radical shift in the political climate that established climate change as the number one political issue in Australia. A quite striking and very wide range of interests began to give voice to their concerns in calling for action on climate change. They ranged from the more popular messages conveyed in the release of Al Gore's documentary *An Inconvenient Truth*, which garnered a large audience in Australia, through to the serious economic scenario assessments in Nicholas Stern's *Climate Review* (2006) commissioned by the United Kingdom, which was echoed in the 2007 climate change report and follow-up 2008 reports prepared by the economist Ross Garnaut and commissioned by the Labor Party, to the study prepared under the editorial leadership of the CSIRO, *The Heat Is On: The Future of Energy in Australia*. This report encapsulated the extent of the shift in the mood of the nation and the preparedness to break with historical antipathies and establish common cause in the call for action on climate change. The CSIRO-led report envisaging an energy future for Australia brought together prominent corporations, peak union and community organisations, non-government environmental organisations and many others. The distinctive feature of these interventions was the incorporation of emissions management as an integral feature of energy market scenarios. In the words of the *Future of Energy in Australia*, the solution to the challenge required a holistic approach that demanded much bolder intervention by the state.

S. Rosewarne, *Contested Energy Futures*,
https://doi.org/10.1007/978-981-19-0224-6_3

This coincided with a heightened level of interest by prominent international energy corporations in investing in the energy industry. Several international energy corporations bought into the industry through partnerships with or takeovers of established energy companies, and for a couple this included the acquisition of generating capacity. In addition to these energy companies there was a flurry of investment by independent energy retailers who captured a relatively small consumer base.

The corporate sector was becoming more proactive in declaring the need for policy shifts. In April 2006, the Business Roundtable on Climate Change was established by BP, Australian Insurance Group, Origin Energy, Swiss Re, Visy Industries, Westpac and the Australian Conservation Foundation. Australian Local Government (AGL) joined Frontier Economics and WWF-Australia to release *Options for Moving Towards a Lower Emission Future* (2006). State governments adopted greenhouse gas abatement schemes that set short-term goals, and the centrepiece of a range of policy measures being canvassed was the introduction of measures to price emissions.

The broader Australian community was also very supportive, many having participated in climate rallies and pursued personal steps themselves to lobby for change, joining or setting up community climate action groups and participating in campaigns to frustrate the operation of the coal-industrial system, such as with the anti-coal camps (Rosewarne et al. 2014; Pearce 2017). State governments introduced feed-in tariffs which provided incentives for households to invest in the installation of rooftop solar panel systems, and this marked a critical turning point in the energy market.

The canvassing of emissions trading schemes was a clear measure of the significance in the shift in the nation's climate political discourse. The shift was mirrored in the receptiveness of the Business Council of Australia, the peak council for corporate Australia, and the Australian Bankers Association to the idea of an emissions trading scheme. Financial institutions had become an ascendant force within the Council and fostered a change of thinking within the organisation by embracing the need for a proactive stance on emissions governance and explicitly supporting an Emissions Trading Scheme (ETS). Within the Coalition cabinet there was a softening of its stance, giving qualified endorsement for the introduction of an ETS.

The shift in the BCA's environment policy outlook was conveyed to the Prime Minister and quietly backed by senior public servants. In December 2006, Prime Minister Howard established a Prime Ministerial Task Group on Emissions Trading. The Task Force included representatives from banking, the energy sector and mining.[1] It handed down its report in May 2007, recommending the introduction of a hybrid emissions trading scheme. The Prime Minister accepted the recommendation at what at first glance could be interpreted as a capitulation save for the fact that Howard announced that the scheme would not commence until 2012.

However, events soon overtook the Coalition government's resistance to supporting renewable energy. There was a surge in investments in renewable energy technologies, and this had an immediate and discernible impact in containing the rate of increase in the nation's emissions from 2007 as we have observed in Fig. 3.2. It was evident, nonetheless, that following this initial decline, cutting emissions any further would prove quite a challenge.

Increased investment in renewable energy technologies, and particularly wind farms, soon translated into increased renewable energy generating capacity, the extent of which can be observed in Fig. 3.1. The increase was initially gradual, but the investment continued to grow into the following decade. The interest in investing in coal-fired generation flat was abandoned, such that all new generating capacity was attributable to renewable sources.

The more telling story of the unfolding energy transition is captured in Fig. 3.2, which maps the increase in the different sources of renewable energy. The increase in electricity generated by wind and solar really took off from the mid-2000s and accelerated from the mid-2010s. All new generating capacity from 2009 was attributable to renewables sources. The relative significance of the different sources of renewable energy is depicted in Fig. 3.2. Wind and solar are the technologies that have really been driving the transition.

Distinguishing the contribution of these technologies according to scale also highlights an intriguing story. The relative significance of small-scale rooftop solar panel generating capacity, mainly household rooftop solar panels although an increasing proportion of small enterprises have been installing panels, and large-scale solar and wind farms is mapped in Fig. 3.3. This highlights the ever-growing significance of household investment in solar alongside the continuing cumulative increase in solar voltaic panel farms. As the contribution that solar farms make to

[1] Most of the business representatives invited to participate in the Task Group were signatories to the CSIRO report.

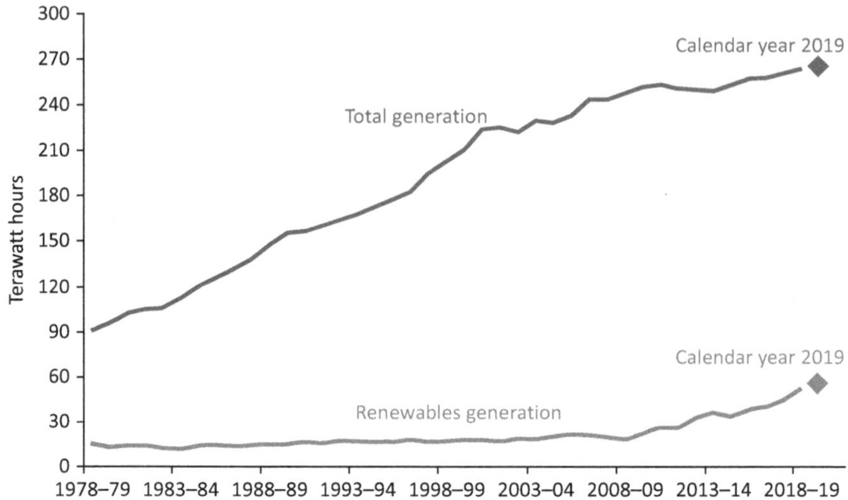

Fig. 3.1 Australia: electricity generation sources. (Source: *Australian Energy Update 2020*: Fig. 3.3 Australian electricity generation Sources: Department of Industry, Science, Energy and Resources (2020c) Australian Energy Statistics, Table O, International Energy Agency (2020), World Energy Balances)

generating capacity increases, so the proportion of electricity generated from coal-fired power stations has been decreasing.

The positive effect of the investment in renewable energy technologies on energy-related emissions was apparent almost immediately, and most obviously from 2007. The level of emissions recorded in 2007, when the first dip in emissions occurred, was 60 per cent higher than 1990 figures, as can be observed in Fig. 3.2 (Department of Industry, Science, Energy and Resources, 2020c). The Clean Energy Regulator reported in 2020 that emissions in the electricity generation sector had decreased over the last four years with 'notable reductions of 4.8 million tonnes of CO_{2e} (2.9 per cent) in 2018–2019 and 7.5 million tonnes of CO_{2e} (4.6 per cent) in 2019–2020 on the previous reporting year' (Clean Energy Regulator 2020).

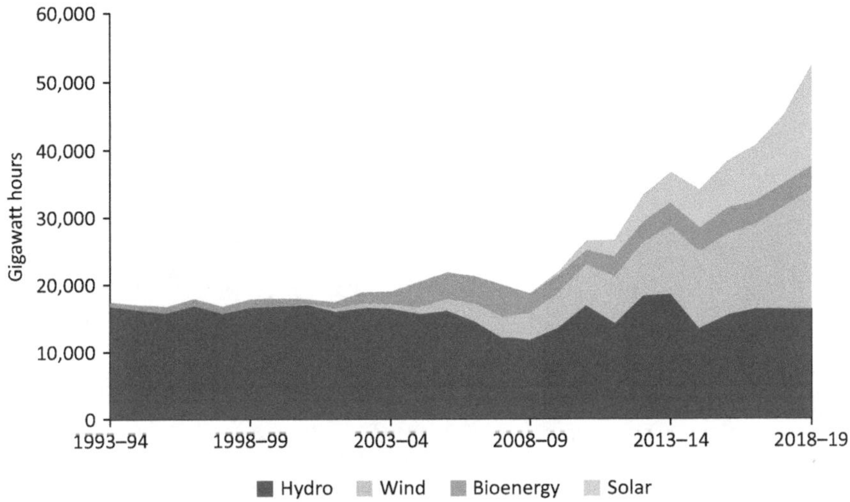

Fig. 3.2 Australian electricity generation from renewable sources. (Source: Department of Industry, Science, Energy and Resources (2020c) *Australian Energy Update 2020*)

A Watershed in Australia's Climate Politics

The irony of this turnaround coincided with Howard's acquiescence to those interest groups and more particularly industry calling for action on emissions. Howard's acquiescence was little more than a fait accompli, an acknowledgement that climate politics had gained traction and that there was overwhelming support for the federal government to develop an emissions mitigation policy. The proposed 2012 introduction date created the political space that would provide opportunity for conservative political forces to try to block any initiatives to reduce emissions that could contain the development of coal mining and expansion of coal exports. The November 2007 election, however, dulled any momentum they could muster. Howard's popularity in his electorate had plummeted over 2006 and he lost his parliamentary seat at the 2007 federal election. The Liberal-National Party Coalition lost government to Labor, although the Labor Party did not secure a majority in the Senate.

Following the swearing-in of the Rudd-led Labor government, the Prime Minister lost no time in moving on climate change. Within the

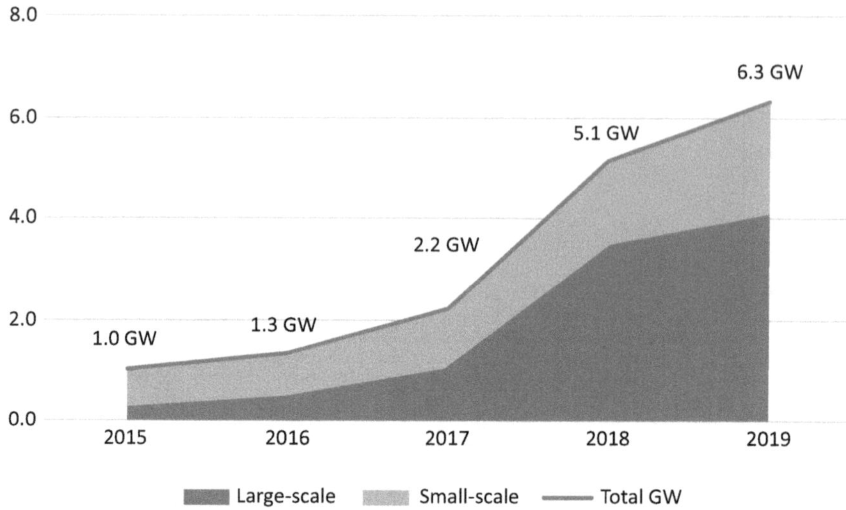

Fig. 3.3 New renewable capacity installed in Australia, by calendar year. (Source: Clean Energy Regulator (2020) National Energy and Greenhouse Reporting 28 February)

month, Rudd attended the December COP meeting held in Bali to announce that the new government's first act would be to ratify the Kyoto Protocol. A policy template had been commissioned by the Labor Party, and the author of that review, economist Ross Garnaut, was assigned to refine the policy frame, and this formed the basis of the Labor government's proposed *Carbon Pollution Reduction Scheme* (CPRS).

The 2008 *Garnaut Review* drew upon the principle of tradable property rights to advocate an emissions trading scheme based on trade in carbon permits that could be adopted with the object of meeting one of two possible emissions reduction scenarios, one aiming for limiting global concentrations of emissions to 450 ppm and a fallback position of 550 ppm.[2] The *Review* proposed that emission permits be auctioned. The government would aim to sell permits at a low price during the transition phase. Mindful of equity considerations, the government committed to

[2]The targets were based on a normative judgement of what was politically possible and were not informed by reference to any goals regarding the assimilation of the greenhouse gases.

providing assistance to low-income households to compensate for the system which would result in higher energy prices. It also envisaged providing some support for areas in which coal-fired power stations were located. Emissions-intensive, trade-exposed industries would be shielded from international competition through the provision of compensation in the event that a comprehensive global emissions trading scheme was not adopted.

This set in train a veritable free for all as those who could be hurt by an emissions trading scheme also began lobbying for compensation. The Prime Minister's Taskforce charged with reviewing and refining the CPRS came under considerable pressure to accommodate the compensation claims. The Taskforce recommended adopting a more liberal method for compensating the energy-intensive trade-exposed industries by grandfathering the issue of permits, recommending that free permits be granted to cover up to 90 per cent of emissions produced by establishments that generated 2000 CO_{2e} tonnes or more and 60 per cent for establishments producing between 1000 and 2000 CO_{2e} tonnes. This, it was maintained, would shield emissions-intensive trade-exposed industries from being subjected to a competitive disadvantage. The CPRS Taskforce also resolved to support assistance for coal-fired power generators through the *Electricity Sector Adjustment Scheme* and supported targets being met through carbon offsets and sequestration initiatives. This in turn resulted in the government agreeing to support other measures, appeasing farmers with the introduction of the Carbon Farming Initiative that rewarded land-based sequestration initiatives and adding another cost to the CPRS. Funds for carbon capture and sequestration research were also increased.

The *Scheme* was presented as the basis for Australia playing a proportionate role in the global effort to reduce emissions and the cost of doing so was argued to be manageable. In fact, as bargaining over the terms of the CPRS unfolded, the notion that the costs were manageable became less and less realistic. The state was being called upon to act as a bank to appease those constituents dissatisfied with the CPRS. In the end it did not matter. Despite having radically recast the proposed emissions trading scheme from that initially recommended by Garnaut, changes which, incidentally, Garnaut accepted as being politically necessary, the CPRS failed to win sufficient support in the parliament to be legislated. Rudd's leadership of the Labor Party was challenged in the lead up to the next federal election. Julia Gillard was installed as a leader and took the Labor Party to the election. Upon being re-elected, a redesigned emissions management proposal was tabled. The *Clean Energy Future Package* (*CEFP*) won the backing of the Greens and was legislated. An emissions trading scheme was to be established. The value of permits would be fixed for three years after which a market-driven system would operate.

The Policy Breakthrough: Reinstating a Role for the State

The *CEFP* laid the foundations for the state engineering a potentially radical transformation in electricity generation in Australia. However, in response to criticism that the *Clean Energy Future Package* was essentially a revenue-raising exercise, the government committed to returning all revenues raised through the sale of permits. It established the Energy Security Fund, which would oversee the issue of free carbon emissions permits and make cash payments to affected electricity generators. The Fund would meet the cost of the closure of up to 2000 megawatts of the dirtiest power stations by 2020. The Fund would decouple emissions reduction from the goal of economic growth ensuring that the cost of emissions reduction would not be a drain on the budget. An amount of $5.5 billion was allocated to meet the cost of issuing free permits and cash compensations and additional funds would be made available to meet the costs of power plant closures (Pearce 2017). The commitment to compensation invited businesses to lobby for their share. The concessions in response to business claims, alongside the issue of free permits, and other financial assistance made for massive windfall gains. One billion was paid to power generators in the first year of the Fund. EnergyAustralia, which had bought the privatised Yallourn Power Station, estimated that it would receive permits in 2012 and 2103 with a market value of over half a million dollars. Four companies that used a disproportionate share of electricity and which were already being subsidised through State-based long-term below-market price power supply contracts—Tomago, Rio Tinto, Alcoa, Norsk Hydro and Alumina—would be awarded a disproportionate share of the compensation.[3] International Power, the British Corporation which

[3] The Victorian Labor government and the Coalition government have recently re-negotiated a four-year $155 million support and subsidies for Alcoa's smelter, including a $19.2 million a power supply agreement with AGL, Origin Energy and Alinta that would require it to free up power to avert a State-wide blackout. Alcoa is in the final year of a $240 million price subsidy to support continued production (Toscano 2021; Macdonald-Smith et al.

had bought the Hazelwood power plant in Victoria, one of the most polluting in the world, sought compensation to close the plant, which was an audacious power play, given it had completed due diligence on the prospective purchase and was aware that the terms of the purchase could entail closure without compensation (Rosewarne 2012). One estimate (Perry et al. 2013) suggests that the magnitude of compensation the government had committed to substantially exceeded the likely net revenue and that householders would end up bearing a disproportionate share of the cost of the *CEFP* in higher costs.

Perhaps the most critically significant institutional development to be established as a result of the deliberations with the Greens over the *CEFP* was an agreement to cement a more active role for the state in supporting the uptake of renewable energy. The Greens demanded that the government adopt a proactive role for the state by setting up two agencies. The Australian Renewable Energy Agency (ARENA) would provide seed funding for renewable energy projects. The Clean Energy Finance Corporation (CEFC) would be established as a debt financier with a bank of $10 billion. The CEFC could draw on the fund to provide an advance that a renewable energy project developer could leverage against to raise additional funds to support a planned project. The CEFC legislation explicitly precluded the Corporation from investing in any ventures other than renewable energy, low-emissions technology and energy-efficiency projects. The *CEFP* also included funding for establishing a Climate Change Authority that would provide an independent voice, conduct audits and provide advice on policy. These agencies heralded a paradigmatic change in the emphasis of government climate change policy, although the real significance of the role that ARENA and the CEFC would play in the transitioning of the energy economy was probably not fully appreciated at the time. The two agencies marked a radical departure from the longstanding conservative opposition to building an emissions mitigation programme on the edifice of the state rather than the 'free market'.

The CEFC has proved a remarkably successful story. As a renewable energy investment catalyser, since it began operating in mid-2012, it has

2021). Alcoa accounts for 10 per cent of electricity in Victoria. Tomago, the single largest user of electricity in New South Wales, has announced that its international competitive advantage is being eroded because of rising power prices and that, if it is to continue smelting, it expects comparable support for its aluminium smelter (Kirkwood 2020; Williams 2021).

committed $8 billion in lifetime investment commitments that have activated almost $30 billion of additional investment supporting the launch of some 150 mostly large-scale projects (Learmouth 2020). These projects have been established across a range of industries and spread across the Australian landscape. The CEFC has also generated above market returns on its investments, averaging over 6 per cent each year, well above the 2.94 per cent government bond rate and debunking any suggestions that it is a drain on the public purse (CEFC Annual Reports, 2012–2013–2019–2020).

ARENA has supported many innovative projects from early stage innovation to later stage demonstration projects and their commercialisation. To date, over the course of its life it has dedicated $1.63 billion to backing 566 projects involving a broad range of technologies (ARENA, 2020).

The continuing work of ARENA and CEFP has had a real material impact on the energy industry. While not specifically incorporated into NEM governance, the two agencies have contributed to underwriting the transformation of the NEM in backing investments that are spearheading the transition to greater reliance on meeting the nation's power needs through renewable energy technologies. Figure 3.1 indicates that from the late 2000s all the increase in generating capacity is attributable to renewable technologies and this is associated with the decline in the rate of increase in emissions observable in Fig. 3.2. Unlike previous initiatives to mitigate emissions as a discrete process managed without systematic reference to the energy market, ARENA has financed several projects in conjunction with the Australian Energy Market Operator, including enhancing the capacity of the grid network to accommodate intermittent sources of energy generated from renewable technologies with a view to strengthening the integrity of the NEM and supply reliability.

The Fossil Fuel Kickback

The CEFC and the ARENA have had an enduring impact on the transformation of the energy mix in Australia, but this has not gone uncontested. On the contrary, when the CEFP was introduced in 2011 it was confronted by what must go down as perhaps the most vicious and dishonest political campaign in Australian history and the opposition to an emissions reduction policy framework has persisted. The *Clean Energy Future Package* was attacked for having allegedly misled the voting population in the campaign leading up to the 2011 election. Prime Minister Gillard was accused of lying, having gone into the campaign promising to introduce an emissions trading scheme based on carbon permits, but instead imposing a fixed price for three years which the opposition leader claimed was 'one great big tax'.

The 'anti-tax' campaign was waged through to the 2013 federal election when Labor was defeated and the leader of the Coalition, Tony Abbott, was elected to office. Abbott immediately set about trying to repeal the *CEFP* architecture. He moved to dismantle the ARENA and the CEFC but was blocked because he could not muster the necessary numbers in the Senate to pass the legislation (Murdoch and Mills 2021). Abbott abolished the emissions trading scheme and froze out the board of the Climate Change Authority considering them too closely aligned with the Labor Party. The Authority's mandate was more narrowly defined to concentrate on monitoring the Carbon Farming Initiative and National Greenhouse and overseeing the Energy Reporting system, and Labor-appointed members were gradually forced to retire.[4] Abbott did succeed in having Labor's expanded emissions reduction target reviewed, although the agreed reduction in the target was relatively marginal. The one substantial policy change was an expansion in funding for the Emissions Reduction Fund (ERF) which promised greater reward for farm-based sequestration, a policy designed to strengthen electoral support for the Coalition in rural areas.[5]

Reflecting its negative approach to policy, the Abbott-led Coalition government released the *Energy White Paper* in 2015 to supposedly mark a new era in which free market forces would prevail. The *White Paper*'s one distinguishing feature was its failure to develop any new initiatives. It basically reiterated the well-aired objectives of 'deliver[ing] competitively priced and reliable energy supplies by promoting competition in energy markets, increasing energy productivity and facilitating investment in energy resources development' and thus defending the future for fossil fuels (Macfarlane 2015).[6] An emissions reduction policy was to be

[4] Following the demise of the Climate Change Authority, a public fund-raising exercise resulted in the establishment of the Australian Climate Council in 2013. The Council provides an independent voice on climate change.

[5] This was as much about pork barrelling in the attempt to win support of farming communities, but it has not been overly successful. There has been minimal call on the funds set aside to reward the farming sector for sequestering CO_2 because the measurement of sequestration is still the subject of testing.

[6] Macfarlane, the then-Minister for Industry, retired in to take-up the chief executive officer position with the Queensland Resources Council.

designed, but the government's priority was focused on strengthening competition to lower energy prices and it looked to the ACCC to take the lead on this by investigating gas pricing.

In 2015, Abbott committed Australia to sign onto the Paris Accord. This may have seemed as though the Prime Minister had either discovered the road to Damascus or was somewhat overwhelmed and disorientated in mixing with the leaders of the world. Truth be told, agreeing to a timetable that committed signatories to reduce emissions by 2050 to keep climate change within a 1.5 °C threshold was not going to be onerous and was three decades away, and besides, Abbott had only recently successfully negotiated the reduction of Labor's expanded emissions reduction target to 26–28 per cent by 2030 based on 2005 levels. However, the failure to develop a meaningful policy response that would map the pathway to meeting the emissions reduction generated disquiet within the Coalition government, and in 2015 Abbott was displaced as leader of the Coalition and as Prime Minister. The more liberal-minded Malcolm Turnbull was elected as the new leader, and Prime Minister, and this removed the immediate political obstacles for a serious reconsideration of energy and emission reduction policies. A severe cyclonic storm that wiped out much infrastructure, resulting in a system shutdown and widespread power outage in South Australia on 28 September 2016, provided the occasion for the reappraisal of policy. The federal Coalition government led by Turnbull sought to make political mileage out of the event by launching an attack on South Australia's Labor government led by Jay Wetherill. The attack criticised the government for having been overly precipitous in supporting the transition to renewable energy while not attending to the potential risk of jeopardising energy security by relying on one or two sources of intermittent energy.[7] The criticism did not stand the test of time as other extreme weather events were being reported and testing supply reliability, and this was increasingly being viewed as the consequence of climate change affecting weather patterns (McGreevy et al. 2021).[8] More point-

[7]The Minister for Energy and Emissions Reduction, Angus Taylor, also took the opportunity to criticise the Victorian Labor government for accelerating an energy transition and the problems that the increase in intermittent renewable energy would present for the transmission and distribution networks (Garvey 2020).

[8]There have been a number of extreme weather events affecting reliability of supply. For example, on 31 January 2020, six transmission towers in south-west Victoria collapsed in a severe convective wind downburst, separating the Mortlake Power Station, the Portland aluminium smelter and South Australia from the rest of the NEM. South Australia was

edly, the hysterical attacks on the Wetherill government indicated that even the more liberal elements in the Coalition were beholden to the coal-industrial complex.

An Energy-Emissions Policy Breakthrough?

Not to be sidelined, the Council of Australian Government's Energy Council sought to move beyond the caustic debate. With a view to identifying potential courses for action, in 2016 the Energy Council commissioned a review into the future security of the energy system to be chaired by the government's chief scientist, Alan Finkel. The *Review*'s primary tasks were to chart a way forward to support the transition to a low-emissions economy while ensuring supply reliability and energy security as well as to do so 'at the lowest cost'.

Having signed the Paris Accord—which was itself a surprising about-face given that it was the then-Abbot-led incumbent government that had signed the Accord—the federal government also had to consider policy developments that would enable Australia to meet the nation's international obligations. The Review was tasked with formulating a national policy that held out the prospect of bringing energy policy and emissions policy under the one banner and into the future. The parameters for emissions reduction had been set in 2015 by the Abbott Coalition government having committed to a 26–28 per cent reduction on 2005 levels by 2030, and meeting the targets was not going to be a real test of the nation's capacity to cut emissions because the Paris targets were decidedly unambitious and even if met the world could expect world temperatures likely increasing by more than 1.5 °C. Notwithstanding this conservative target, the government has refused to make any further commitments to either increasing Australia's target or signalling a timeline that is comparable to that set by other nations that had signed the Paris Accord and were committed to net-zero emissions by 2050 (Murphy and Morton 2021). This intransigence hamstrung the *Review*.

Notwithstanding the Review's emissions reduction objective conforming to that set by the government, commissioning the review created a storm within the Coalition. The Finkel *Review* presented the Coalition

'islanded' for a time from the National Energy Market. However, a greater proportion of interruptions to supply are the result of temporary shutdowns and maintenance of coal-fired generators (Mcardle 2018).

government with an energy plan that heralded a potentially transformational moment, one that acknowledged that the nation had to confront the realisation that coal-fired power within a generation would unlikely figure in the energy mix. The *Independent Review into the Future Security of the National Electricity Market* released in 2017 provided a blueprint for the future development of the energy system that was organised around a 'clean energy target', setting an emissions reduction timetable consistent with Australia's international obligations. The plan would remove the uncertainty over emissions reductions, enhance security and, by drawing on a broad range of technologies that included gas, guide an orderly transition to a low-carbon economy. The Review's recommendation, encapsulated in the *Clean Energy Target*, recognised that this would require bringing new generation capacity into the market, which was already happening, but a reliable system also required better planning for the exit of existing, and mostly, coal-fired capacity. This, the Review contended, could be better managed by requiring energy generators to sign onto a 'Generator Reliability Obligation' that would apply in each region. Generators would be required to provide three years' notice of an intention to shutter a plant and this arrangement would provide sufficient time to make good the loss of generating capacity with the closures. In seeking to allay concerns that the governance architecture be strengthened, the *Review* was also concerned that successfully engaging participants in the market would have to be driven by incentives and not penalties.

The *Review* was framed around two broad energy system operation and governance imperatives—reliability of supply and security—that played out in a particularly acrimonious debate within the Coalition party rooms and in the public domain. The *Review* recommended that the whole governance structure be strengthened and that the Council of Australian Governments (COAG) Energy Council should assume the lead in policy design and implementation. The *Review* called upon the COAG Energy Council to develop some sense of leadership and recommended the Energy Council adopt a strategic vision that would entail both tiers of government, State and federal, committing to a 'nationally consistent approach to energy policy'. It recommended funding should be increased to support the industry regulator, 'enhanced market monitoring capabilities and an operator with a broader planning role … [to] help drive better overall system outcomes'.

Adoption of the plan would also necessitate more intense oversight. In the *Review*'s assessment of governance arrangements, there was good reason to establish an independent institutional force to shepherd the *Clean Energy Target*. The *Review* was in fact more strident in recommending that a new agency be established to drive the process. The recommendation was adopted with the formation of the Energy Security Board (ESB) spearheading policy development and implementation, and Dr Kerry Schott, former chief executive officer of Sydney Water, banker and on the boards of several companies, was appointed to lead the Board and was to be joined by the heads of the Australian Energy Market Commission (AEMC), the Australian Energy Regulator (AER) and the Australian Energy Market Operator (AEMO), on a three-year contract to implement the recommendations of the *Finkel Review*.[9] The ESB set to work immediately, focusing on the implementation of the proposed policy (Finkel et al. 2017: 6).

In addition to exploring various options to accelerate emissions reduction, the other principal objective of the Review was to defend and enhance the integrity of the NEM and the grid system. The *Review* recommended strengthening infrastructure capacity to better position the industry operator to accommodate the declining significance of synchronous generators and the increase in Distributed Energy Resources. This would require the AEMO to assume a more active role in order to enhance greater security and reliability in the transmission and distribution infrastructure. Transmitting power to the grid could be managed more effectively and securely by equipping AEMO with the authority and technological capability to regulate the flow of electricity. New market processes could be explored, such as Demand Side Management, which would entail large users of electricity being compensated for not drawing on electricity during periods of high peak demand. Similarly, the introduction of governor

[9] The Energy Security Board was established under the authority of the COAG Energy Council to whom it reports. The contract has recently been extended to complete the task of designing a 2025 market. The ESB board has the chief executives of the AER, the AMEC and AEMO on its board. It does not share with AEMO a comparable organisational and operational level of autonomy and formally it is more directed in its role, being explicitly established. Given that the majority of the Finkel Review recommendations were never endorsed, the ESB's brief is formally much less ambitious than was originally foreseen, limited to promoting reliability of supply, discouraging over-investment in the grid and networks, and reducing prices. In practice, the ESB has adopted a more ambitious and directing role in promoting the orderly development of the NEM, and while there have been some disagreements and tension arising with the ESB and AEMO occasionally having different agendas, they have more or less mapped similar energy trajectories that do more than what the ESB's brief would indicate.

settings utilising smart technologies installed in households and other distributed energy generators, as well as large users, would equip AEMO with the ability to take control of the flow of power shutting off electricity being fed into the network by householders or other forms of distributed energy in order to stabilise frequency and voltage of electricity flows. AEMO was being called upon to become more proactive in cohering the national market by, for instance, undertaking systematic scoping exercises that prioritised projects. The *Review* envisaged AEMO coordinating with State governments involved in endorsing new generation development projects. State governments would be required to commit and sign-off on a project and, in the event that a private investor of an endorsed project was not going to proceed, the State government would take on the responsibility for financing the project and thereby minimise the possibility of stranded infrastructure that had been constructed to accommodate the projected increase in generating capacity. Exercising oversight of transmission capacity, including increasing the scope for cross-border transmission with the construction of additional interconnectors, was also regarded as a role that could be taken on by AEMO to strengthen the integrity of the NEM. Better management of supply and effective demand management would cost, although this could be met by increasing the impost on energy enterprises, but it would also obviate the need for additional investments in infrastructure.[10]

This sat uncomfortably with the rationale that had informed the energy market liberalisation project, and, if past experience was any indication, this projected retreat from the unfettered role of the free market was not going to be acceptable to the conservatives within the Coalition, and it wasn't (Grattan 2017; Kelly 2017). The *Review* clearly anticipated the criticisms that would be mounted from within the Coalition parties' room.

[10] Strictly speaking, the Finkel Review's recommendation for more robust management of the infrastructure foundations was not an argument for an enhanced role for the state. AEMO, the Australian Energy Market Operator was going to be at the centre of the plans to future secure a critical component of Australia's energy market governance, but it was not formally a constituent part of the state system. It acted in the interests of all energy market operators. Private enterprises accounted for 40 per cent of AEMO board representation with State and federal governments 60 per cent. AEMO was a relatively independent governing entity. The status of the Energy Security Board with respect to its place in the state system is more ambiguous. It was established as an arm of COAG and is a forum that brings together the heads of the AEMC, the AER and AEMO, and these set the agenda which has been more adventurist than its mandate would suggest.

In sanctioning a greater role for state agencies in determining the trajectory of future development, it advocated a low-emissions future that was not tied to a particular technology. This reckoned on the benefits engendered by diverse energy generation and storage forms, though this still raised the spectre of coal's finite future, and the cyclonic storm that caused the South Australian blackout provided the platform for advocates of coal to critique Finkel's *Clean Energy Target* proposal and the leadership of Malcolm Turnbull.

The three issues framing policy debate—energy supply reliability, power prices and cost implications, and emissions reductions and investment in renewable energy technologies—were brought into sharp relief following the cyclonic storm in South Australia. The South Australian Labor government had been backing the accelerated uptake of renewable energy and had become the largest per capita generator of clean power in Australia. The Coalition government went into overdrive criticising what it deemed to be an over-reliance on a less-than-reliable intermittent source of energy and a costly one at that.[11] The intervention exposed the level of antipathy towards renewable energy within the Coalition. Several investigations into the blackout were conducted by AEMO, AEMC, the chief scientist Alan Finkel, the South Australian government and the Productivity Commission, and there was no downplaying the severity of the storm causing the blackout, but there were safeguards built into the system to protect it from unforeseen events. Sudden changes in load had caused these to trip and this had cumulative knock-on effects across a network designed to respond by insulating different stages from potential stress. The system-wide shutdown, and the occasional shutdowns occurring elsewhere, focused attention on the vulnerability of renewable energy technologies just as the storm's intensity was a telling reminder of the possible implications of climate change.

The political hysteria that had been aroused by events in South Australia revealed the intensity of the entrenched opposition to renewable energy. The Prime Minister and the Treasurer voiced their concerns about supply reliability given the extent of the reliance on renewable energy, but their displays of concern were very much theatre. The more conservative forces within the Coalition seized the opportunity to not only rail against the South Australia's renewable energy agenda but also ramp up their attack on the Finkel review which advocated a *Clean Energy Target*. It was led by

[11] Electricity prices in South Australia were the highest in the country.

Abbott who maintained that the proposed emissions reduction timetable was too disruptive and costly. He had the backing of the Coalition partner, the National Party. Faced with the prospect of not having sufficient support within the Coalition, Prime Minister Turnbull abandoned the *Clean Energy Target* component of the Finkel *Review* in late 2017.

The Energy Security Board sought to chart a way through the political mire in a pared-back policy proposal, the *National Energy Guarantee* (NEG). Turnbull announced that the National Energy Guarantee was intended to 'lower electricity prices, make the system more reliable, encourage the right investment and reduce emissions' (Turnbull 2017). Subsidies and incentives for renewable energy would also be scrapped under the NEG plan. The two key elements of the National Energy Guarantee would require retailers to ensure reliability of energy supply, drawn from dispatchable sources and rely increasingly on renewable energy to meet Australia's international commitments on emissions reduction. In issuing a Briefing Paper, the Clean Energy Council identified widespread criticism of the NEG for being a costly and complex solution that would have little impact on emissions mitigation (Stock and Brailsford 2018; Bourne 2018). A focus of concern among those providing feedback to the discussion paper was that it would have a negative impact on competition in the NEM, especially affecting small retailers and risk further entrenching the market power of energy companies which own both retail and generation businesses. The emissions reduction target that was set for the electricity sector was considered to be deficient.[12] The targets were less than those State and Territory governments had committed to and there was no long-term target.

While conservative forces within the Coalition were not supportive, it is not clear that State and Territory governments were all that enamoured with the NEG, and their support was crucial given that the proposed policy would require legislative changes by State and Territory governments. These forces were adamant in their opposition to legislation that locked in an emissions target, and Turnbull announced that he would withdraw the bill and pursue the NEG objectives by means of regulation (Knaus 2018). He did. The ESB's Schott despaired at politicians' 'bad behaviour' in simply wanting to score political points, and Schott indicated the ESB's

[12] The Clean Energy Target was the only substantive policy proposal that had been dropped from the COAG Energy Council endorsed Finkel Review (Potter 2017).

determination to proceed with the job the Board had been set up for (Potter 2017).

A Change of Guard and the Resuscitation of the Fossil Fuel Industrial Complex

Prime Minister Turnbull's attempts to establish an energy and emissions policy based on the Finkel Review's recommendations to secure the future of Australia's energy market failed to make any headway. The Coalition Party room rejected the proposed Clean Energy Target and Reliability bill and when the Party rejected a compromise policy proposal in the National Energy Guarantee policy, Turnbull's political career came to an abrupt end. The Treasurer, Scott Morrison, won the support of the Party room, assumed the Prime Ministership and Turnbull resigned. In an intriguing development, the retreat from policy action on emissions was all but absolute and this was overseen by a fresh cohort of Prime Ministerial staff that were handpicked from the resources sector. There was an all but complete transplanting of personnel from the energy sector into the PM's Office. This marked the recovery in the reassertion of the power of the coal-industrial complex. The personal leverage that could be exercised within the Prime Minister's office could not be better represented in the inner-most circle of federal government. Three of the most senior staff in the Prime Minister's office were recruited directly from the sector's peak councils: the PM's chief of staff from the Minerals Council of Australia (MCA); a senior advisor who had been ousted from the chief executive officer position at the MCA following pressure from BHP and the PM's executive officer who had previously been a partner for Crosby Textor and had headed the advisor brief for Glencore.

Paradoxically, at a moment in history when the political authority of the fossil fuel industrial complex appeared to be on the way to becoming a spent force, the fossil fuel industry seemed to gain a new sense of purpose to revive and to try to capitalise on the political turmoil. While capital in general had determined that the future lies in renewable energy, the Morrison-led federal Coalition government was reasserting the Coalition's confidence in a fossil fuel-based future. It affirmed its defence of fossil fuels and moved to limit the extent to which ARENA and the CEFC could provide financial support for renewable energy generation projects, although it failed in passing the legislation required to do so. The government had, for instance, proposed that the CEFC dedicate resources to supporting the development of gas-fired generation plants and carbon capture and storage (Mazengarb 2020c).

The reassertion of government support for fossil fuels in the energy mix is being underwritten by a shift in the business focus on winning government support to push down the price of gas and increase the volume of gas being supplied into the Australian market.[13] The resources sector has been working quietly behind the scenes to restore their political stocks. The sector has been advocating a role for gas as an energy source. Greater recourse to gas, it contended, would solve the supply reliability problem of solar and wind farms as intermittent sources of electricity. Gas is being promoted as a fuel that can provide firming capacity as well as facilitating the transition to a cleaner energy future. Interestingly, it has been the resources peak industry associations that once again have been leading this debate, and the refashioning of the policy debate to support gas has been setting the terms of the national policy debate. The public sales pitch to promote gas has been less vigorous, more circumspect than was the case with the coal lobby.

The Australian Petroleum Production and Exploration Association has been a leading donor to the Coalition and had been providing input into the design of policy and development approvals (Marsh 2017; West 2018; Brissenden 2021; Foley 2021). Since assuming office, Morrison and his Energy and Emissions Reduction Minister have been striving to reconnect with the business leaders in the resources sector, which seems to be engendered by the continuing preoccupation to sustain Australia's reputation as the world's leading energy superpower (McHugh 2011; Evans-Pritchard 2013; Macfarlane 2020).[14] Morrison has appointed several individuals who are closely aligned with the coal-industrial complex to ARENA including his close friend, Justin Punch, and Anna Metysek, former advisor to Angus Taylor, and made appointments to the Emissions Reduction Assurance Committee, which oversees the integrity of the Emissions Reduction Fund. Other new appointments to the ERF included the economist Brian Fisher who led Australia's Kyoto negotiations, David Byers,

[13] Australia has developed into the largest exporter of LNG in the world. Gas producers had locked in long-term supply contracts which has created a shortfall in supply into the Australia market and resulted in substantial inflation (Climate Council of Australia 2017).

[14] In the past couple of years, the rhetoric has shifted to the potential of Australia to become a global green energy superpower (Dargaville et al. 2018; Carroll 2020; Garnaut 2019).

former executive officer of the Minerals Council of Australia and former cement industry executive Margie Thomson (Mazengarb 2021a, b, 2020b; Morton 2019). The Morrison government quietly assembled a group of energy and engineering leaders to formulate a post-COVID economic recovery programme. It commissioned Grant King, the former chief executive officer of Origin Energy, which has substantial investments in gas, and Business Council of Australia President, to head the team. He was joined by Clean Energy Regulator chair David Parker, Andrew Macintosh, head of Emissions Reduction Assurance Committee, and the chief executive of the Australian Industry Greenhouse Network, to review participation in the government's Emissions Reduction Fund and how to incentivise voluntary industry participation in emissions reduction (Department of Industry, Science, Energy and Resources, 2020a).

Within the year, the rapport between the Coalition and resources industry leaders was on full display. The federal government established the National COVID-19 Advisory Board to advise, from a business perspective, the government's management of COVID-19 and its plans for economic recovery. Nev Power, former Fortescue CEO and chair of Strike Energy, was appointed to chair the Board. Catherine Tanna, the chief executive officer of EnergyAustralia, joined the committee as did Greg Combet who had worked as a consultant for the gas industry after leaving parliament. The committee recommended that an ammonia plant be established in Narrabri in the vicinity of the contentious gas project. It also recommended revisiting the west-east pipeline from the North-West Shelf, a proposed gas line designed to increase supply to the east coast, which had been considered but rejected by the federal government in 2018 (Davies 2020). The former chief executive officer of Dow Dupont, director of Saudi oil giant Aramco, and board member of engineering and resource infrastructure company Worley, Andrew Liveris, was also appointed to the committee. Liveris chaired a manufacturing subcommittee that recommended plans for public ownership of new gas pipelines, underwriting gas supply projects and a national gas reservation policy, although Power advised a Senate Committee the report was not endorsed by the Commission (Long 2020; Murphy 2020; McIlroy 2020; Stott and Aedy 2020).[15] Liveris chaired the Northern Territory's COVID-19 Economic Reconstruction Committee. In an unusual arrange-

[15] A report commissioned by the National COVID-19 Co-ordination Commission that recommended plans for public ownership of new gas pipelines, underwriting gas supply

ment, fellow Worley board member, Martin Parkinson, joined him on the Commission (Wilkins 2020). The Committee recommended cutting red tape to fast-track development projects and endorsed Darwin's multi-billion-dollar Inpex gas project as well as the Beetaloo Basin shale rock gas project (Roberts 2020; Thompson 2020).

If one needed any further evidence of the Coalition's re-embrace of the fossil fuel industrial complex, then the appointments in April 2021 of Grant King to chair the Climate Change Authority and Susie Smith, chief executive officer of the Australian Industry Greenhouse Network to the Authority's board should suffice. But there was more. Within a week of the announcement of King's appointment the Environment Minister in the New South Wales Coalition government announced the establishment of the Net-Zero Emissions Clean Economy Board, what seemed to be envisaged as the State equivalent to the federal Climate Change Authority. The Minister announced the appointment of former Prime Minister Turnbull to head the board, but within the week, and following Turnbull speaking out against opening more coal mines in New South Wales, the appointment was abruptly terminated, and government leaders declared that coal did have a continuing future in the State.

COVID-19 has provided an opportunity for refashioning one arena of the political landscape. An old alliance has been reformed, albeit with a somewhat different emphasis based on the reconstitution of the coal-industrial complex to incorporate gas. This is playing out in terms of State governments being pressured to lift bans on exploration licences and support the development of gas as a back-up source for power generation (Macdonald-Smith 2021). The reforging of the links between industry leaders and government, and the inner-government Ministry in particular, appears to have steeled the Coalition government's defence of fossil fuels as the platform from which to reignite the momentum of the Australian economy.

However, this has not been without some fundamental contradictions. In defending the paramount place of coal in the energy mix, the government has emphasised the importance of coal as a source of dispatchable electricity, whereas solar and wind are held to be intermittent and an unreliable source of energy and, as inclement weather events have demonstrated, more vulnerable to extreme weather events. Renewable energy

projects and a national gas reservation policy, was not according to Power endorsed by the Commission (McIlroy 2020).

can be subject to supply uncertainty and unpredictable price spikes. One corollary of this is ensuring the continuity of coal-fired power, and this has placed the Coalition government in the awkward position of having to afford consumers protection from sudden cuts in supply. The Finkel *Review* recommended a reliability of supply strategy that required energy generators to give advance notice of any plans to shutter plants. In the case of AGL's announcement in late 2017 that it intended to close the no-longer economically viable Liddell power plant in 2022, and upon reconfirming this decision, the Minister for Energy and Emissions Reduction, Angus Taylor, commissioned a working party to consider if there were any means of prolonging the life of the power station. Following some quite public heated exchanges initiated by the Minister who was set on maintaining the primacy of coal as the energy source, AGL resolved to postpone the closure by one year but reaffirmed the planned closure much to the chagrin of the Minister. In an extraordinary departure from its historical role as the party representing business, and with the withdrawal of the NEG, the government moved to shore up reliability by proposing a policy of divestiture that would subject energy generators to the divestment of assets should they close a power plant without the imprimatur of government and provide evidence that energy supply would not suffer (Grattan 2018).[16]

This was clearly overreach. The thought of a belligerent state dictating to energy corporations sent off alarm bells within the corporate world (Remeikis 2018; Murphy 2018). Minister Taylor ratcheted up criticisms of AGL and the future of the Liddell plant and endorsed proposals advanced by his National Party colleagues that the government should dedicate resources to meeting a need for firm, coal-based generating capacity. The Coalition indicated that if it could not secure a commitment from a private venture to take on the task of building a coal- or gas-fired power plant, preferably in northern Australia, it would consider constructing a state-owned power plant contradicting its ideological free market

[16] The advance notice of closure recommendation was made in an attempt to reconcile the Minister for Energy and Emissions Reduction's criticisms of AGL that followed the company's announcement in December 2015 that it was considering the future of its Liddell plant. The sudden closure of the Engie-owned Hazelwood brown coal-fired plant in March 2017 had come as a shock to the government and resulted in substantial wholesale price increases. The Minister for Energy and Emissions set up a taskforce to consider measures to keep the Liddell plant operating and brought considerable pressure to bear on AGL to extend the life of the plant. The Finkel *Review* looked to the closure notice requirement as a means of effecting a more orderly transition from coal.

philosophy. The government subsequently awarded $4 million to a company that had not had any experience in the energy industry, and whose office was a small beachside shack, to conduct a feasibility study of a coal-fired generating plant at Collinsville in Queensland (Coorey 2019; Smee 2020a, 2020b). An audit report has since determined that awarding the grant did not comply with federal grant rules (McIlroy 2021).

The States to the Rescue: Emissions Reduction Leadership

In the meantime, while the federal government's policy intransigence and resistance to committing to an emissions reduction programme reflected the intensity of the pro-coal faction within the Coalition and its preparedness to bring down the government, State and Territory governments have taken hold of the climate change baton and jettisoned a fossil fuel-intensive future. Without exception they have committed to ambitious emissions reduction targets. In addition to supporting household installation of rooftop photovoltaic panels, they have been backing a whole host of renewable energy projects; the most exciting and politically potent of the State projects has been the establishment of regionally located renewable energy zones. These are radically transforming the energy landscape.

In 2018, the COAG Energy Council invited the AEMO to map a plan that outlined various options for coordinating investment in the generation and transmission sectors. The interim *Integrated System Plan* was released in mid-2018, but the involvement of the Energy Council in charting the orderly development of energy policy all but evaporated with no meetings scheduled for a year following the Prime Minister's decision to dissolve COAG (Filatoff 2020). The National Cabinet was established to focus on responding to the COVID-19 pandemic while a National Federation Reform Council was established under the auspices of the Cabinet and committees convened to focus on prioritising reforms in six areas, one of which was energy. In May, and reflecting the Energy and Emissions Minister's policy priorities, the Energy National Cabinet Reform Committee identified its three key policy development objectives: ensuring reliability and security of the electricity grid; redesigning the NEM to take effect from 2025, a responsibility that fell to the ESB; and developing a package of reforms to unlock new gas supply (Department of Industry, Science, Energy and Resources 2020b).

The striking omission from the Energy Reform Committee was, once again, any reference to emissions reduction, and the Coalition remains resistant to an emissions reduction target beyond the existing 2030 goal. Avoiding any direct reference to emissions, in 2019, the Minister for Energy and Emissions Reduction cobbled together the draft of a policy proposal which he claimed was technologically agnostic or neutral and which holds out the promise of transitioning from reliance on fossil fuels.[17] The *Technology Investment Roadmap* is, in fact, overwhelmingly concerned with maintaining reliability of supply and delivering energy at a low price. Yet to be enacted in legislation, it is an underdeveloped concoction that has been subject to widespread criticism (Mazengarb, 2020a; Matich 2020). It claims to steer the nation's transition to reducing carbon emissions by earmarking five priority technologies for investment: 'clean' hydrogen, energy storage, 'low carbon' steel and aluminium, carbon capture and storage and soil carbon sequestration. The rationale for this selection has not been articulated and is obviously at odds with the notion of technological neutrality. Gas was not included in Taylor's original list of priority technologies, but it is now being promoted as a low-emissions fuel that will drive the post-COVID-19 economic recovery and is regarded as an energy source that will be used in the production of clean hydrogen.[18] And, with a view to underwriting the *Technological Roadmap*, the Minister proposed to redefine the mandates for the CEFC and ARENA to reduce their support for renewable technologies and redirect financing to the technologies identified in the *Roadmap*.

The *Technology Investment Roadmap* is a risky venture, contradicting the Coalition's public stance on the dangers of governments dedicating funds to picking winners when this is exactly what the *Roadmap* intends. The record of carbon capture and storage sucking up considerable financial support with next to nothing to show for it is clear. Likewise, there is not much to show for the decade-long experiment in soil sequestration (Lam et al. 2013). The *Roadmap* presents coal and gas as mature technologies that no longer require government support, as it argues is also the case with solar and wind technologies. Quite obviously, this positing

[17] The Coalition is promising to finalise an energy and emissions policy sometime in the first half of 2021. Some elements of this have been flagged, but one persistent criticism that industry has made is that the policy uncertainty has endured for too long.

[18] The policy Underwriting New Generation Investment scheme is also about picking winners.

of coal and gas as no longer in need of state assistance is at odds with the history and continuing large scale of subsidies provided to support fossil fuels extraction as it is with the decision to provide seed funding for the Collinsville coal-fired power station study. Likewise, the Coalition's proposed amendments to the Clean Energy Finance Corporation Act, which would remove the restriction on fossil fuel projects being able to apply for CEFC funding (Murdoch and Mills 2020), and the intense lobbying to pressure State and Territory governments to lift restrictions on exploration and development of gas deposits make a mockery of the ostensible downgrading of the place of coal and gas in the Australian economy.[19] The proposed amendment would also affect the future standing of solar and wind technologies. Australia signed onto the Paris Agreement emissions reduction target, yet the Coalition refuses to formally endorse the goal of zero emissions by 2050 beyond signalling that it is an aspiration. The *Technology Roadmap* is a hastily contrived plan without a target and a target with an ill-defined plan.

Conclusion

The Australian energy market has been an experiment in the liberalisation of markets. Based on an ideal construct of the 'free market', the unbundling, corporatisation and then privatisation of state-owned and controlled utilities was represented as a project that would ignite a competitive impulse, that would attract new investment, engender greater efficiencies in the generation and distribution of electricity and, most importantly, drive down prices and thus the cost of power for households and industry. Dismantling the monopoly state power utilities would invite new investment and as a decidedly incidental consideration provide innovative ways of confronting the challenge of reducing greenhouse gas emissions and doing so in an efficient or least-cost manner. As the market has evolved, the story has proven to be more complicated than the liberalisation narrative presents. Firstly, privatisation of a state enterprise did not necessarily command the political support required to accomplish the sell-off of public property. Communities and unions resisted the privatisation

[19] The climate sceptics within the National Party are pursuing this agenda. The Environmental Defenders Office has sought the intervention of parliamentary committees to block moves by the Minister for Energy and Emissions Reduction from lodging an amendment to the CEFC Act that would redefine what constitutes a 'low-emissions' technology that would permit CEFC funds to be deployed to fund investment in the Federal Government's 'gas led Covid-19 recovery' (Murdoch and Mills 2020).

programmes, and some State governments continue to refuse to entertain privatising utilities because of the financial benefits, the dividends, the utilities generate.

Secondly, while privatisation attracted fresh investment to the industry, this was not always motivated by intentions to invest in new technologies that would make for a more dynamic industry. Some investments were simply opportunist, capital seeing an opportunity to buy into the industry to profit from quick capital gains or by rent-seeking (Rosewarne 2012). The industry has become dominated by three energy corporations—AGL, EnergyAustralia and Origin Energy—which, through backward and forward linkages in the energy supply chain, gives them a competitive edge and places them in the position to exercise market power. Other, smaller ventures are also adopting the vertically integrated model to secure their place in the liberalised energy market. Price inflation has been the recurring concern and there have been a plethora of inquiries over time into why the so-called free market has failed to deliver competitive prices. Such inquiries now seem to have become an enduring feature of the industry putting pay to the notion of the 'invisible hand' ensuring efficiency and reliability in supply.

A constant criticism of governments has been the failure to develop and implement policies that provide a clear and well-defined regulatory frame that provides certainty for planning and operating. The lack of policy consistency and the absence of policy durability have made for considerable uncertainty (Scales et al. n.d.). Industry has raised the volume of its concerns that this is affecting investment. But it has been the inability to forge strategies to incorporate the cost of the environmental externalities associated with emissions produced in the course of generating electricity, as well as the fugitive emissions associated with the extraction of coal and gas, that has proved to be the most outstanding manifestation of policy failure and investment uncertainty. The talk of pricing emissions has perhaps been the most hotly contested issue in the definition of the liberalised energy market yet does not appear to be any closer to resolution.

As we have observed, the exercise of power is not restricted to the market. Emissions reduction policy design has been frustrated by the exercise of extra-market power. The lobbying efforts of the emissions-intensive industries have proved a defining feature of policy design, the force of the carbon club, and Morrison is well acquainted with this (Wilkinson 2020). The projection of industry's political power is exemplified in the Prime Minister's staff appointments, with disproportionate numbers being recruited fresh from resource sector peak bodies and who have helped reinstate gas on the energy agenda. Similarly, the selection of individuals who have managed resource industry-related companies to lead teams developing post-COVID economic recovery programmes highlights how structural and personal links can define the prioritising of the fossil fuel-based pattern of accumulation, giving substance to the form of the coal-industrial complex.

In reasserting that the energy sector was not 'a free market nirvana' but rather 'a bastardised market with a legislated rate of return', Morrison defended the state and 'law [that] was targeted at situations where sweetheart deals came at the expense of consumers' (Coorey 2018; Dean 2017; Grattan 2018). Yet, he is not above striking sweetheart deals himself and hiding behind the illusion of a free market that is turning to gas as a means of driving the energy system to an ostensibly cleaner future as well as underscoring Australia's status as a global energy superpower.

The anarchy that Schott contends characterises the energy market reflects her frustration with the opposition of the conservative Coalition government to developing a policy agenda that does more than simply appeal to a fast-unravelling legacy status quo. The decision by Prime Minister Turnbull to commit the Commonwealth government to backing and financing the expansion of Snowy Hydro well before any evaluation, let alone an environmental impact assessment was undertaken, appears to have been motivated by his desire to leave a legacy marking one success story in his failure to break the climate policy impasse that has defined Coalition governments. And a costly legacy it is going to be, with the initial estimated $2 billion blowing out to twice that amount before any estimate is made of the likely environmental costs of the additional dams, pump stations and transmission lines that will cut a swathe of destruction through the Kosciuszko National Park (Williams 2020). Clearly Turnbull was keen to settle on something other than nothing—his legacy if you like—in this policy space in order to try to inject a level of investment certainty that has been missing. The conservatives within the Coalition, and the rump of the National Party, in particular, fail to appreciate that the transition from the coal-fired driven power days is rapidly being supplanted by the scaling up of investment in renewable energy technologies. Angus Taylor's decision not to convene a COAG Energy Council meeting after the federal election held in May 2019 suggests that behind that mask of self-confidence is someone who knows he is on the back foot but doesn't have the maturity to acknowledge it (Ludlow and Macdonald-Smith 2019).

References

Australia (2008) *Carbon Pollution Reduction Scheme.*

Australia (2013) *The Clean Energy Future Package.*

Australian Renewable Energy Agency (2020).

Bourne, Greg (2018) "'Botched' NEG must still pass series of legislative hurdles' Climate Council 14 August. n.12

Brissenden, M. (2021) 'Fired Up' *Four Corners* ABC 12 April.

Carroll, D. (2020) 'Report: Australia shapes as green "energy super power"' *PV Magazine* 27 November.

Clean Energy Finance Corporation (2020) *Annual Reports.*

Climate Council of Australia (2017) *Pollution and Price: the cost of investing in gas.*

Coorey, P. (2019) 'PM concedes to Nats: pledges to consider new coal plant in Qld' *The Australian Financial Review* 26 March.

Coorey, P. (2018) 'Scott Morrison's government and business drift apart' *The Australian Financial Review* 4 December.

CSIRO (2006) *The Heat Is On: The Future of Energy in Australia*, Canberra.

Dargaville, R., C. Wang and C. Hamilton (2018) 'Can we make Australia a renewable energy exporting super power' *Renew Economy* 5 December.

Davies, A, (2020) 'Why is the Covid omission backing a fertiliser plant as it top recovery project?' *The Guardian* 3 May.

Dean, L. (2017) 'Banking sector not a 'free market nirvana' Treasurer' *Mortgage Business* 1 September.

Department of Environment and Energy (2017) *Australia's emission projections 2017*, December.

Department of Industry, Science, Energy and Resources (2020a) 'Expert panel identifies opportunities to reduce emissions', 19 May.

Department of Industry, Science, Energy and Resources (2020b) 'Energy National Cabinet Reform Committee' 24 December.

Department of Industry, Science, Energy and Resources (2020c) *Australian Energy Update 2020,* Canberra.

Evans-Pritchard, A. (2013) 'Australia to be an energy 'superpower'' *The Sydney Morning Herald* 16 December.

Frontier Economics and WWF-Australia (2006) *Options for Moving Towards a Lower Emission Future.*

Filatoff, N. (2020) 'COAG Energy Council in the time of Coronavirus: The COAG Energy Council met on 20 March 2000' *PV Magazine* 23 March.

Finkel, A., C. Munro, T. Effeney and M. O'Kane (2017) *Independent Review into the Future Security of the National Electricity Market: Blueprint for the Future*, Canberra: Council of Australian Governments (COAG) Energy Council June.

Foley, M. (2021) 'Political donors 'undermining' gas-led recovery' *The Sydney Morning Herald* 13 April: 12.

Garnaut, R. (2008) *The Garnaut climate change review : final report* Cambridge University Press.

Garnaut, R. (2019) *Superpower: Australia's low-carbon future.* Black inc.

Garvey, J. (2020) "State to blame' for every *logjam' The Australian* 24 February: 4.

Grattan, M. (2018) 'Malcolm Turnbull and his NEG continue to haunt the government' *The Conversation* 5 December.

Grattan, M. (2017) 'Critical back benchers push back on Finkel clean energy target plan' *The Conversation* 13 June.

Kelly, J. (2017) 'Reject Finkel and end subsidies, demand Nats' *The Australian* 11 September: 3.

Kirkwood, I. (2020) 'Tomago shareholders Rio Tinto and CSR hoe in over "too high" power prices' *The Newcastle Herald* 20 July.

Knaus, (2018) 'The Liberal Party is self-destructing over energy' *The Guardian* 20 August.

Lam, S.K., D. Cen, A.R. Mosier and R. Rush (2013) 'The potential for carbon sequestration in Australian agricultural soils is technically and economically limited' *Scientific Reports* 3, 2179: 1–5.

Learmouth, I. (2020) *Financing the Energy Transition.* Clean Energy Finance Corporation.

Long, S. (2020) 'Government's COVID Commission manufacturing plan calls for huge public gas subsidies' *ABC News* 21 May.

Ludlow, M., and A. Macdonald-Smith (2019) 'State renewable targets put pressure on coal-fired power' *The Australian Financial Review* 13 July: 3.

Macdonald-Smith, A. (2021) 'Coal exit threat to investment' *The Australian Financial Review* 12 March, 1.

Macdonald-Smith, A., P. Ker and p. Durkin (2021) ''Jobs saved in $155m Portland smelter deal' *The Australian Financial Review* 20 March, 29.

Macfarlane, I. (2015) 'Resources can help QLD become an energy superpower' Queensland Resources Council 27 November.

Marsh, S. (2017) 'The great gas con: how Australia got sucked dry' *Michael West Media* 22 May.

Matich, B. (2020) 'Taylor bets $18 billion on emissions technology, ignores own hot air' *PV Magazine* 23 September.

Mazengarb, M. (2021a) 'Taylor slammed for "stacking gas lobbyists" on Climate Change Authority' Renew Economy 9 April.

Mazengarb, M. (2021b) 'Morrison and Taylor continue to stack government bodies with fossil fuel allies' *Renew Economy* 25 January.

Mazengarb, M. (2020a) 'Anything but wind or solar: Taylor's technology plan is "road to nowhere"' *Renew Economy* 22 September.

Mazengarb, M. (2020b) Taylor appoints his former advisor and a prominent RET critic to ARENA board' *Renew Economy* 24 July.

Mazengarb, M. (2020c) 'Morrison to redirect climate funds to carbon capture and big emitters' *Renew Economy* 19 May.

Mazengarb, M. (2020d) 'Australia won't budge on 2030 climate targets, keeps mum on longer term intentions' *Renew Economy* 12 May.

Mcardle, P. (2018) 'Both QLD and SA islanded from the rest of the NEM on Saturday 25th August 2018' Wattclarity 25 August. n.23

McGreevy, M., C. MacDougall, M. Fisher, M. Henley and F. Baum (2021) 'Expediting a renewable transition in a privatised market via public policy: the case of South Australia' *Energy Policy* 148: 1–14.

McHugh, L. (2011) 'Australia to be an Energy Superpower' *Future Directions* 22 June.

McIlroy, T., (2021) 'Power station feasibility grant broke funding rules: Auditor-General' *The Australian Financial Review* 19 March.

McIlroy, T. (2020) 'Nev Power distances COVID-19 commission from gas-led recovery plan' *The Australian Financial Review* 4 June.

Morton, A. (2019) 'Coalition quietly adds fossil fuel industry leaders to emissions reduction panel' The Guardian 23 January.

Murdoch, C., and J. Mills (2021) 'Clean Energy Finance Corporation Amendment (Grid Reliability Fund) Bill 2020' Bills Digest No. 41, 2020–21 Parliament of Australia.

Murphy, K. (2020) 'Australia's Covid commission downplays 'green recovery'' *The Guardian* 11 August.

Murphy, K., (2018) 'Coalition revises 'big stick' policy following backbench objections' *The Guardian* 3 December.

Murphy, K., and A. Morton (2021) 'Scott Morrison refuses to budge on climate target as Biden pledges to halve US emissions by 2030' *The Guardian* 22 April.

Pearce, R, (2017) *Pricing carbon in Australia: Contestation, market failure and the state* London: Routledge.

Perry, N., S. Rosewarne and G. White (2013) 'Australia's Clean Energy Policy: taxing carbon and the illusion of the equity objective' *Ecological Economics* 90: 104–113.

Potter, B. (2017) 'ESB chairman Kerry Schott slams governments 'behaving badly'' The Australian Financial Review 20 November.

Remeikis, A., (2018) 'Business community unites to urge Coalition to abandon energy 'big stick'' *The Guardian* 3 December.

Roberts, L. (2020) 'Andrew Liveris appointed co-chair of NT's post-coronavirus taskforce Economic Reconstruction Commission' *ABC News* 26 May.

Report of the Prime Ministerial Task Group on Emissions Trading (2007).

Rosewarne, S., (2012) 'Right to compensation was folly from the start' *The Conversation* 7 September.

Rosewarne, S., J. Goodman and R. Pearse (2014) *Climate Action Upsurge: The Ethnography of Climate Movement Politics* London: Routledge.

Scales, B., D. Carmody, G. Swift and A. Rattray (n.d.) (Energy Reform Implementation Group) *Energy Reform -The way forward for Australia -Final Report* Council of Australian Government Energy Council.

Smee, B. (2020b) 'Collinsville coal-fired power station is backed by 'big mining companies' Queensland LNP says' *The Guardian* 24 July.

Smee, B. (2020a) 'Shine invited to apply for Collinsville grant two days after securing it' *The Guardian* 24 July.

Stock, P., and L. Brailsford (2018) *Briefing Paper: What are stakeholders saying about the National Energy Guarantee* Canberra: Climate Council.

Stott, E., and R. Aedy (2020) 'Keeping the focus on manufacturing after coronavirus' *ABC News* 30 June.

Thompson, J (2020) 'Territory Economic Reconstruction Commission releases blueprint for NT coronavirus recovery' *ABC News* 1 December.

Toscano, N. (2021) 'Alcoa's struggling Portland smelter secures lifeline with power deal, $160m in subsidies' *The Age* 19 March.

Turnbull, M (2017) 'National Energy Guarantee to deliver affordable, reliable electricity' *Malcolm Turnbull Media Releases* 17 October.

West, M. (2018) 'Revolving doors: How the fossil fuel industry has the government snared' *Michael West Media* 21 November

Wilkins, G. (2020) 'Two for one: company gets two directors on Covid commission' *Crikey* 29 June.

Williams, P. (2020) 'Power grid shake-up to boost supply' *The Australian* 30 April: 18.

Williams, P. (2000) 'Snowy Hydro secures $3.5bn for expansion, stays 'on track, on budget' *The Australian* 6 April: 15.

Wilkinson, M. (2020) The Carbon Club Sydney: Allen and Unwin.

CHAPTER 4

The Uneven Path to a Renewable Energy Future: From the *Laissez Faire* Market Rhetoric and Beyond

The Energy Market Liberalisation Project: Reconstituting the State and Economistic Trajectories

The speed with which these reforms proceeded was remarkable, especially when one considers that there has been no substantive emissions mitigation policy developed since the Abbott government dismantled the Labor government's *Clean Energy Policy* in 2014. The foundations for the liberalisation projects were set when State and Territory heads of governments met in 1990 to consider industry reform, and from that point on a reasonably well-executed plan to establish a national energy market was put in place. Given the constitutional authority for energy rested with the states, a coordinated legislative change agenda was spearheaded by the State governments that put in place an agreed National Electricity Code that laid out the rules for the market and established the National Grid Management Council to coordinate the construction of an interstate grid that would provide the interconnector infrastructure that would provide some substance to the development of a national market. The market would be based on setting up a wholesale electricity supply pool based on available generating capacity, with a view to supply being matched with demand. Spot prices would be set at ten-minute intervals and contract settlements would be based on average prices over half-hour time frames. The National

S. Rosewarne, *Contested Energy Futures*,
https://doi.org/10.1007/978-981-19-0224-6_4

Energy Market Management Company (NEMMCO) was established in 1997 to take over the role of the National Grid Council and manage the wholesale market on behalf of participants. The South Australian and Victorian governments began privatising state-owned utilities.

The first trial run of cross-border trading took place in 1997 and the National Electricity Market was formally launched in late 1998 with New South Wales and Victoria its founding members, and other States joined in the following years. Following a review of market management arrangements for electricity and (in some States) gas, the Australian Energy Market Commission was established by legislation in each State and Territory in 2004 to set the rules and regulations to govern and to guide how best to develop the sector. The Australian Energy Regulator was formed as a division within the Australian Competition and Consumer Commission to enforce energy market rules and regulations. This included regulating rules pertaining to the natural monopoly transmission and distribution sectors as well as determining rates of return for the enterprises in these two sectors and monitoring the wholesale electricity market. A review of the new market resulted in responsibility for oversight of gas trading being incorporated into the national energy market. The Australian Energy Market Operator was formed in 2009 to take over the role of NEMMCO. In a novel organisational arrangement, private energy operators, which were being levied to support the cost of market management, were given representation on the board of the Australian Energy Market Operator. The federal and State governments retained control of the AEMO holding 60 per cent of voting rights.

The energy markets' governance arrangements were defined by rules set out in the *National Electricity Law*, and the rules were subject to continuous review, additions, modifications and removal. The flexibility in the rules was designed to accommodate the development of the industry and the various challenges that became evident and calls for rule changes could be suggested by any party engaged in the industry. This meant that the energy market and its governance institutions are best regarded as 'works-in-progress'. Initially, the rules and regulations were guided by a broad range of objectives. This was evident in the guidelines that informed the role of the National Grid Management Council. The Council's mandate required it to 'encourage open access and free trade in bulk electricity and coordinate the efficient, economical and environmentally sound development of the electricity supply in eastern and southern Australia' (Rann 1988). However, a distinguishing feature of the governance arrangements as they have evolved is that, notwithstanding the Commission's responsiveness to new challenges, there has been increasing emphasis on economic governance and the deepening of competition as the paramount market objective.

The Instrumentalist Emphasis of Energy Market Governance

As the governance arrangements have developed, so governance has become more instrumental in its orientation. The National Electricity Objective, which was the original instrument guiding the responsibilities of the national energy market governance agencies, formed the basis of *National Electricity Law* that was legislated in 2004. The Law, which was first legislated in South Australia with other States following suit in 2004, incorporated the Objectives which articulated in quite precise and limiting terms the market governance arrangements. These were specifically defined as economic objectives: 'to promote efficient investment in, and efficient operation and use of, electricity services for the long-term interests of consumers of electricity with respect to: price, quality, safety and reliability and security of supply of electricity'. Absent from this fine-grained specification of energy market governance was any reference to the impact of electricity generation on the environment, let alone any mention of emissions. This erasure of emissions from the purview of energy governance was an extraordinary omission given that the international community was still wrestling with ways in which to foster emissions mitigation and doing so contemporaneously as was the case with the European Union which was on the cusp of introducing an emissions trading scheme. Of equal significance was how this affected the priorities and the attitudes of the governance agencies. For instance, the Australian Energy Market Commission declared in its 2011 *Strategic Priorities for Energy Market Development*:

> While social and environmental objectives have a direct bearing on Australia's energy market, these issues are in the jurisdiction of policy makers. The scope of the AEMC's work is limited to pursuing economic efficiency in electricity and gas markets for a given policy framework … it is also important to recognise that the AEMC is not directly responsible for addressing climate change or developing climate change policies but does have a role in

> advising governments on the impact of such climate change policies on the NEM and gas markets. (AEMC 2011: 12; 2019a)

This did not preclude consideration of climate change concerns, but the focus was asymmetrical; the agencies' interest in climate change concerns would be aroused only when mitigation or adaptation was impacting on the delivery of energy services. This translated into a reluctance to consider the possibility that emissions or climate change might have a direct and real material bearing on economic performance. In its submission to the Finkel Review, the 2017 *Independent Review into the Future Security of the National Electricity Market*, the Australian Energy Regulator queried: 'whether we should include an environmental objective in the National Electricity Objective (NEO)' (AER 2017: 8). There was some pressure from other stakeholders for environmental factors to be incorporated in the Electricity Rules, but these were opposed by the AEMC which had a quite narrow interpretation of its responsibilities. The AER rejected the inclusion of an environmental objective because: 'The problem we face is that there is currently not a national commitment within and across governments to energy and emissions reduction policy integration'. The observation is quite specious because the argument could have just as equally have been made about economic objectives, and it was anticipated by an AER board member, Ed Willett, that planned reviews of the industry would address this omission although the AER remained wedded to the *laissez faire* market script opining 'It is better to let the market 'decide' (Willett 2008; AER 2017: 11).

The AER was emphatic in declaring the paramount objective was economic:

> The NEM is founded on a principle that reliance on competitive markets, where feasible, will deliver the best outcomes for consumers in terms of price and innovation. (AER 2017: 5)

In fact, the AER regarded any attempt to estimate the value of environmental benefits, or conversely and more importantly externalities, as misplaced because these 'may only be quantified if there is an identifiable tax, levy or other payment associated with environmental or health costs which producers are required to pay' (AER 2021). This was a position shared by the agencies that made up the NEM governance architecture. In the summaries of submissions to the Finkel Review, it was noted that the NEM governance agencies, the Australian Energy Regulator, the Australian Energy Market Operator and the Australian Energy Market Commission, were at one in arguing 'against the inclusion of any reference to environmental considerations in the NEO [National Electricity Objective]' (Finkel et al. 2017: 176).[1] AEMO's position was to be expected given that it represented the private energy operators that could be expected to be resistant to the idea of a reckoning being made of industry emissions. In fact, this was no whimsical act. As AEMO (2019–2020) advised, this position was prescribed in the *National Electricity Law* and the National Electricity Rules (AEMC 2020a) that defined the regulatory framework and the objectives of energy market liberalisation. However, the National Electricity Objective does not actually proscribe reference to or consideration of emissions and climate change (AEMC 2019b). The AEMC, the AER and the AEMO have quite simply elected to not engage environmental concerns.

The reason for elaborating in such detail on the omission of emissions from the fine-grained specification of energy market governance is that benefit-cost analysis will be of limited value when a reckoning of the cost of environmental externalities associated with generating electricity is not made. Such oversight leaves out a critical element from the *laissez faire* construct, namely, the economic costs associated with emissions. In conventional economic theory, some consideration of such costs has to be made in order for producers and consumers to have a complete understanding of the benefits and costs that face them to optimise their economic wellbeing. To do otherwise makes for second best solutions. As we have noted, there is a well-worn path within economic theory that has sought to come to terms with this market failure, the failure of the market

[1] The AEMC (2019b: 8–9) does recognise that it cannot simply ignore the impact of climate change: 'The Commission makes its decisions on rule changes with reference to the national energy objectives. These objectives do not specifically require the Commission to have regard to the long-term interests of consumers with respect to climate change or the environment. … However, to make decisions that meet the national energy objectives, the Commission considers whether its decisions are robust to any impacts on price, quality, safety, reliability and security of supply of energy or energy services, if these matters are impacted by mitigation or adaptation risk that manifests due to the issue of climate change.' See, also AEMC (2011: 12). The AER has acknowledged that there is a need for closer coordination of energy and emissions policy (AER 2017). The Australian Energy Council rejects the idea of incorporating an environmental objective into the National Electricity Objective because it contends that this would compromise the efficiency goal (Donoghue 2016). AGL was more supportive (Australia Senate Standing Committee 2016).

to put a price on externalities. The standard-bearer for dealing with market failure is Coase's 1960 'Social Cost' essay which argued the case for establishing property rights over an amenity that was affected by an economic enterprise and which was resulting in costs that impacted on other economic agents that were not accounted for in the market. This provided the logic behind emissions trading schemes: the level of emissions could be capped and $CO_{2\text{-}e}$ generating enterprises would have to purchase the right to emit or, at least, cut back on their polluting activities to reduce their carbon footprint to engender an optimal outcome.

Proposals to introduce emissions trading schemes have faced determined and persistent opposition. This is reflected in the AER's position on the issue of market failure: should there be some measure of the hidden costs associated with emissions produced as the corollary of coal combustion in generating electricity? The AER acknowledged market failure but did not regard this particular form of market failure as a systemic concern. In the AER assessment there were more pressing issues that warranted addressing through market interventions than emissions. 'We consider that some of the issues we face in the market at the moment are not a result of the failure of the NEM but have been caused by a retreat from this principle of relying on competition and markets' (AER 2017: 11). This clearly compromised the integrity of the liberalised energy market model that is being produced supposedly as a mirror image of the *laissez faire* model. This has critically significant consequences for evaluating the relative merits of coal as an energy source and as the preeminent producer of emissions. Coal's ostensible competitive (cost) advantage has to be reframed in this context. As the Productivity Commission noted: 'climate change policy in Australia has been developed largely independently of energy policy' and this has been the success story of the coal-industrial complex narrative (2017: 16).

Insofar as the market design outcome has been based on a partial and selective application of the *laissez faire* model, the omission of emissions from the economic calculus raises the question as to how much the advocacy of energy market liberalisation has played to the advantage of coal as an energy source. The privileging of the market—'It is better to allow the market to "decide"' as the AER claims (AER 2017: 11; AEMC 2019b: 13)—amounts to an appeal to depoliticise the decision-making process. This, of course, has been one of the objectives of the carbon club, to expunge debate on the reliance and future of coal. It also highlights one dimension of the fictional account of the energy market liberalisation.

The Myopic Conviction in the Free Market and the Perfect Competition Fiction

The AER's passing remarks on market failure demonstrate a conviction in the energy market liberalisation project based on the presumption that dismantling the system of state-owned power utilities has by and large led to the emergence of a market characterised by a level of unparalleled competition. Notwithstanding the fact that those stages of the electricity supply chain involving the generation of electricity, the transmission of power via the grid and the 'poles and wires' distribution networks supplying power to households are natural monopolies. The AEMC and the AER have confidence in the energy laws and regulations to provide sufficient monitoring oversight and governance powers to enable any evidence of rent-seeking to be reined in to establish a quasi-perfectly competitive market. The establishment of the NEM was regarded as providing more flexibility and enhanced supply options to diminish the dominant position that these enterprises experienced in the State-defined markets. The competitive spirit was also envisaged driving technological developments that would broaden the choice of supply options that would help erode profiteering in the market (AER 2017: 11).

Much of this confidence in the operation of the market rests on the belief that, once unleashed, competition would become all pervasive and self-reinforcing. Given that the liberalisation project was establishing an entirely new market, there was a presumption that competition would be automatically inscribed in market relations. However, apart from several inquiries into the reasons for the sustained inflation of electricity prices, there has been very little critical appraisal of the structure and organisation of energy business in the NEM. The one considered reflection has sought to explain the lack of competition in terms of the limited competition in the retail sector, an issue that is taken up in the following chapters. This chapter explores further the extent to which the promise of the liberalisation project has succeeded in establishing competition as a ubiquitous feature of the NEM. The focus will concentrate on five relatively distinct issues: state underwriting of the coal-industrial complex; energy companies reinstating vertical integration to underpin competitive advantage, and the ACCC sanctioning this anti-competitive development; the contradictory effects of market deepening; the influence of international capital in defining the industry's development trajectory and the circumscribed nature of the NEM and how this is impacting on the development trajectory.

Underwriting the Coal-Industrial Complex and Competitive Advantage

The energy market had become organised around energy corporations prosecuting their respective interests. One example of this that gave the coal-fired generators a contrived competitive advantage has been in the form of state financial support (Campbell 2015; Market Forces 2021). Through a range of subsidies and development approvals, Federal and State governments have contributed substantially to the development of an industry that has been among the most profitable in the nation. While debate raged on the link between fossil fuels and climate change, according to one conservative estimate, Australian governments were subsidising research and development on fossil fuels to the tune of between $2.9 and $3.5 billion a year (Bast et al. 2014). A 2020 *IMF Working Paper* estimated that Australia provided fossil fuel subsidies of as much as US $29 billion a year (Coady et al. 2019; UNEP et al. 2019; Mazengarb 2019a). Depreciation and research and development write-off concessions added to the magnitude of support (McDowall 2021). As much as 60 per cent of royalties from coal mining in Queensland was being ploughed back into the sector in the early 2010s (Peel et al. 2014; Reidy 2003; Reidy and Diesendorf 2013). The Australian Taxation Office identified coal mining as the single most significant beneficiary of the fossil fuel tax credits (ATO 2017). Such support contributed to the 'competitive' advantage of coal as an input in energy generation.

The Gillard Labor government's *Clean Energy Policy*, which introduced a carbon price that was designed to incentivise the transition from reliance on coal-fired power, provided a raft of 'compensation' packages to offset the carbon impost. This was despite the architect of the *Carbon Pollution Reduction Scheme* advising that compensation was not necessary. Prior to the commencement of the *Clean Energy Policy*, the government had provided subsidies to the value of some $1 to $1.8 billion to boost energy production (Spratt 2013; Vorrath 2015). When the government introduced the *Clean Energy Policy* both the coal mining industry and coal-fired generators were provided with substantial compensation packages, with promises to deliver compensation of between $2.3 and $5.4 billion to the generators depending on the carbon price, and the generators had already made substantial windfall profits by including the carbon impost in wholesale electricity prices before the impost commenced. The generators also sought to lock in compensation for closure of power plants despite the fact that, in undertaking due diligence when negotiating the purchase of privatised power plants, they would be proceeding on the basis that there would be no right to compensation for shuttering the plant.[2]

For as long as they have lasted, the scale of subsidies underwriting the extraction of coal contributed to coal-fired power's competitive advantage over renewable energy fuel sector. This was not necessarily going to last because just as investment in renewable energy generating capacity was taking off there was increasing international pressure being brought to bear to 'rationalise and phase out inefficient fossil fuel subsidies' (Vorrath 2015; Treasury 2019). Certainly, the magnitude of subsidies, and particu larly direct subsidies to encourage production, has been in decline from around 2010 although the rate of decline has been diminishing (IEA and OECD 2019). The support has, however, continued. This has been evident in a number of instances. It takes the form of prospective royalty relief, in effect, an off-balance sheet subsidy. This is the case with the Queensland government's endorsement of the proposed Galilee Basin coal mine project being developed by Adani (Zillman and Horn 2020). The support is based on the first-mover principal that supporting the development of the Adani venture would open the Basin for more extensive mining (Slattery 2019). It takes the form of the changes proposed to planning laws in New South Wales that would preclude consideration of 'Scope 3 emissions' in development applications, thereby limiting a fuller appreciation of market failure and the costs of emissions (Cox 2019; Mazengarb 2020c). Tax minimisation arrangements have proved to be another aspect of the behind-the-scenes state support for development provided by the state as well the ability of the energy corporations' ability to influence the design of the fiscal system. More than any other industry, the resources and energy sector has benefited from the broad range of tax concessions that aim to encourage exploration and development and foster energy-intensive accumulation.[3] Taxation rules have worked to the

[2] The proposal to provide compensation for closing plants, under a planned 'Contract for Closure' scheme to accelerate emissions mitigation was dropped from the Clean Energy Bill.

[3] The resources sector had launched a well-funded campaign to successfully block the Gillard Labor government project to introduce the Resource Super Profits tax, designed to

financial advantage of fossil fuel companies as well as a number of energy companies (Bagshaw 2018). Several leading energy companies have reported robust returns at the same time as they have been able to minimise tax. Some have been able to avoid paying any taxes at all (Market Forces 2021; Turnbull 2020; West 2015, 2021; Aston 2015; Greenpeace 2015). Some of the biggest emitters have been permitted to increase their emissions under the terms of the Emissions Reduction Funds Safeguard Mechanism and not penalised for failing to meet scheduled baseload emission reduction targets (Australian Conservation Foundation 2020; Morton 2020; Jotzo 2020).[4]

The coal-industrial complex has sought to deflect criticism of state support for coal by arguing that the call for support for renewable energy is more substantial than for coal. The contention that renewables have a competitive advantage over coal because of subsidies is one that the Minister for Energy and Emissions Reduction has been prosecuting in defence of coal having a place in a technologically neutral emissions mitigation policy. It is an argument that has been forcefully criticised as being a misrepresentation based on distorting the nature of support (Parkinson 2018). And the notion that there has been a significant claw back of subsidies needs to be considered in the context of the dramatic increases in subsidies underwriting the exploration and extraction of gas. The ACCC has also recommended that subsidies provided under the MRET should be abolished in 2021 (ACCC 2018a).

The Valorisation of Market Power

The several inquiries into what factors had been driving the inflation of electricity prices all identified the checks on competition as the key. Four factors have been noted, each canvassing distinct correcting remedies. Firstly, the advocates of energy market liberalisation recognised that a potential obstacle to the competitive impulse driving down power prices was the absence of substantive competition in the generation, transmission

capture a share of the high profits being declared by iron ore and coal mining companies, which was dropped following immense lobbying and replaced by the Minerals Resources Rent Tax that would be levied on profits generated from the mining of non-renewable resources.

[4] The Coalitions to reward Santos by paying the gas company using carbon capture and storage technologies to bury its emissions makes an even greater farce of the Safeguard Mechanism (Thornhill 2021).

and distribution of electricity because of these sectors' characteristically natural monopoly form. It was presumed that the market power derived from the monopoly position could be contained by expanding the reach of the market, increasing competition across State borders with the formation of the NEM. In practice, this has not eventuated. Secondly, the more direct measure to counter the market power associated with natural monopolies was for the state to do what the market could not, and this was to set the retail prices that consumers could be charged. In the context of the NEM governance architecture, the authority vested in the AER empowered it to determine reasonable rates of return on an energy enterprise's investment and to adopt this as a benchmark for setting 'efficient' prices in the wholesale market. Incidentally, the AER was not the only authority charged with setting prices because the governance remnants of state-owned utilities meant that some State authorities set prices, and this was largely framed in terms of ensuring minimum rates of return to foster investment.

Thirdly, the AER's price-setting role was not without its challenges because it had to reconcile the effect that any recommendation on rates of return would translate into prices that consumers would be charged, and this was a pressure that limited price increments and minimised the scope for corporate rent-seeking. Conversely, in an industry that was under pressure to increase investment to augment generating and distribution capacities and cement supply reliability, more generous rates of return were necessary to induce investment. The state and age of the generator facilities and transmission and distribution network infrastructure had to be taken into account as did the ambition to foster investment to upgrade the system. The record indicates that, on balance, energy corporations have been rewarded with high rates of return and expanded capacity beyond what was believed to be necessary. The frequently expressed criticism of the AER was that it sanctioned the 'gold plating' of corporate energy's profits (Dimasi 2015; King 2015; Lasker 2017; Ludlow 2019). This remains an ongoing concern leading to calls to better resource the AER and enhance its governance authority (Macdonald-Smith 2018).

The Electricity Rules and funding constraints limited the options available to the AER. While too much state intervention in the energy market is eschewed by the energy market liberalisation stalwarts, the AER has defended its determinations arguing that there was a need for increased investment and the Electricity Rules severely limited the options available for incentivising investment (AER 2012). Some argue that more resourcing would place the AER in a better position to be able to make more considered judgements on rates of return required to attract investment. Others contend that enhancing the role of the AER would only serve to worsen the profiteering problem, while an alternative approach has emphasised the potential of utilising the power of the market to act as a check on the magnitude of returns. Empowering consumers to give voice to their preferences, responding to different pricing signals, is advocated as one possibility (Dimasi 2015; AER 2012).

Fourthly, price inflation has also been attributed to the ability of some of the larger energy corporations to exercise market power. Much to the chagrin of the Business Council of Australia, vertical integration of enterprise has become a pronounced feature of the National Energy Market, a structural arrangement that is argued to compromise the intensity of competition (ACCC 2018b: 123). Indeed, there has been significant vertical integration of energy corporations in Victoria, South Australia and New South Wales. AGL purchased a 35 per cent stake in the Victorian Power Station Loy Yang A in April 2004 and in 2007 acquired Torrens Island, South Australia's largest power station to dominate the industry. CLP, Hong Kong's Light and Power Company and operating as EnergyAustralia, has expanded through acquisition, purchasing TRUEnergy's retail and generation assets in mid-2005. Origin Energy when formed as a result of the Boral demerger had investments in generation and retail, including interests in gas, and has expanded its interests in both spheres of the energy supply chain through a series of takeovers commencing in 2007, including the Eraring power station in New South Wales.

A second tier of vertically integrated enterprises now features in the NEM. Red Energy/Lumo Energy (Snowy Hydro) is integrated with substantial portfolios of generation assets. ERM Power-ERM Power Retail (and recently acquired by Shell), Engie-Simply Energy (Engie) and Alinta also own major generation assets. With AGL, Origin and EnergyAustralia, these vertically integrated businesses account for nearly 70 per cent of residential customers across the NEM and some 85 per cent in New South Wales (ACCC 2018a: 135). In NSW, Victoria and South Australia, they account for around 90 per cent of generation capacity. A third layer has been built around a smaller scale and involves Hydro Tasmania (Momentum), Meridian Energy (Powershop) and Pacific Hydro (Tango).

The high degree of ownership concentration across each stage of the electricity supply chain creates a noticeable competitive advantage, especially for AGL, EnergyAustralia and Origin Energy. These three benefit from a range of efficiencies. Having control of the supply chain enables them to reduce transaction costs and build reliability in supply. The ability to adjust prices across the different stages of the chain can result in a competitive price advantage in the retail end of the business and deliver taxation benefits (ACCC 2018a: Ch. 5). They have also not been averse to engaging in 'opportunistic bidding' in the wholesale market taking advantage of 'dispatch and settlement prices [in the wholesale market which] are determined over different time frames' (ACCC 2018a: 107).

The differences in the bargaining power that the diverse cohort of enterprises bring to the market are discernible in the structural asymmetries that distinguish the NEM. The supplanting of state utilities by private enterprise has not engendered a free market in which all participants have comparable bargaining power. Yet the stalwarts of the NEM continued to express their conviction in the force of the perfectly competitive market despite the NEM bearing little resemblance to the idealist *laissez faire* free market construct in which all market participants compete on equal terms. The market is dominated by the large energy corporations, and the logical corollary would be for some form of third-party intervention to contain the effects of the inequities that can or do compromise efficient outcomes, but those best placed to do so, the very institutions that have been established to govern the energy market, the AEMC, the AER and the ESB, resist intervening. The AER observed that while 'government intervention can help manage an identified market issue, its wider market impacts are complex' because intervention can distort market signals, affecting private sector investment decisions (AER 2020: 63). The Energy Security Board has endorsed this position: best to leave the market structure as it is and not compromise its operation with state intervention.

This has played to the advantage of the coal-fired generators. The coal-industrial complex can explain its market advantage as the consequence of the force of competition, in fact hiding behind the rhetoric of a perfectly competitive model that remains an idealised construct.

The *laissez faire* market stalwarts have looked to market innovations as a means for overcoming some of the structural hurdles that block perfect competition in the energy market. One such innovation is the futures market. It functions as a parallel domain to the actual physical market in which the competitive impulse can be ignited. The vibrancy of this parallel market is reckoned to be greater than the energy market because it introduces another sphere of competition that can influence pricing and volumes by engaging other investors who are not directly involved in trading in energy trading through their participating in the speculative trade in energy futures (AER 2020: 102–103). Engaging in futures trading and options swaps provides a means for hedging against volatility in prices and supply and can instil confidence in trading in supply. Generators, for instance, can hedge against the possibility of future price volatility by negotiating for a quantity of energy being supplied at an agreed wholesale price at some future designated time. This insures against a potential decline in prices and shores up confidence, but, on the other hand, it also makes the generator vulnerable to the possibility that the contract is executed when future electricity prices are increasing, and the generator misses out on what would be a higher return. By the same token, retailers who potentially insulate against future price increases by locking in a futures contract to purchase electricity at a set price also risk the possibility that future prices could fall and they end up paying a higher price than would otherwise be the case. This opportunity to engage in the futures market is argued to amount to a deepening of the market, bolstering market liquidity and introducing additional competitive pressures because of the possibility to trade actual supply contracts or derivative contracts. Futures trading has become a more significant feature of trading in energy, with trading in electricity futures on the ASX up 43 per cent 2018–2019 on the previous financial year, and the trades in futures represent 243 per cent of underlying NEM demand (AER 2020: 104).

However, this market deepening can also provide another disproportionate advantage for the vertically integrated enterprises (ACCC 2018a: Ch. 5). Vertical integration provides the means to hedge within the business entity and this ability places the big three in a stronger position when negotiating hedge contracts. This can be to the double disadvantage of small retailers. Stand-alone generators and retailers do not have the capacity to hedge against price fluctuations through accounting adjustments within the enterprise. Because small retailers have to purchase power from the pool in which spot prices are being set, they are purchasing power at rates that are likely to be less competitive than the prices that the big three are charging for supplying power into their retail arms as the ACCC discovered. Small retailers are more vulnerable to price volatility and having settled on a fixed price that they supply electricity into the retail market, hedging becomes a more pressing safeguard, but as fewer substantial enterprises, they face a futures market in which their negotiating power is generally not as robust as the big three. Interestingly, demonstrating its unquestioned faith in the market's ability to quash such possibilities, the

ACCC has argued that compelling the larger energy corporations to participate in hedging would increase liquidity and open up the futures markets to expand the opportunities for the smaller enterprises to engage in trading.

There are also some potential market-distorting effects in linking the trade in electricity supply contracts with futures contracts. One of the formidable challenges unfolding in the industry arises with the determination of market prices. Historically, energy prices for coal-fired generators were based on average costs, given the operational nature of the plants, operating on a 24/7 cycle. By way of contrast, energy generated from renewable sources, such as wind and solar farms, tends to set prices equivalent to marginal cost, which is negligible when these plants are generating. This can have devastating consequences for the competitive position of the coal-fired generators. Hedging provides some scope for dealing with this anomaly, but it can also add to the distortion in pricing because hedging is essentially a process of calculating what the market price will be at some point in the future, setting a so-called discovery price. Ideally, for this to reflect the actual costs of generating power, it should gravitate around the likely marginal cost of the power, the 'efficient price', which in conventional economic discourse is associated with the 'least cost point of production'. The process of anticipating future prices is based on predicting the future technology mix and possible policy frameworks, levels of supply and demand compared to the current market situation. This can result in the process of price formation being inverted with future price expectations affecting current price formation and establishing a disconnect between the cost of generating and the reward for generating. This particular form of deepening the energy market can engender more instability and inequities in the energy market.

Too Little or Too Much State Intervention in the NEM: Whither the *Laissez Faire* Market?

A recurring theme of the inquiries into the state of the industry has been the appeal for more robust governance of the industry at the same time as criticisms claim that there is far too much intrusion into the energy market. State intrusion has frustrated industry efficiency and contributed to the inflation of prices according to the dominant second narrative (Harper et al. 2015; Vertigan et al. 2015; ACCC 2018a). Better governance is held up as the means to underwrite the force of the competitive impulse at the same time as a seemingly contradictory argument is advanced that regards too much intervention as obstructing competition.[5]

What is missing from these contradictory reflections on the role of state is the need to consider the structure of the industry and the uneven power relations that give the large corporations the ability to capture a disproportionate share of the revenue being generated. The energy governance agencies are very much wedded to the conception of the energy market that is founded on competition and the hegemony of private enterprise and this informs the logic in the appeal for minimising the role of the state in the functioning of the market. The rationale for the minimalist state is critical to framing the energy governance agencies' policy actions in reproducing or modifying the patterns of accumulation. This is clear in the ACCC being convinced of the notion that state intervention crowds out the opportunities for private enterprise. This informed the ACCC's taking issue with and chastising the New South Wales government for belatedly privatising the NSW grid and its recommendation that the Queensland state utility should be dismantled. Somewhat ironically, it does not rail against the suggestion that the vertically integrated energy corporations should be broken up. Competition between the big three was regarded as 'vigorous' (ACCC 2018a: 136). A decade earlier, the head of the AER had expressed a similar level of confidence in the competitive force of the market overriding the oligopolistic market, a position that is clearly at odds with the conventional economic wisdom (Willett 2008). As Katharine Murphy neatly summed the apparent contradiction: 'The ACCC does not believe it would be appropriate to intervene to unwind the way in which the market has evolved across the national electricity market' (Murphy 2019).[6] As if this evolution was regarded to be an organic process, one effected through the 'invisible hand', and not the product of the pressure brought to bear by the coal-industrial complex.

The Commission's ideological disposition to endorsing the sanctity of the *laissez faire* market sits uncomfortably with the plethora of market failures and recommendations that call upon the state to shepherd the free market into existence, to do what it has failed to do. But this tells only part

[5] Reflecting the increasing scale of governance, the AER noted that its staff numbers had doubled since 2015 (AER 2019), while the increasing role of coordination had resulted in AEMO making a larger claim on resources (Macdonald-Smith 2020a).

[6] Enova's chair, Alison Crook, has alleged instances of anti-competitive behaviour with evidence of generators failing to switch on gas peaking plants to meet peak demand, driving up and profiting from wholesale price spikes (O'Malley 2021).

of the story. It overlooks the compromised formation of the market, given the omnipotence of the vertically integrated energy enterprises and just how much the reality of the market falls short of the ideal construct of the perfectly competitive market. The ACCC and the energy market governance agencies charged with the responsibility for promoting the integrity of the energy market have become trapped by their own ideological zeal, reluctant to recommend intervention for fear this contradicts their mandate but having to do so in order to rectify the instances of systemic market failure.

In keeping with the dominant conventional wisdom, the Minister for Industry reasserted in the *Energy White Paper 2015* the principle on which the energy industry was founded:

> Our guiding principle is that markets should be left to operate freely, without unnecessary government intervention. Competition, productivity and investment will deliver reliable and cost competitive energy to households and business.

In the context of the determination of the Minister for Energy and Emissions Reduction to drive investment to maintain the stock of coal-fired generators and his threats to divest energy corporations of their assets if they failed to abide by this government directive, it had become obvious that the conviction in the *laissez faire* market was little more than hyperbole. The rhetoric of the free market as the preferred, the 'efficient', mode of organising the market could be sustained for only as long it underwrote the fossil fuel-based regime of capital accumulation. A less sanguine tone was also being uttered by the then-Treasurer and subsequently Prime Minister, Scott Morrison. Morrison responded to questions about how he could reconcile the threat to divest AGL conflicted with the Coalition's idealist 'free market government':[7]

> Well, hang on a tick, though. You are assuming that AGL—and any of our energy companies—are operating in some sort of free-market Nirvana. They don't.

[7] The Coalition government had threatened AGL with divestiture if it proceeded with its plan to shutter the ageing Liddell power station in New South Wales.

> They operate in a highly regulated market. And particularly when you get into transmissions companies: they have guaranteed, legislated rates of return. (Sales 2019; Coorey 2018)[8]

Elaborating on the question of market power in the energy market, Morrison conceded that state intervention was necessary with 'the law … targeted at situations where sweetheart deals came at the expense of consumers' (Grattan 2018). The standout to be drawn from his observation could not have been blunter. It acknowledges that the competitive impulse that is supposed to define the NEM is not as pervasive and intrinsic as is claimed, notwithstanding the fact that the government simultaneously holds onto the fiction of the free market as the foundation of the energy economy.

The National Energy Market as Fiction

In a withering report into governance and regulation arrangements in the NEM, the Productivity Commission noted that: 'The "National" in the NEM is progressing too slowly' (Productivity Commission 2013: I: 7). In fact, there is good cause for questioning the veracity of the NEM as a concrete construct rather than an idea or work in progress and we can attribute this disconnect to the construction of the NEM being a discrete phenomenon whose advocates have sought to restrict its oeuvre to a narrow conception of energy that eschews any consideration of emissions and commends a coal-fired accumulation regime.

In the first instance, the extent to which a cohered energy market incorporating the States of Queensland, New South Wales, Victoria, South Australia and Tasmania and the Australian Capital Territory has been established is greatly exaggerated. Describing this organisational

[8] Speaking in his capacity as Treasurer at a Bloomberg event the previous year, he had made a similar observation: 'over time' large businesses (such as banks) which depend on a regulatory environment can 'get very good at seeing those regulations worked to their advantage'. They are highly regulated sectors for very good reason, and, over time, you need to recalibrate that regulation to ensure the customer is at the centre of the universe:

> The actions I've been taking … whether it's in the energy sector or whether it's [the bank levy] in the banking sector are really about trying to restore greater competitive pressures in these markets by increasing the power of the customer. (Dean 2017)

arrangement as a National Energy Market is misleading. In the first instance, the economics of the geographical limits the reach of the transmission network. Western Australia and the Northern Territory are not participants in the National Energy Market. The notion of the NEM being organised around the pooling of generating capacity in the five States and the ACT which is drawn upon to feed demand sits uncomfortably with the reality. In practice, the pooling of capacity and the supply of power occurs in five regions that are primarily organised with respect to State markets, the ACT being part of the New South Wales' pool, with wholesale prices still mostly set on a regional basis (AER 2017: 27). This focus on State markets is reinforced by the Australian Energy Regulator with the emphasis of its regulatory oversight concentrated on the State markets as discrete spheres. AEMO oversight is framed by a similar region-based form of organisation (AEMO 2019).

This approach to the physical dynamics of the NEM as a composite of discrete State-based energy systems carries through to retail market governance. Certainly, the NEM has been formed on the basis of some common regulatory principles and practices that must be met across these energy markets, but the paramount requirement has been that competition is the driving force in energy retail markets, and this is premised on the respective State-based retail markets complying with the 'competitive neutrality' obligation. Once this was put into effect, it was taken as a given that the state should intervene in competitive market practices as little as is organisationally necessary. Relevant State pricing authorities continue to set price caps, but these must be in accord with the 'common national framework for regulating retail energy markets' (AER 2020: 236). State and Territory governments for the most part continued to set reliability standards (Productivity Commission 2013: I, 7). The federal Coalition government and the NEM governance agencies have sought to assert some authority over retail price setting by highlighting the different rates that apply. South Australian governments have been the subject of much criticism for having had the highest retail prices of all States in the NEM, with much of the blame for this being attributed to the State's overzealous commitment to renewable energy (Doran 2016; Bickers 2017). South Australia now has the lowest retail prices, the direct result of its reliance on renewable energy.

In 2019, after years of criticism that the industry had failed to halt inflation, often under the cover of deceptive and opaque marketing practices, the Federal government set a default price that would apply in most retail markets. This marked the potential for a convergence of retail prices across the NEM, or at least it did momentarily because, not to be outdone, the Victorian State government adopted a default price. Notwithstanding this assertion of federal authority over the operation of the power sector, and State moves that undermined the federal policy, one consequence of the adoption of a default price has been that it strengthened the regulatory role of the Australian Energy Regulator in exercising greater oversight of pricing across the national market.

The equally, if not more, significant reason for calling into question the integrity of the national electricity market project has been the State and Territory governments determination to formulate energy policies that mitigate emissions and support the transition to renewable energy sources. Each State and Territory government has launched renewable energy investment policies independently of each other and with respect to their place in the NEM. The policies contrast markedly with the Coalition's policy based on the paramountcy of coal-fired power, and this cuts across the notion of a national energy market. As the Energy Security Board has observed in the *Post-2025 Market Directions Design Discussion Paper*, it has had to confront the reality that each jurisdiction has become more assertive in defining its vision of the future shape of the energy economy, and the policies to build greater reliance on renewable energy are being developed with little or no reference to the development of the national energy market. Charged with the responsibility for advising on a long-term, fit-for-purpose market design for the National Electricity Market, the ESB has recognised that each State and Territory continues to exercise their sovereign authority in adopting policies that chart pathways for hastening the transition to renewable energy, and which are quite distinctive of each other's and any notion that there should be a common design (ESB 2021). In the absence of the Coalition government demonstrating any leadership on policy design, the ESB argues that it is important that the State initiatives should be linked and embedded in a common set of principles that would assist its efforts to develop a coherent NEM in a federated political system and align with the AEMO's *Integrated System Plan*. These jurisdictions have endorsed the establishment of a competitive market, have responded to popular community support for renewable energy and have embraced the need for action to reduce their carbon footprints. Most, but not all, have demonstrated a heightened determination to reduce reliance on coal-fired electricity placing emissions mitigation at the front and centre of their energy policies.

The decisions of State and Territory governments to promote the turn to renewable energy also exposed another weakness in the liberalisation project. Supplanting state involvement in generating and distributing electricity through privatisation and the sale of disaggregated energy utilities to establish the idealised competitive market did not automatically ignite forces that would impel the transition from coal to renewable sources of energy. There was no automaticity in the 'invisible hand' being aroused from its slumber to recast the energy market. The State interventions in the energy markets have contrasted with the inability of successive Coalition governments to resolve a comprehensive energy-emissions mitigation policy. Two initiatives distinguished how the States and Territory governments incorporated climate policies as core features of their energy policies. Defying federal government inaction on climate change policy, State and Territory governments set even more ambitious renewable energy targets and introduced feed-in tariffs that radically altered the pace of change towards building a more carbon-benign national energy industry.

Riding on the wave of popular support for action being taken to mitigate the threat of climate change, the ACT became the first State or territory in Australia to announce a renewable energy target. In 1997, the ACT government proposed a target for stabilising greenhouse gas emissions attributable to the ACT at 1990 levels by the year 2008 and reducing these emissions by a further 20 per cent by 2018 (Logan 2002). In 2000, the ACT launched its greenhouse strategy which outlined various measures that were being pursued to achieve its 2008 target. The New South Wales government had committed to reducing greenhouse gas emissions to 2000 levels by 2025, and by 60 per cent by 2050. In 2005, the government introduced its mandatory NSW Renewable Energy Target (NRET), relating to all electricity consumed in NSW. The government announced that it would provide ongoing support to the renewable energy industry, imposing a target on electricity retailers that would include renewable energy certificate trading and an enforceable penalty for non-compliance in those instances where retailers failed to meet their targets (New South Wales, Department of Energy, Utilities and Sustainability 2005). Trade-exposed, energy-intensive users were to be exempted from bearing the costs of the scheme. The Victorian government responded to the Tambling Review becoming the first State in 2006 to mandate the generation of 10 per cent of electricity from renewable sources by 2010. In July 2007, South Australia's Labor government passed the *Climate*

Change and Green House Reduction Act. This set a State renewable generation target of 20 per cent by 2014. This target was soon exceeded and the target subsequently revised upwards. In 2000, with the issue of the *Queensland Energy Policy: A Cleaner Energy Strategy* the Queensland government endorsed the Commonwealth government's Renewable Energy Target, indicating that it would not exceed that target, although it subsequently adjusted its target for retailers to commit to providing up to 15 per cent of energy from renewable sources by 2011 and 18 per cent by 2020. New South Wales considered a similar scheme at the same time as the government was establishing well-defined objectives to frame energy policy (New South Wales 2005).

Feed-in tariffs have proved more contentious as a possible incentive for households to contribute to emissions mitigation. The Greens had sought to get the federal government to introduce a scheme that provided uniform feed-in tariffs across the nation, but this was rejected. The federal government was opposed to introducing feed-in tariffs to reward households installing rooftop solar panels and copying a scheme that had proved successful in Germany. Indeed, federal Labor's Minister for Energy had made clear his aversion to feed-in tariffs. In his opening address to the Clean Energy Council's *Solar Flagship Forum* held in Brisbane in September 2009, he took issue with the: 'technologically-prescriptive and ideologically-based feed-in tariff. … Feed in Tariffs are not the solution some make them out to be' (Ferguson 2009).

This did not discourage the South Australian State government from following the German feed-in tariff model. South Australia became the first of the States to announce in 2006 that it was planning to introduce a feed-in tariff to encourage the uptake of household solar photovoltaic rooftop panels as a key element in the objective of becoming carbon neutral by 2020. The legislation was passed in 2008, and other States and the Territory soon followed with their feed-in tariff legislation. Victoria introduced a feed-in tariff in 2009, adjusting it in subsequent years. By 2009, most had commenced paying feed-in tariffs. To encourage the take-up of solar panels most governments offered feed-in tariff rates that exceeded the prevailing commercial retail supply rates.

So successful were these schemes, within a comparatively short time frame, fiscal pressures prompted State and Territory governments to reduce the rate of tariffs for new applicants and for the most part they have since ended their programmes. Energy retailers also began offering households feed-in tariffs, although their rates have steadily declined over time, with several no longer contracting feed-in tariffs with households. More recently, in the endeavour to avoid energy retailers abandoning feed-in tariffs altogether, some States have permitted retailers to introduce time-variable feed-in tariffs, allowing rates to be reduced during those periods of the day when panels were most productive. The establishment of virtual power stations by several of the larger energy companies has offset this retreat, and other enterprises have launched local aggregation initiatives. Households are being enlisted to install rooftop solar panels, connect to the grid, transmit power and be rewarded with feed-in tariffs to play an active role in transforming the energy mix.

Conflicting Trajectories of Energy Market Formation

The National Energy Market is now characterised by the increasing misalignment of energy policies as the States and ACT have embarked on energy industry accumulation models that are based on investment in renewable energy. The policies place emissions reduction at the core of energy policy and, paradoxically, this is radically transforming the energy landscape. Private investment in renewable energy technologies, underwritten by State and Territory support, belies the Coalition government's determination to maintain fossil fuels as the premier force in electricity generation. The NEM reflects a fractured policy horizon in which the form of the energy system is being pulled in different directions. In contrast with the Coalition government that is holding onto the old industrial-scale fossil fuel-powered economy, State governments are constructing more flexible, decentralised renewable energy systems. This has forced the hand of the Australian Energy Market Operator as it has become progressively more concerned with the lack of correspondence between federal energy policy and State and Territory energy policies that incorporated emissions mitigation. According to the NEM agencies and the Energy Security Board, the absence of a national policy to deal with the mismatch of new generating capacity and transmission infrastructure was contributing to the growing inefficiencies across the market and the high price Australian households and businesses were having to pay for their power. The Australian Energy Council expressed its concern that the State initiatives were compromising the integrity of the NEM (Costin 2020).

The practical implications of the State initiatives have been the high uptake in renewable energy generating capacity that could not be readily accommodated by the established transmission and distribution grid which had been constructed to dovetail with the industrial-scale coal-fired power generators and the one-way flow of energy. This disconnect has been causing increasing concern within AEMO, with the CEO, Audrey Zibelman, calling for greater coordination of investment in renewable energy technologies to avoid instability across the NEM. AEMO's *Integrated Systems Plan* is designed to accommodate the ever-growing significance of distributed resource energy, including the increased generation of renewable energy from regional locations (AEMO 2020; Potter and Macdonald-Smith 2018). In an angry response to the Coalition party room's rejection of the *National Energy Guarantee*, the chair of the ESB, Kerry Schott (2020), characterised 'the general state of affairs at the moment as being one of anarchy' (Potter and Macdonald-Smith 2018; Latimer and Hannam 2018). Meanwhile, the lack of synergy in policy objectives across political jurisdictions was being amplified as State governments ratcheted up their support for the installation of household rooftop solar panels, community renewable energy programmes and encouraged the development of commercial wind and solar farms.

The adoption of emissions reduction targets by each of the States and Territory and the Coalition government's refusal to follow their lead has effectively brought an end to what had been a unique experiment in political jurisdictions demonstrating unanimity of purpose in establishing the National Energy Market (Parkinson 2019b, 2019c; Mears 2021). Victoria announced its intention to formally break away from the rules governing the NEM and passed legislation that modified the *National Electricity Rules* removing restrictions on its plans to proceed with the construction of large-scale wind and solar farms, and they are doing so with the backing of the AEMO (Maisch 2021; Towell 2020). The New South Wales government launched its energy infrastructure plan in 2020, initially focusing on the development of three renewable energy zones, which roughly correspond with zones determined in conjunction with AEMO, but it has since incorporated additional zones into its future energy programme to help meet its renewable energy target and curry political favour in a couple of electorates (Letts 2019; Parkinson 2019c; Fowler 2020; Mazengarb

2020a; Warren 2020; Macdonald-Smith 2021a).[9] Queensland established the state-owned CleanCo, a state generator and retail utility, in 2019 to spearhead meeting its 2030 50 per cent renewables energy target (Ludlow 2019a), and this created some pressure for Queensland to abandon the NEM grid (Vorrath 2019). South Australia had pressed ahead with fostering renewable energy generation and investing in battery storage capacity in several innovative initiatives and effectively abandoning any connection with national benchmarks. The ACT has achieved its goal of 100 per cent electricity from renewable sources by negotiating several Power Purchasing Agreements with wind and solar farms in New South Wales and South Australia (Mazengarb 2020b). Even Tasmania's Premier announced that the State was contemplating quitting the NEM (Vorrath 2018).

The distinctive feature of the State and Territory policy initiatives is that they are charting a quite different energy future to that held onto by the Coalition government. Each has formulated an energy development strategy based on reducing emissions that they consider will engender a new era of sustained economic growth which will be distinguished by the greater diversity in the sourcing of renewable energy. The high uptake of rooftop solar panels by households is another manifestation of the firming of an alternative development path. In this instance the phenomenon of the household assuming more responsibility for meeting their energy needs is represented as an empowering process that is making the coal-fired generation model look even more redundant. These developments are transforming the locus of the energy market dynamic away from one based on the NEM, industrial-scale generation and the embedding of coal to State-based, flexible and household-focused renewable energy dynamics.

There is a still further dimension to the shifting locus of the energy dynamic with the internationalisation of the energy industry. The privatisation of state owned utilities resulted in international capital assuming an unprecedented role in the industry's development. Indeed, attracting such investment was regarded as one of the potential benefits of privatisation.[10] With one exception, there has been negligible interference by the federal

[9] In its determination to include gas in the energy mix, the Coalition government was courting support to lift bans on exploration and extraction. It agreed to provide $2 billion to NSW for agreeing to ensure that gas would be included in the energy transition mix.

[10] The courting of international investors in the energy and resource sectors continued through the 2010s. See, for instance, Australian Trade Commission (2015) *Investment opportunities in Australian Resources and Energy.*

government in limiting foreign investment in energy.[11] The early investments in the industry, such as the purchase of Victoria's brown-coal fired power plants, were proof that privatisation would attract foreign investment and reinforce the pivotal role of coal in the energy system. The first tranche of investments seemed to be more motivated by the prospect of capitalising on acquiring an under-valued asset than by any vision of assuming a more active role in reshaping the energy market. The entities investing in the acquisition of established assets were not necessarily interested in the long term. Engie's acquisition of the Hazelwood power station is a case in point. The French utility invested very little in upgrading the plant, letting it run down and hoping to be compensated when it closed in 2017. Engie sold its Loy Yang B power station in 2018. Some investments involved international-national consortia that were interested in profiting from short-term asset inflation, buying into the system and then on-selling their holdings.

From around 2008 onwards, however, the focus of foreign investment in Australian energy radically shifted just as there was a dramatic increase in the magnitude of investment. This has partly been the consequence of State governments inviting global energy corporations to invest in renewable energy technologies and this has been linked to a more conscious endeavour to make their mark on the energy landscape.[12] The South Australian government's overtures to several global energy corporations, such as the installation of the Elon Musk's Tesla 100-megawatt battery, are having a substantial material effect. Some prospective investors regarded Australia as a testing ground for new technological applications, and the success of such projects proved that the possibilities could be realised: enhancing energy generation capabilities (Parkinson 2017; Taylor 2018; ARENA 2019; Clean Energy Regulator 2019; McCullough 2017). These endeavours are having a knock-on effect as international sources of capital ratchet up investment in utility-scale wind and solar farms in

[11] The Coalition government blocked bids for the 99-year lease of Ausgrid by the Chinese Government-owned State Grid Corp and Hong Kong-listed Cheung Kong Infrastructure on the grounds that they already had substantial interests in Australian energy infrastructure (ABC News 2016; Grattan 2016). The decision to block the State Grid and CKI bid cost the New South Wales taxpayers some $5 billion (Williams and Foley 2016).

[12] The low-interest world of COVID-19 has no doubt increased the appeal of investing in renewable energy ventures. By contrast, EnergyAustralia's 2013 bid for the Mt. Piper (1400 MW) and Wallerawang (1000 MW) power stations and associated infrastructure, in partnering Delta was approved (EnergyAustralia 2013).

Australia and be on the ground floor of expanding renewable energy generating capacity in what many see as one of the most attractive countries in which to invest. This has seen a snowballing of international investment in renewable energy projects, described by the Climate Council quite simply as a 'bonanza' (Taylor 2018).

The relatively less significant role of Australian-domiciled capital in renewable energy is partly the consequence of the cultural milieu that favoured the fossil fuel energy system. It can also be attributed to the lack of ready access to global capital markets that could be drawn upon to finance investment. Australian banks have historically backed the resources sector, including coal, and it is only very recently that new ways of financing investment in renewables have been established. This is the case with corporate capital, which has recently entertained long-term electricity supply contracts at agreed price contracts through the negotiation of Purchasing Power Agreements (Clean Energy Council 2021).

With a relatively stable political system, the identification of new development zones, the potential for funding support from ARENA and the CEFC and the acceptance of international investment and, crucially, the backing of State governments, Australia has become a country that some of the biggest energy corporations feel compelled to invest in, and these investments constitute a dominant position in each and every facet of the electricity and gas value supply chains. For instance, international energy corporations that have invested in the transformation of Australia's power industry include French energy companies Engie, Neoen and Total, the French state-energy nuclear power company EDF is now investing in renewables, Spain's ACCIONA and Iberdrola and Infigen Energy, Hong Kong's Cheung Kong Infrastructure (with interests in Citipower, Powercor, SA Power Networks, United Power, APA Group) and CLP (Energy Australia), the Chinese state-owned State Grid (with interests in ElectraNet, AusNet Services and Jemena), Singapore Power (with interests in AusNet Services and Jemena), Canada's ATCO, Italy's ENEL, New Zealand's Tilt Renewables, Shell (ERM Power), Germany's RWE Renewables and BP are among the many international investments in Australia. The magnitude of investments is bolstered by investment and superannuation and pension funds, such as the Swiss SUSSI and Canada's Optrust, the Caisse de dépot et placement du Quebec. Many are buying into the sector following the approval of development applications in order to not miss out on the opportunity to reap the rewards to be had from first-mover advantage. The investment bonanza has not been without its problems. The enthusiasm with which State governments have established regional renewable energy zones has resulted in some foreign investors being susceptible to investing in a wind or solar farm before AEMO has 'confirmed grid connection requirements' (Potter 2018). Most importantly for understanding the dynamics of energy market formation is that the internationalisation of the energy sector is being forged on the back of State government commitments to transitioning out of coal and doing so in terms of a different spatial frame, the State, to that of the national market proselytised by the energy market liberalisation stalwarts alongside capital investment that is framed by international considerations.

Anarchy and the Precarious State of the NEM

There is a lack of common cause across the federal-State jurisdictions. The determination of the federal coalition to persevere with the idea that the shape of the energy system should be left to the competitive forces of the *laissez faire* market mask what is little more than a rhetorical ruse to defend carbon capital. The splintering of federal and State objectives has given rise to what the head of the Energy Security Board, Kerry Schott, claims is testing the 'commitment of governments to ... [the] national approach to energy policy' (Foley 2021a, 2021b). State jurisdictions in viewing energy and emissions reduction management as one package, as we have noted, are exercising a degree of independence in policy design and implementation that belies the NEM construct. Each is pushing the development priorities of the energy markets in different directions and diminishing the coherence and the integrity of the NEM. This, according to Schott, sees 'the general state of affairs at the moment as being one of anarchy' (Potter and Macdonald-Smith 2018).

The lack of coordination across the two tiers of government has prompted AMEO and the ESB to adopt more oversight of the energy system. Its *Integrated System Plan* is endeavouring to demonstrate how the range of electricity generation technologies can be reconciled and cohered to secure a greater sense of order in the market and to do what the liberalised energy market cannot do, and that is to ensure that distributed energy resources and variable energy resources can be accommodated within the transmission and distribution infrastructure. AEMO has adopted a more assertive role in pursuit of these objectives. In September 2019, it began overriding the power system's normal operation to manage system frequency and voltage issues in north west Victoria and south west

NSW. New rules have been established to sanction such interventions as the norm, as standard operating procedure. AEMO is now vetting new generating facilities connecting to the grid, and management of flows has also been extended to cover household rooftop solar panels with the recent blocking of household power being fed into the network (Harmsen 2020).

The difficulties in reconciling the fundamentally different approaches to climate change governance remain. Whereas in the federal arena the conservative Coalition parties have been resolutely opposed to any initiatives that would privilege investment in renewable energy over other technologies, and most strikingly coal, State and Territory governments have placed emissions mitigation and support for renewable energy front and centre of energy policy. This is playing out quite consistently across the different jurisdictions as State and Territory governments reckon on transitioning their energy economies to supplant reliance on coal-fired power. While the States and Territory are, formally at least, still signatories to the NEM project, they are leading the transformation in the energy economy by exercising State sovereignty as AEMO has acknowledged their constitutional right to manage energy—and emissions—within their respective jurisdictions, although New South Wales and Queensland remain wedded to the extraction of coal for export (Finkel et al. 2017: 167).

Of course, the States and Territory have not been alone in effecting this transition. The residential household sector has been leading the transition, indeed leading the world, with the installation of rooftop photovoltaic panels in a grassroots-driven transition. The transition is also being underwritten by a shift in the dominant discourse. Addressing climate risk has captured the attention of a swathe of Federal state agencies and authorities—the Australian Prudential Regulation Authority, the Australian Securities and Investments Commission, the Australian Accounting Standards Board, the Auditing and Assurance Standards Board and the Reserve Bank of Australia—prompting calls for action. Key peak business organisations have joined in this chorus for movement on emissions policy or formed alliances such as the Investor Group on Climate Change or the Carbon Disclosure Project, and these have been reinforced by shareholder activist groups that expose the failure to take account of the reckoning of emissions in the energy-intensive world of business. To these must be added all those enterprises that have abandoned coal as a source of energy and/or have set emissions reduction targets or, as with banks and insurance companies, are declaring that they will no longer service coal mining or coal-fired power stations. And there are the global resources corporations that are disposing their holdings in coal. This marks a new reality, one rooted in a climate change politics and not a faux appeal to the ostensible perfectly competitive market in which the 'invisible hand' delivers an objective or technological solution to the preferred modes of generating electricity.

Bubbling away, there has been a mounting retreat from engagement in the NEM that gathered momentum following the abandonment of the National Energy Guarantee in September 2018. In late 2020, the ESB released a discussion paper endorsing moves to develop the transmission grid to accommodate the turn to renewables. This coincided with State governments declaring their intention to accelerate and support renewable energy investment in generating capacity and storage. The push towards a renewable energy future has taken hold. Given the scale of international investment in renewable energy technologies in Australia it is apparent that the transition that is underway, while very much a State-led process, is now being underwritten as an international project. The fact that there has been no new investment in coal-fired power generation since 1988 and the recent announcement of the planned 2028 closure of the Yallourn power station, four years earlier than planned and other earlier-than-planned plant closures have since been announced, are unambiguous signals of fossil fuel-based energy's finite future (Climate Council of Australia 2018; Narayan 2019; Fowler and Ker 2021).

The Coalition Party's partnership with the coal-industry industrial complex and the attempt to reinvigorate this alliance has been overshadowed by events. The concerns expressed by some, such as Paul Broad, the Snowy 2.0 chief executive officer, that the ESB policy proposals were moving away from a market-based principles approach, a move which he criticised, exposes the spurious nature of the appeal to the idealist *laissez faire* construct as the most efficient and effective means of energy market governance. Turnbull's announcement to support the expansion of the Snowy hydro scheme was a spur-of-the-moment decision, certainly not one tested in the market or by a posse of economists (Macdonald-Smith 2021a). The Snowy 2.0 proposal seeks to replicate the industrial-scale electricity generation systems of old and to recover its iconic status as a nation-building project as much as it is conceived as marking the legacy of a Prime Minister who capitulated to the coal-industrial complex. Its nemesis, environmental externalities, is going to involve the swathe of destruction it will create as it carves its way through the Kosciuszko National Park.

The rush is on to mount a solution to the increasingly precarious state of energy supply as the energy corporations built on the back of coal-fired power face the reality that they cannot compete with renewable energy. Victoria was the first to formally break from a national approach in 2020, committing to supporting five new renewable energy zones and passing legislation to support battery installation (Evans 2021; Macdonald-Smith 2020a). Other States are following in one form or another. The illusion that the NEM fits the mould of the perfectly competitive market is the subject of review. The AEMO is developing an *Integrated System Plan* that places much greater emphasis on coordinating development rather than leaving development to the unpredictability of the 'invisible hand' of the free market. The Energy Security Board has also been enlisted to lend its intellectual weight to mapping the regulatory framework to shape the post-2025 national energy market. These initiatives represent a dramatic shift in policy formulation, and while a systematic incorporation of a reckoning of the cost of emissions in assessing the relative merits of different modes of generating electricity remains elusive, they do acknowledge that the *laissez faire* market cannot ensure energy reliability and security. In taking on the tasks of extra-market coordination and planning to enhance energy market, they are also focused on reinvigorating the NEM project. There are, however, other futures being explored.

References

ABC News (2016) 'Ausgrid lease: Treasurer Scott Morrison blocks sale to Chinese, Hong Kong bidders' 11 August.

AMEC Vertigan, M., G. Yarrow and E. Morton (2015) *Review of Governance arrangements for Australian Energy Markets – FinalReport* October. Commonwealth of Australia.

Aston, H. (2015) 'Energy giants called to explain billions in tax havens' *ACAPMAg* 15 July.

ARENA [Australian Renewable Energy Agency] (2019) 'Australian Renewables growing at record rates' 7 August.

Australian Conservation Foundation (2020) 'Morrison government climate policy allows big polluters to increase emissions' 13 August.

ACCC [Australian Competition and Consumer Commission] (2018b) *A Fair Deal on Energy.*

ACCC [Australian Competition and Consumer Commission] (2018a) Restoring electricity affordability and Australia's competitive advantage. *Retail Electricity Pricing Inquiry—Final Report* June.

AEMC [Australian Energy Market Commission] (2020a) *National Electricity Rules* Version 158.
AEMC [Australian Energy Market Commission] (2019a) *AEMC Annual Report 2018–19.*
AEMC [Australian Energy Market Commission] (2019b) *Applying the Energy Market Objectives* 8 July.
AEMC [Australian Energy Market Commission] (2011) *Strategic Priorities for Energy Market Development.*
AEMC [Australian Energy Market Commission] (2020b) *Annual Report.*
AEMO (2020) *2020 Integrated System Plan.*
AEMO [Australian Energy Market Operator] (2019) *Electricity statement of opportunities 2019.*
AER [Australian Energy Regulator] (2021) *Draft DER Integration expenditure guidance note*, July.
AER [Australian Energy Regulator] (2020) *State of the market report 2020*
AER [Australian Energy Regulator] (2019) *AER Annual Report 2018–19.*
AER [Australian Energy Regulator] (2017) *AER Submission to the Future Security of the National Electricity Market* 14 March.
AER [Australian Energy Regulator] (2012) 'Submission to Senate Select Committee on Energy Prices' 17 September.
Australia Senate Standing Committee on Environment and Communications (2016) 'Retirement of Coal Fired Power Stations' *Interim Report* 28 November.
Australian Taxation Office (2017) 'Fuel tax credits—business'
Australian Trade Commission (2015) 'Investment opportunities in Australian Resources and Energy'.
Bagshaw, E. (2018) 'Staggering: $90 billion lost in resource tax' *The Sydney Morning Herald*, 12 March.
Bast, E., S. Makhijani, S. Pickard and S. Whitley (2014) The fossil fuel bailout: G20 subsidies for Oil, *Gas and Coal Exploration* London: Overseas Development Institute and Oil Change International.
Bickers, C. (2017) 'Turnbull government criticises South Australian energy plan as a problem for other states' *News.com* 14 March.
Campbell, R. (2015) 'Subsidisation of Abbot Point coal port expansion' *The Australia Institute Briefing note* 16 January.
Clean Energy Council (2021) 'Clean Energy Australia Report 2021'.
Clean Energy Regulator (2019) 'The 2018 Renewable Energy Target report' Canberra.
Climate Council of Australia (2018) *End of the Line: Coal in Australia.*
Coady, D., I. Parry, N.-P. Le and B. Shang (2019) 'Fossil Fuel Subsidies Remain Large: An Update Based on Country-Level Estimates' *IMF Working Paper* 2 May.
Coorey, P. (2018) 'Scott Morrison's government and business drift apart' *The Australian Financial Review* 4 December.

Costin, L. (2020) 'Concern for nation over $32b power plan' *Crookwell Gazette* 9 November.

Cox, L. (2019) 'NSW considers laws to stop courts and planners blocking coal mines on climate grounds' *The Guardian* 2 October.

Dean, L. (2017) "Banking sector not a "free market nirvana": Treasurer' *Mortgage Business* 1 September.

Dimasi, J. (2015) Bringing an end to electricity network gold-plating' *The Conversation* 29 April.

Donoghue, K. (2016) 'Objecting to the objectives' *The Australian Energy Council* 15 December.

Doran, M. (2016) 'S.A. weather: No link between blackout and renewable energy, expert say' *ABC News* 29 September.

EnergyAustralia (2013) 'EnergyAustralia acquires Mt Piper and Wallerawang power stations' *EnergyAustralia News* 25 July.

ESB (2021) 'Post 2025 Market Directions Design Discussion Paper'.

Evans, S. (2021) 'Power suppliers back ESB call for unified energy market reform' *The Australian* 6 January: 13.

Ferguson, M. (2009) 'Solar Flagships Forum' 24 September, Brisbane.

Finkel, A., Moses, K., Munro, C., Effeney, T. and O'Kane, M. 2017, *Blueprint for the Future: Independent Review into the Future Security of the National Electricity Market*, Final Report, 9 June.

Foley, M. (2021b) "Absolute urgency' needed to stop grid failure as renewables take off' *The Sydney Morning Herald* 5 January.

Foley, M. (2021a) 'Energy industry warns disjointed climate and policies risk network failure' *The Sydney Morning Herald* 5 January.

Fowler, E. (2020) 'NSW green plan best for Liddell, report says' *The Australian Financial Review* 27 November: 26.

Fowler, E., and P. Ker (2021) 'Five power stations to 'run at a loss' by 2025' *The Australian Financial Review* 24 February, 3.

Grattan, M. (2018) 'Turnbull and his NEG continue to haunt the government' *CityNews.com.au* 5 December.

Grattan, M. (2016) 'Morrison blocks Chinese bids for NSW power grid' *The Conversation* 11 August.

Greenpeace (2015) 'Some fossil fuel companies could be paying less tax than you. Wait, what?' 17 August.

Harmsen, N. (2020) 'Power granted to switch off household solar in SA to prevent statewide blackout' *ABC News* 19 June.

Harper, I., P. Anderson, S. McCluskey, M. O'Bryan (2015) *The Australian Government competition policy review: final report*, Treasury, Canberra.

IEA and OECD [International Energy Agency and the Organisation for Economic Co-operation and Development] (2019) *Update on recent progress in reform of inefficient fossil-fuel subsidies that encourage wasteful consumption.*

Jotzo, F. (2020) 'Morrison government dangles new carrots for industry but fails to fix bigger climate policy problem' *The Conversation* 20 May.
King, T. (2015) 'How 'Goldplating' the Australian Electricity Grid Is Killing Off Coal-Fired Generation' *Institute for Energy Economics and Financial Analysis* 15 August.
Lasker, P (2017) 'Power prices: Australia has a gold-plated electricity grid that consumers can't afford' *ABC News* 18 July.
Latimer, C., and P. Hannam (2018) "I haven't left anger yet' NEG architect slams policy 'anarchy" *The Sydney Morning Herald* 10 October.
Letts, S., (2019) 'Renewable energy is growing, but NSW and Queensland are still undershooting their targets' *ABC News* 10 July.
Logan, B. (2002) Standing Committee on Planning and the Environment, Legislative Assembly for the Australian Capital Territory, 11 October.
Ludlow, M. (2019) 'Fears new "gold-plating" will push up power prices' *The Australian Financial Review* 25 November.
Ludlow, M. (2019a) 'Queensland's CleanCo targets 'duck curve' *The Australian Financial Review* 28 October.
Macdonald-Smith, A. (2021c) 'Solar households face tougher rules' *The Australian Financial Review* 24 February.
Macdonald-Smith, A. (2021a) 'Energy Australia flags challenges in pricing, competition' *The Australian Financial Review* 23 February, 15.
Macdonald-Smith, A. (2020a) 'Power prices, intervention worries investors' *The Australian Financial Review* 14 December: 20.
Macdonald-Smith, A. (2020b) 'States splinter market' *The Australian Financial Review* 1 December: 36.
Macdonald-Smith, A. (2018) '"Gold-plated" energy market seen as huge risk of National Energy Guarantee' *The Australian Financial Review* 26 February.
Maisch, M. (2021) 'Victoria decides to go it alone on transmission to unlock more large-scale renewables and batteries' *PV Magazine* 19 February.
Market Forces (2021) 'How your tax dollar subsidises fossil fuels' 21 January.
Mazengarb, M. (2020b) 'ACT bucks trend to include emissions in new planning decisions' *Renew Economy* 26 August.
Mazengarb, M. (2020a) 'NSW clears the way for 17GW of wind, solar and storage capacity with new renewable zones' *Renew Economy* 16 March.
Mazengarb, M. (2020c) 'Canberra records 18 per cent fall in emissions as renewables kick in' *Renew Economy* 5 May.
Mazengarb, M. (2019a) 'Global fossil fuel subsidies reach $5.2 trillion, and $29b in Australia' *Renew Economy* 13 May.
McCullough, Robertson (2017) Renewable Energy in Australia: Market and Industry Overview.
McDowall, W. (2021) 'R&D tax credits can be a significant source of taxpayer support for fossil fuel innovation' *Environmental Research Letters* 16(6).

Mears, A. (2021) 'The end of the NEM as we know it' *Renew Economy* 28 February.
Morton, A. (2020)'A 60% rise in industrial emissions points to failure of Coalition' safeguard mechanism' *The Guardian* 12 February.
Murphy, K. (2019) 'Australia's tortured energy debate: What is the state of play?' *The Guardian* 13 October
Narayan, M. (2019) 'Energy Insider' Energy Networks Australia.
New South Wales, Department of Energy, Utilities and Sustainability (2005) *New South Wales Energy Target: Explanatory Paper* November.
O'Malley, N. (2021) "Power generators caused price spikes" *The Sydney Morning Herald* 21 June: 7.
OECD [Organisation Economic Cooperation and Development] (2019) 'Fossil fuel support data and Country Notes'.
Parkinson, G. (2019c) 'States go it alone on energy after Coalition buries talk of emissions' *Renew Economy* 25 November.
Parkinson, G. (2019b) 'NSW announces 3,000MW regional energy zone, and energy security target' *Renew Economy* 22 November.
Parkinson, G. (2018) 'The rapidly disappearing subsidies for wind and solar in Australia' *Renew Economy* 14 May.
Parkinson, G. (2017) 'Australia leaps up global renewable energy rankings' *Renew Economy* 17 May.
Peel, M., R. Campbell and R. Denniss (2014) Mining the age of entitlement: State government assistance of the mining and fossil fuel sector' *Australia Institute* 23 June.
Potter, B. (2018) 'AEMO imposes tough conditions on new wind and solar in Victoria's 'full' grid' *The Australian Financial Review* 15 October.
Potter, B., and A. Macdonald-Smith (2018) 'Rooftop anarchy' *The Australian Financial Review* 13 October, 17.
Productivity Commission (2017) *Shifting the dial: Energy Five year productivity review*. Supporting Paper No. 11, Canberra.
Productivity Commission (2013) *Energy Network Regulation.*
Rann, A. (1988) 'Electricity industry restructuring: A chronology' *Background Working Paper 21*, Parliamentary Library, Canberra: Parliament of Australia.
Reidy, C. (2003) 'Subsidies that Encourage Fossil Fuel Use in Australia' *Institute for Sustainable Futures* Working Paper CR2002/1.
Reidy, C., and M. Diesendorf (2013) 'Financial subsidies to the Australian fossil fuel industry' *Energy Policy* 31(2): 125–137.
Schott, K. (2020) 'Calm down and carry on with the energy revolution' *The Australian Financial Review* 25 November: 36.
Sales, L. (2019) 'Scott Morrison says he might run for the Liberal leadership when job becomes vacant' *ABC 7.20 Report* 9 April.

Slattery, C.H (2019) 'Fossil Fueling the Apocalypse': Australian Coal Subsidies and the Agreement on Subsidies and Counter Measures' *World Trade Review* 18(1): 109–132.

Spratt, D. (2013) 'New study reveals Labor will hand $2.3–5.4 billion profit bonanza to Australia's dirtiest power stations from carbon price compensation' *The Generator* 20 February.

Taylor, A. (2018) 'Renewable energy investment 'bonanza' is coming to Australia, experts argue' *ABC News* 15 January.

Thornhill, J. (2021) 'An Australian oil giant will be paid to bury its own emissions' *Bloomberg* 2 November.

Towell, N. (2020) "Energy laws have let us down' Victoria pulls the plug' *The Age* 17 February.

Treasury (2019) 'G20 Pittsburgh Summit 2009: Fossil fuel subsidies'. *Attachment A*.

Turnbull, N. (2020) 'Should taxpayers get tax-payer funded COVID-19 bailout? *Michael West Media* 7 May.

UNEP, OECD and IISD (2019). *Measuring Fossil Fuel Subsidies in the Context of the Sustainable Development Goals* UN Environment, Nairobi, Kenya.

Vorrath, S. (2019) 'Crazy gang: Coalition MP wants Queensland to leave main grid' *Renew Economy* 21 August.

Vorrath, S. (2018) 'Tasmania wants to "quit" broken NEM' *Renew Economy* 5 February.

Vorrath, S. (2015) 'Coal production subsidies cost Australians $1.8b a year' *Renew Economy* 17 September.

Warren, M. (2020) 'What the NSW energy plan really means' *The Australian Financial Review* 16 November.

West, M. (2021) 'Revealed: Australia's Top 40 Tax Dodgers for 2021' *Michael West Media* 15 January.

West, M. (2018) "Revolving Doors: how the fossil fuel lobby has governments ensnared' *Michael West Media* 21 November.

West, M. (2015) 'Tax strategies may distort power sales' *Michael West Media* 23 March.

Willett. E. (2008) *Commissioner, ACCC and Member, AER APP Energy Regulatory and Market Development Forum Inaugural Forum Conference Meeting* 27 June 2008.

Williams, P., and B. Foley (2016) 'Australia discovers cost of blocking China in Ausgrid sale' *The Sydney Morning Herald* 21 October.

Zillman, S., and Horn (2020) 'Queensland Government signs deal to allow Adani to defer royalty payments for unspecified period' *ABC News* 1 October.

CHAPTER 5

Households' Embrace of Renewable Energy: The Birth of the Prosumer and Pathway to Empowerment?

Households and, to a lesser extent, small business enterprises have led the investment surge in renewable energy technologies and especially in solar voltaic rooftop panels in Australia. While other countries have experienced substantial investment in solar panels, the scale of investment and the comparatively short-time frame in which this investment has been concentrated, mostly over the last decade, distinguishes the Australian community's engagement with decarbonising electricity generating capacity. The surge of investment in renewable energy has resulted in the largest penetration of residential rooftop solar panels in the world.[1] A record number of rooftop installations were completed in 2020 (Clean Energy Council 2021a, 2021b). Three million households have rooftops capturing solar

[1] The Energy Supply Association of Australia reported in 2015 that 'Australia clearly leads the world in the installation of household scale distributed solar PV … double the residential solar PV penetration rates of the next country (Belgium), and more than three times the penetration of Germany and the UK' (Clean Energy Council 2019). Australia is the ninth in terms of total solar PV capacity (Sheth 2017). The uptake of rooftop solar PV generating capacity in Australia has been dramatic, increasing from 424 GWh in 2009–2010 to 1530 GWh in 2010–2011 to 8512.4 GWh in 2017–2018, and the increase in capacity accelerated in 2020 and the level of installations has continued to increase (Department of Industry, Science, Resources and Energy 2020; Australian Energy Statistics Table O; Hannam 2020). Still, small-scale rooftop solar panels account for only 3 per cent of total generating capacity.

S. Rosewarne, *Contested Energy Futures*,
https://doi.org/10.1007/978-981-19-0224-6_5

radiation that have become components in the nation's energy infrastructure. These now account for more than 10 per cent of Australia's renewable energy generating capacity.

The scale of this investment and households' contribution to the National Electricity Market is a uniquely Australian phenomenon (Energy Security Board 2019). No other country matches the magnitude of household involvement in energy generation, and the level of engagement in the energy system has intensified, with new installations in 2019 exceeding 2.2 GW of additional capacity (Clean Energy Council 2019b, 2020). Moreover, investment in solar panel installations has been entirely voluntary, encouraged by some assistance provided in the form of rebates and feed-in tariffs. The average installation size has been increasingly complemented by record increases in household battery installations as the cost of this technology falls, extending the scale of the household power-generating machine. It has been predicted that rooftop solar will provide a quarter of electricity generated in Australia by 2040.

This investment is transforming the energy system. Alongside the expansion in the distributed electricity being generated by the utility-scale wind farms and solar PV installations, the transformation is creating energy market uncertainty. The established social relations and political economy of energy are being disrupted. In a market that was historically defined by state utilities that possessed a monopoly over the generation and distribution of electricity, and which more recently has been dominated by three major energy corporations that successfully bid for the utilities' business, households and small businesses have, in effect, had next to no say in the functioning of the system.[2] The long history in Australia of households as consumers purchasing electricity generated by state-owned or corporate utilities is being broken at the same time as governments have been constructing the National Electricity Market. The historical passivity of

[2] The Council of Australian Governments Energy Council established the consumer advisory panel, which was subsequently incorporated as the Energy Consumers Australia to provide input to the COAG Energy Council and liaise with the Australian Energy Market Commission and the Energy Regulator Australia and represented the interests of residential customers and small business customers with regard to the price, quality, safety, reliability and security of supply of energy services. The Morrison government dissolved COAG soon after the declaration of the COVID-19 pandemic. The COAG Energy Council disappeared in the process, replaced by the Energy National Cabinet Reform Committee, and in the process the voice that the Energy Consumers Association had in the COAG forum was not extended to the National Cabinet.

households in electricity markets is giving way to a more active agency as they reduce their reliance on purchasing energy through the market, and they have become more aware of services offered by competing retailers, more confident in seeking services that include the supply of green energy. In investing in the capacity to generate electricity themselves, residential households are challenging the old energy order, reconstituting the social relations of energy relations, engendering new ways of organising the energy system as they lead the transition to a clean energy future and thus more able to 'take control of their bill' (Australian Energy Regulator n.d.-a; Brakels 2019).

This transition is being celebrated as one of empowerment. Viewed through a conventional economics lens, this is embodied in the shift in power relations within the realm of the market as contractual relations between households and energy retailers are being recalibrated. A more significant form of empowerment is envisaged when households invest in renewable energy technology to become more energy self-reliant, and even self-sufficient, enabling them to give voice to the form of their engagement with the market. Such investments are envisaged as holding out the potential for households joining forces in clusters or as communities to transcend the limits of their bargaining power when acting individually. The increased significance of household generating capacity, and the effect this has on enhancing bargaining power, is becoming apparent with the growing imperative for energy retail businesses to capture this energy in order to help secure their share of the market. The offer of feed-in tariffs is one manifestation of this empowerment, which becomes rooted in a recasting of the exchange relations that define the organisation and structure of the market. State regulations that impose supply reliability requirements on corporate electricity retailers as a condition of their operating licence, and the representation of these arrangements as constituting the makings of *virtual power plants*, suggest a sense of cooperation between households and energy retailers. This masks a more radical shift in the residential sector's bargaining position that can potentially disadvantage the corporate retailers.[3]

[3] Given that households entering such commercial arrangements engage with the retailer rather than other households, it is somewhat misleading to refer to the virtual power plant as resembling a community of like-minded citizens. There is no necessary spatial or social contiguity that draws together the participants as a community. It makes more sense to describe the participants as a cluster of households.

The increasing employment of renewable technologies is dissolving the passive demeanour of the household and providing the means for residential Australia, for households and communities, to contribute to constructing a zero-emissions future. Insofar as this is manifest in different forms of empowerment, the transformation in social relations and the political economy of the energy system that is being wrought, or proposed, is a highly contested process (Hicks et al. 2014). This is not only evident in how energy is being generated but also manifests in the distribution or supply of electricity, at what cost, and who are, or will be, the main beneficiaries of any transformation. While the allusions to household empowerment have entered the lexicon of the energy sector to capture what is regarded as the progressive evolution of the energy market, this is a characterisation that is invested in the defence of the liberalisation project, the *laissez faire* market and the neoliberal mantle on which it is constructed.

This chapter illuminates the importance of the shifting place of the household as a critically significant agent of change, an agent combining forces with others, which is forcing the reconsideration of the viability and the future of the established infrastructure, the industrial-sized scale power plants and the transmission grid that has become so central to the integrity of the National Electricity Market. The analysis focuses on developments in Australia because of Australia's preeminence in household investment in rooftop solar voltaic panel installation and some of the innovations that have developed to marshal household generating capacity.

If the promise of the engagement with renewable energy technology is held up as being empowering for individuals, households and communities, then this also has implications for the current dominance and roles of the energy corporations and the state in energy governance. We return to consider how capital and the state are responding to these challenges in Chap. 11. Here we turn attention to focus on the motivations and catalysts driving households' investment in renewable energy technologies, firstly examining the rationale for deciding to invest in the technologies and the calculus that shapes investment. From this interest in the sovereign realm of the household, the analysis then considers how the take-up of renewable technologies opens a range of options for households to explore their empowerment in the market and the varied institutional arrangements this engenders. In the subsequent chapter, we interrogate the reach of empowerment.

The *Cri de Coeur*: Unleashing the Force of the Market?

In 2001 the federal government launched the Australian government's Mandatory Renewable Energy Target (MRET), a target that mandated the power sector to generate at least 2 per cent, or 9500 Gwh, of electricity from renewable sources by 2010 from 1997 levels. The MRET was introduced by the conservative Coalition government as a policy gesture aligned with the broad objectives reflected in the Kyoto Protocol, notwithstanding its refusal to ratify the Protocol. As was the case with the Protocol, voluntarism rather than compulsion was regarded as the preferred policy frame for engaging commitments to increasing reliance on energy from renewable sources, and the MRET introduced a market-based incentive in the form of a rebate on investment in renewable energy technologies. Investors in renewable energy technologies would be rewarded with the issue of renewable energy certificates' which could be traded to provide a financial incentive. A distinctive feature of the MRET scheme was that it was organised around two approaches to supporting investment in renewable energy technologies: the issue of a tradable renewable energy certificate for each 1 MW produced by a power station from clean energy sources above a baseline figure, and which could be power generated from solar, wind, wood or bagasse waste; secondly, rebates for small-scale investments that generated up to 10 kW for solar photovoltaic panel installations, up to 10 kW for wind turbines and hydro systems up to 6.4 kW. The small-scale scheme provided an incentive for households and small businesses to invest in renewable energy technologies, and especially in rooftop solar photovoltaic panels. In 2009 the federal Labor government increased the target to deliver 20 per cent of electricity from renewable sources, the equivalent of 45,000 GWh, by 2020, signalling that the incentive for investing in renewable energy would have a longer future. The election of the conservative Coalition government in 2014, and its determination to pull-back from the Labor government's climate change policies resulted in the reduction of the target in 2015, to generating 33,000 GWh of renewable energy in 2020. For a short time, this discouraged interest in both large-scale and small-scale investment in renewable energy technologies.

Initially, the MRET scheme had little effect in exciting much interest in investing in small-scale renewable energy technologies. The market value of the rebate was insufficient to offset the high costs of such technologies.

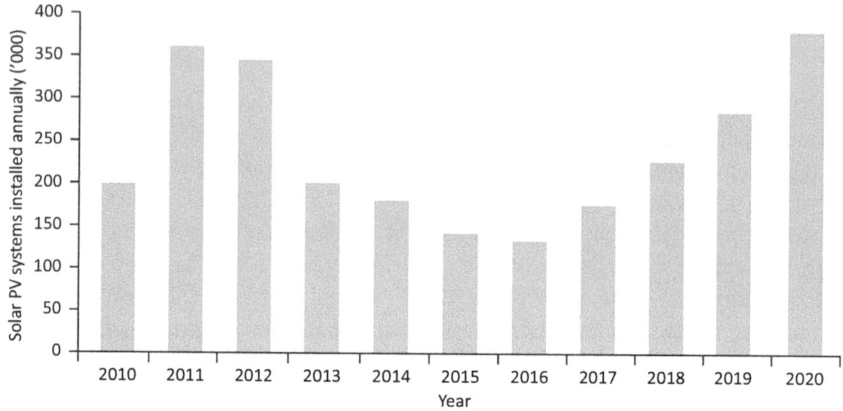

Fig. 5.1 Solar annual installed capacity for household and commercial up to 10 KW. (Source: Clean Energy Council (2020) *Clean Energy Australia Report for 2020* (68))

However, within the decade several factors combined to drive a dramatic shift in household behaviour when investment in rooftop solar voltaic panels, and to lesser extent wind turbines, took off. The magnitude in the shift is clearly depicted in the annual rooftop solar panels generating capacity being added to the system. Figure 5.1 provides an indication of the extent to which the election of the Labor Party to federal office in 2007 appears to have had a positive effect on household decisions to invest in rooftop solar panels and convinced State and Territory governments that their emissions mitigation policies could articulate with national policy. Investments in small-scale solar power increased quite markedly during the Rudd Labor government and then tapered off in the face of growing political turmoil and failure to secure sufficient parliamentary support to pass the *Carbon Pollution Reduction Scheme* bill and remained so even though the Gillard-led Labor government succeeded in having the Clean Energy bill legislated. While successive Coalition governments offered no additional support for household panel installation, the pace of installations remained somewhat flat. Either way, despite the state subsidies, the uptake of photovoltaic panels by households was claimed to reflect the success of the energy market liberalisation project, a claim that had more substance from 2016 when household investment in rooftop panels ratcheted up at a quite significant and escalating rate. Post-2016 trends can be said to reflect household dissatisfaction with the lack of policy debate with the Coalition and the significant increase in retail electricity prices. Once again, the surge in household investment in rooftop solar is a more emphatic display of users calculating the savings from generating power from the grid and avoiding some of the costs associated with the reliance on drawing power from the market.

The conventional economic discourse on the empowerment of households reckoned on the market becoming more of a force for advancing the economic wellbeing of households, and this was based on at least five key developments: ending state-owned utilities' absolute monopoly of the generation, distribution and supply of electricity; the introduction of market-based incentives that brought within their economic reach the range of options for households to more effectively exercise their preferences for low-emissions energy, including installation of rooftop solar voltaic panels; State and Territory government initiatives to introduce market-based instruments in the form of federal rebates and State-funded feed-in tariffs; the evolution of the household as an informed and active economic agent in an energy market that bore the traits of a *laissez faire* market to foster the exercise of the freedom to choose; and the expectation that, as the market matured, the ongoing innovation and development of what was a relatively new technology would drive down the cost of the technologies to expand choice options, of choosing between continuing to rely on energy provided by a commercial retailer or investing in solar voltaic rooftop panels and other technologies to become more self-reliant in energy.

There was nothing natural about these developments; they did not unfold of their own accord. The effective functioning of market processes rested on the imprimatur and proactive initiatives of the state. Indeed, state intervention underwrote the liberalisation project. The neoliberal agendas of corporatisation and privatisation of state-owned utilities involved the state vacating its role in the generation, transmission and supply of a public good, although this was not a comprehensive process that all States signed onto. The establishment of the National Electricity Market was regarded as pivotal to the goal of inscribing the competitive spirit in the energy market to drive increased efficiency and to empower households in their energy relations. On the other hand, the setting up of the National Electricity Market as the basis of competitive market formation, underpinned by the corporatisation and privatisation agenda, necessitated the seemingly contradictory act of establishing several state energy market agencies to design the rules and governance processes and to regulate competition. These agencies have played a crucial role in the efforts to remove obstacles that have impeded the flourishing of market competition.

The two critical and foundational catalysts in the liberalisation project rested on bringing an end to State control of energy utilities and dismantling State borders to create a 'national' market. The process of privatising state-owned utilities was based, firstly, on unbundling the three main functions of generating, transmitting and retail distribution and supply, then corporatising the entities prior to their sale to private bidders. This promised to create a number of competing electricity retailers from whom energy could be purchased and introduced a level of competition that had been historically absent from the sector, at least with respect to the distribution and supply stages of the energy supply chain. The institutional architecture established to regulate and maintain the reliability of energy supply as system underwent transition and brought another layer of the state to the energy stage. Faced with new energy generators, distributors and retail suppliers competing for a share of the energy business, households were presented with unprecedented freedom to choose with whom they could contract to supply their energy needs. For the first time, they could demonstrate their personal preferences for minimising their energy costs by switching from one retailer to contract with another, although there were not a lot of options to choose from.

A further and more groundbreaking dimension to the developing competitive environment was based on the state drawing on its redistributive powers to offer rebates on the installation of solar panels and feed-in tariffs on electricity fed into the grid. These two measures provided incentives for households to invest in the installation of rooftop solar panels to generate low- or zero-emissions electricity themselves and to take the benefits of the freedom to choose and the meaning of competition to a whole different level. The National Electricity Market enabled consumers to voluntarily express their preferences and, if they so desired, influence the pace at which coal-fired power was replaced by renewable energy sources.

There have been, however, several obstacles that have impeded the promise of the competitive spirit and which stand in the way of the market-driven empowerment of household consumers. The establishment of the National Electricity Market did not automatically mirror the ideal of the perfectly competitive *laissez faire* market. There were several ways in which the National Market diverged from the classic economic construct, several instances of what the conventional economic method nominates as market failures. The state energy market governance agencies were charged with the responsibility for addressing these market failures, assisted by other key state apparatuses responsible for challenging instances of uncompetitive practices, and most notably the Australian Competition and Consumer Commission (ACCC), to better align energy market practices with the ideal construct. So, the state was never removed from the functioning of the national competitive energy market.

Firstly, the breakup, unbundling and then privatisation of the State-owned utilities, the power plants, the transmission infrastructure and retail businesses, as occurred in Victoria, South Australia and New South Wales, failed to ignite much competition.[4] Privatisation did not fundamentally change the organisational structure of the sector. Post-privatisation, the sector is characterised by high rates of concentration of ownership, and the exercise of oligopolistic and monopoly power. Twenty years after the launch of the NEM, three energy corporations have reconstituted the vertically integrated structure of the sector. The three service two-thirds of the Australian market, and the extent of their dominance is even greater in the 'competitive' markets of New South Wales, South Australia and Victoria.[5] The Queensland state-owned utilities dominate market share in that State, consolidating State-owned utilities under a single umbrella, although it has established some competition by forming retail enterprises that service higher populated areas of the south-east area of the State. The energy market in Tasmania is dominated by the State-owned Hydro Tasmania. Although this dominance in retailing is declining very gradually, a development courtesy of state authorities' relatively recent and concerted efforts to encourage households to shop around and consider shifting suppliers, the energy corporations draw on their dominance in the generation and distribution of electricity to underscore their significant bargaining, as well as policy lobbying prowess in the sector as a whole (Australian Energy Regulator n.d.-a).

Secondly, the structure of the industry has acted as an impediment to any substantial price competition as has the regulation of the energy

[4] Geographic distance precludes the incorporation of Western Australia and the Northern Territory in the National Energy Market

[5] Ironically, the Victorian conservative government, which had led the push for privatisation of state-owned utilities, advertised the claim that the privatised Victorian energy market was among the most competitive in the world at the same time as retail prices were increasing faster than the prices in Queensland and New South Wales (Collier 2013).

companies operating in the market. This is evident in the inflation of wholesale prices, which has been reinforced by the increasing demand for electricity threatening to outgrow base-load generating capacity. Prices for services provided by the corporations operating the transmission and distribution infrastructure are set by the Australian Energy Market Commission, and the administered prices are set at a level that guarantees a minimum rate of return on investments, a level that is set above normal rates of return by community standards. The guarantee is designed to underwrite maintenance and to incentivise new investment in 'poles and wire' to strengthen the grid network. This has resulted in the 'gold-plating' of returns that resulted in a 70 per cent increase in real network revenues over a decade resulting in as much as a 30 per cent annual rate of return on capital (Australian Energy Regulator 2020; Australian Competition and Consumer Commission 2018). The sector dominated by the three private and the state-owned corporations reports high, above average profit rates. The consequences are manifest in the trend increase in retail prices, which have increased by 67 per cent in the decade through to 2017. Companies hedging against fluctuations in spot prices in the wholesale market and trading in derivatives have served to compound this inflationary spiral as wholesale price adjustments general lag behind falls in the market price.[6] Wholesale prices have been resilient to the deflationary pressures that are being driven by the rapid increase in supply of distributed energy resources.

While the administered pricing practices that secure market advantage for those enterprises engaged in electricity generation and transition are not in the interests of electricity consumers and would seem to be at odds with the promise of consumer empowerment, the exercise of oligopolistic power has had a contradictory effect. Sustained price inflation has acted as a powerful incentive prompting households to consider the merits of investing in rooftop solar panels to generate more of their own power

[6]The wholesale electricity prices are determined by averaging the spot prices that are recorded at ten-minute intervals in half-hour time frames. This sets the wholesale price and price at which supply and demand contracts are settled. This has given energy corporations the opportunity to game the market by withdrawing the lower valued offer and replacing it with an offer to supply electricity when wholesale prices are trending upwards. Those enterprises with the resources can also combine this strategising with hedging. There has been considerable resistance to collapsing the time frame of the ten-minute-spot-price averaged to determine the wholesale price largely because it has yielded robust gains for the large retailers, but the AEMC is in the process of ten-minute wholesale prices.

needs, the cost of which is more predictable (Murphy 2017; Ovo Energy n.d.).

A third obstacle that has frustrated the competitive spirit in the *laissez faire* market has been the low rates of households switching suppliers. The Energy Regulator has explained this in terms of a culture of 'loyalty' towards the large corporate retailers. In some respects, the presumption that households would suddenly transform into animated consumers in the market was always overly ambitious. Given that the corporate bidders generally carried on trading under the same name as the corporatised state-owned utility that they had successfully bid for, which included the transfer of the utility's customer account, continuity of supply went hand in hand with households continuing to contract with the private supplier. This was hardly surprising given that households did not have any experience of competitive markets supplying electricity. Consumers' reticence to investigate the various competitive offers can thus be equally attributed to the impression that little had changed, that energy supplies were a continuum, a hangover from the not-so-long-ago past when state-owned utilities exercised an absolute monopoly and provided a service at prices that were non-negotiable, a time when there was quite literally a complete absence of choice. On the surface, privatisation of the utilities did not appear to change much, apart from the fact that electricity became much more costly.

The notion of consumer loyalty is probably better understood as a feature of a more general characterisation of consumer passivity, at least with respect to switching, and this highlights a fourth obstacle. The reluctance to change longstanding practices is more sensibly attributed to information asymmetry. The lack of real competition and preparedness to switch suppliers reflects the great difficulty in making sense of competing energy supply offers. Opaque pricing offers have made it extremely difficult and time-consuming to deconstruct the details of competitors' supply contracts. Instances of standard offer contracts appearing to be very attractive when coupled with discount deals become much less so when the discounts are of limited duration. The transaction costs in trying to unravel the precise nature of contract offers have been extremely high and this has discouraged switching. Indeed, given the lack of transparency in contract option details, there has been some risk and uncertainty in doing anything other than transferring accounts from the State utility to one of the three dominant energy corporations (Mountain 2018). Consumer passivity or complacency was an understandable reaction to the attenuated nature of pricing competition among retailers. Responding to industry concerns, especially those canvassed by the Australian Competition and Consumer Commission, the Regulator has moved to address this by sponsoring websites that deconstruct and provide details of the range of competitors' price and service options to make comparisons of prices and services easier (ACCC 2018; AER 2020). It took the intervention of the ACCC to prompt the energy regulator to take action to boost the momentum of competition signalling. This amounted to an exercise in selling the liberalised electricity market as a market levelling process, one that was framed in terms of energy democracy secured through the market.

According to the Australian Energy Market Commission, increased advertising has had a positive effect. There has been an uplift in the rate of switching, with one in three households leaving the big retailers to contract with second-tier retailers. However, the claim that these initiatives have met with success overstates the achievements because information still tends to be partial and difficult to decipher, especially as the exercise of choice options remains complicated by the lack of clarity around price options vis-à-vis contracting for the supply of electricity generated from renewable sources. In 2019, the Coalition announced that it was planning on stamping out profiteering by introducing a default price, a reference price that would be the price against which discount offers could be compared. The State and Territory governments that had been considering measures to promote greater transparency in pricing also introduced default prices, setting the regulated price that reflected an 'efficient' price that would prevail in a contestable, perfectly competitive market (EnergyAustralia n.d.). These interventions in the market did result in an upswing in consumers switching suppliers. Nevertheless, making sense of pricing and supply offers is still not straightforward and the government is planning to introduce new competition rules which will give customers access to data detailing their energy pattern use that can be used to assess other retail competitor's pricing (Toscano 2020).

The difficulties in navigating the market are even more exaggerated when trying to contract the supply of clean energy. In the first place, there were initially no retailers who could guarantee that households could be assured a supply of clean energy.[7] Switching to a government-certified

[7] While electing to choose a retailer that could supply green or zero-carbon energy might seem like a reasonably clear proposition, ensuring that what is being purchased is actually green is complicated by the fact that most retailers source power that is generated from a range of technologies. Some offer packages based on different proportions of carbon-free

'Greenpower' retailer did not necessarily guarantee 100 per cent green power because even the 'greenest' of the retailers have been shown to not be delivering 100 per cent carbon-free power-raising concerns about the veracity of the energy nomenclature 'being green'.[8] For the household making the decision to purchase zero-emissions electricity from a retailer, there was no way of knowing that that was what was what was being supplied. The physical properties of zero-emissions electricity were no different to electricity generated from the combustion of fossil fuels and the only assurance that the householder could get the electricity contract was being honoured was if the retail demonstrate that an at least equivalent amount of green power was coming from zero-emissions sources, so the supply arrangement was very much a matter of trust. Those households

electricity, so 'greenness' can be a fluid construct and, when it all boils down, when the switch is turned on 'green' electricity cannot be distinguished from carbon-intensive electricity (Mihm 2018).

> —To understand his hang-up, you need a bit of Electricity 101. If you have solar panels on your roof, the electrons they produce flow across the electric grid like water, following a path of least resistance. As they whiz around, electrons are impossible to track and look identical, whether they're coming from solar panels, a coal plant or whatever. But there is value in keeping tabs on the renewable ones, so energy wonks came up with renewable energy credits (RECs), a tradable financial instrument that corresponds to a certain amount of energy produced by a certain renewable source like solar or wind.
>
> Because RECs have value—ranging from under a penny to a buck or two for each hour's worth of electricity your roof produces, depending on the state—companies like SolarCity can sell them and thus help justify giving you the solar panels for little to nothing. The biggest buyers of RECs are power companies looking to satisfy state-mandated clean-energy requirements, known as renewable portfolio standards. In effect, the power company pays for the right to claim the climate benefits of the panels on your roof and the company uses this feed-in to meet its renewable energy targets, so it has value.
>
> *Mother Jones*—https://www.motherjones.com/environment/2016/01/green-energy-rec-rooftop-solar-panels/

[8] Some energy providers have sought to cash in on the shifting political sentiment by advertising their enterprise as supplying green power and as being active supporters of climate community endeavours. Snowy Hydro advertises its subsidiary energy retailer, Red Energy, as being of the community, Australian-owned and having a 'deep and genuine commitment to local communities'. It partners with various community organisations and, according to the company's chief executive officer, is 'at the forefront of this [renewable energy] transformation' (Australian Made 2020). Yet, the bombastic tone of its advertisement cannot reconcile the intention to promote an image of being of the community at the same time as commercial imperatives frame this being 'really important ... to our customers' irrespective of the fact as an energy retailer that supplies electricity and gas, its 'green' credentials are not absolute.

that have switched to renewable electricity sources, and there are growing but not huge numbers that have, have done so at some cost to themselves, paying a premium for the privilege (Mihm 2018). This premium acted as a disincentive at a time when one new utility, having purchased for a peppercorn price what was supposed to have been a shuttered coal-fired power plant, refurbished the plant to offer electricity at substantially discounted prices (Hendy 2020).

In contrast with the efforts to remove these obstacles were the proactive initiatives based on the introduction of market-based incentives. The most notable for households was the introduction of the small-scale renewable energy certificates. These provided a rebate on the cost of solar panel installation, and the scheme for claiming the rebate was quite straightforward. Panel installers could write the value of the Small Technology Energy Certificates into the contract with the household, which thereby avoided them having to take responsibility for having to claim the rebate. This method has proved to be an enduring encouragement and continues to be so as the cost of panels continues to decline.[9] An equally important market-based incentive that has bolstered the narrative that market liberalisation has aroused consumer empowerment has been the introduction of feed-in tariffs. From 2008, State and Territory governments progressively introduced tariff schemes modelled along the lines of those operating in several OECD countries, with Germany the stand-out model. The promise that, by investing in the installation of solar panels and generating power themselves, households would over the longer term save on their power bills or even avoid having to pay for energy, and these economic benefits could be augmented by payments for exporting power into the grid energy. The feed-in tariffs were one of the strategies deployed by these governments to demonstrate their conviction in hastening the transition from fossil fuel-based energy to renewable energy (Byrnes et al. 2013).

Most State governments had either stopped paying feed-in tariffs or substantially reduced them by 2017. The probability that this would crush interest in installing panels was allayed by retailers starting to offer feed-in tariffs. While the value of feed-in tariffs paid by retailers have in all but a very few cases been much less than the prices they charge for the electricity

[9] The rebates offered under the Renewable Energy Target Scheme are scheduled to continue through until 2023, although the ACCC has argued for the rebate to be removed on equity grounds, which is discussed in the following chapter.

they supply, and this difference is increasing as they reduce the value of feed-in tariff, the escalating costs of importing electricity has reinforced the appeal of investing in panels, and this is more especially so as the cost of panels continues to decline. In Australia, the cost of rooftop solar panel dropped dramatically, falling by 30 per cent over 2012–2016, a fall which has since accelerated. International Renewable Energy Agency (IRENA) estimated there has been a 50 per cent drop in system installation costs in Australia over the four years 2013–2017 (Best et al. 2019; IRENA 2018; Chandler 2018). The combination of these factors seemed to substantiate the rhetoric about the benefits of a competitive market.

Reinforcing this transformation in the organisation of households' place in the energy system has been the concerted efforts of retailers to hold onto and expand their customer base by offering financial advances to assist households install rooftop solar and battery storage systems alongside smart technologies that provide the means for more effective and efficient management of their energy use. This is reorientating investments in rooftop solar panel systems towards much greater generating capacity (Australia PV Institute 2018). This is effecting a more pronounced transformation in the organisation of the energy system as the residential sector becomes a more important component of the nation's electricity generating capacity. Households are morphing into active energy agents and retailers have been underwriting this transformation by convincing those households with panels to entertain partnering in a virtual power station. The retailers' motive in setting up virtual power station partnerships is based on the retailer capturing the household's generating capacity by negotiating with the household to feed electricity to the retailer who aggregates the various sources of power into a virtual pool or reservoir which manifests as an addition to the retailer's generating capacity. In setting up virtual power stations the retailer not only expands the power that falls under its control, it does so without itself having to invest in the physical means of expanding capacity while simultaneously retaining the household custom. This has become a more significant business strategy following AEMO and ESB prescribing the obligation for retailers to meet supply reliability standards by pressuring retailers to build capacity to meet reliability of supply standards.

Household Buy-in to the National Energy System: *Homo Economicus* or Climate Aspirant?

The establishment of the National Electricity Market has not been able to replicate the classic form of the model *laissez faire* market. It has been, and continues to be, a work in progress, and a messy one at that. The promise of competition and greater efficiencies in the generation of electricity, reflected in falling prices, has failed to live up to the promise that market liberalisation would deliver. However, the electricity market has engendered a radical transformation in socio-ecological relations as an increasing proportion of the energy-consuming population invest in solar panel and battery storage systems. Investment in solar panels has strengthened the bargaining position of households by providing a degree of autonomy from reliance on energy retailers to supply households' electricity needs as well as inviting the possibility of a return on the investment. This has clearly injected a different dynamic into the place of the household in the political economy of energy. Likewise, this also permits households to assume a more engaged role in the efforts to decarbonise the energy system, and herein lie the multiple ways in which empowerment can be seen to be unfolding.[10]

The extent to which household empowerment in the energy market has contributed to a dramatic transformation in the place of residential Australia in the energy system is clearly illustrated in Fig. 5.2. The significant impact of the feed-in tariffs provided by State and Territory governments could not be more evident. In conjunction with the rebate offered through the federal MRET and small-scale renewable energy scheme, there was a substantial investment in solar panels from the moment the feed-in tariffs were introduced.[11] The combination of the feed-in tariff and

[10] There is, of course, a contrary argument to be made here that some household solar installations are underwritten by retailer which affects the capacity of households to have on the setting of feed-in tariffs as well as opening up the potential for being subject to surveillance by retailers through the provision of behind-the-meter smart technologies that are represented as providing households with the means to better manage their use of energy but which can also permit retailers to more effectively monitor and manage the energy being generated, used and stored, and controlled. This is discussed in the following chapter.

[11] The rebate provided under the terms of the MRET and the feed-in tariffs is neutral with respect to the renewable energy technologies that are installed. The State and Territory feed-in rebate initially excited some interest in the installation of wind turbines, but there was a

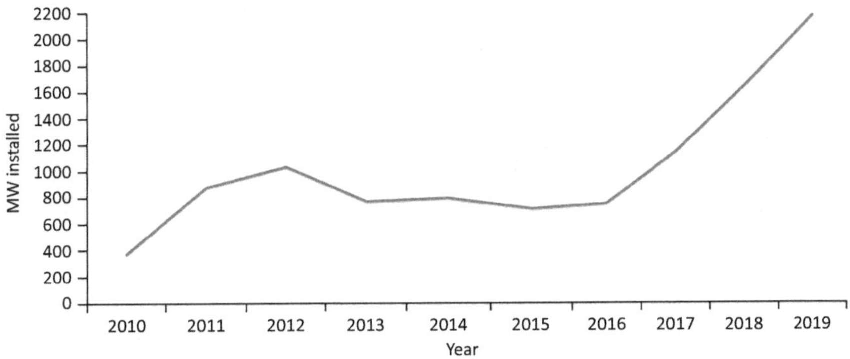

Fig. 5.2 Solar annual installed capacity for household and commercial up to 10 KW. (Source: Clean Energy Council (2020) *Clean Energy Australia Report for 2020*)

the rebate sparked unprecedented growth in household investment in rooftop solar panels towards the end of the 2000s. No doubt, Labor's expanded goals legislated in the Mandatory Renewable Energy Target, and the failure of the Abbot-led Coalition government to completely dismantle the MRET assured the energy-consuming public that the incentive to invest in renewable energy was not just some whimsical gesture. By world standards, this was an extraordinary development. Despite some periodic declines in the level of investments, the wind-back of state feed-in tariffs in 2017 has not daunted the high rate of solar panel installations. The continuing initiatives of State and Territory governments to step up efforts to increase reliance on renewable energy by consolidating household energy generation capacity have added further impetus to household investment in the installation of solar panels. Energy retailers have also introduced schemes that provide financial advances to households to

significant decline in the magnitude of investment in small-scale wind turbines after 2010. There is extremely little residential investment in wind turbines, with next to no rebates being made, which is a clear measure of the relative competitive cost advantage of solar panels that occurred over the course of the 2010s and the overwhelming dominance of the household take-up of solar panel under the small-scale renewable energy rebate (Australian Photovoltaic Institute Ltd. n.d.).

install solar panels and battery storage systems to partner the retailer as a measure for shoring up reliability of supply.

By 2019, 25 per cent of Australian free-standing households had installed solar photovoltaic rooftop panels, establishing Australia as having the highest per capita residential generating capacity in the world (ABS 2017; Egan 2019).[12] Clearly, not all households have invested in the installation of solar panels. Geographic location and the measure of solar irradiation are crucial factors in explaining the high concentration of rooftop solar panels across the country. The south-east metropolitan area of Queensland, the so-called Sunshine State, has the highest proportion of households with solar power followed by South Australia, where over a third of residences have solar panels (Australian Photovoltaic Institute 2020). The significance of this upswing cannot be overstated.[13] As the cost of installing panels has dropped, so the rate of new generating capacity has expanded. As Fig. 5.2 demonstrates, the pace of installations has been quite sustained.

Moreover, while the continuing climate change policy uncertainty has discouraged investment in large-scale renewable energy technologies, such as solar and wind farms, the residential sector remains a dynamic force in the transition from reliance on fossil fuel-sourced energy. Several initiatives are underwriting this. Australian Renewable Energy Agency (ARENA) and Clean Energy Finance Corporation (CEFC) are providing funding support for new housing developments that will be more self-reliant on rooftop solar and battery storage based on a minigrid (ARENA

[12] By way of comparison, California has the largest proportion of households with solar panels in the United States. Some 15 per cent of new residences in California have rooftop solar panels installed. The state mandated that, from the beginning of 2020, all new residential households of less than four storeys were required to install panels, although developers are permitted (Roberts 2020).

[13] By the same token, there is cause for exercising some caution and to not overstate the significance of household investment in solar panels as a means of reducing emissions. Some sense of this can be gained from referring to Figure 3.1, which charts the principal sources of renewable energy being generated. With just 11 per cent of power generated from solar and wind, the renewable energy target has a long way to go, (Clean Energy Council of Australia 2019a). Much of the increased demand for electricity over the post-2000 decade was met by the expansion of coal-fired power, which grew by a robust 10 per cent, and it is only recently that energy-based emissions have begun to decline (Climate Change Authority 2012; Digital Eagles 2020; Hannam 2020). The MRET, in effect, insulated the expansion of coal-fired energy generating capacity from any possible check that could be linked to the commitment to expand renewable energy generating capacity.

2015b; CEFC 2018). Some State governments are partnering with residential property developers to lock in household generating capacity. The South Australian government, for instance, is partnering with a housing project developer to mandate the installation of solar panels and battery storage systems and the ACT has agreed to provide a storage battery for a new housing estate (Mannheim 2020).

State and Territory governments continue to support efforts to promote residential solar panel installations with a view to consolidating State-based generating capacity and progressing their ambitions to meet emissions reduction and renewable energy targets. Equity considerations, through extending access to low-income households and to renters, are another factor contributing to enhancing the empowerment of residential Australia. Some State and Territory governments have supported the installation of solar panels in public housing (Energy Matters 2018b; Lucas 2019). Victoria's 'Solar Homes' programme has a goal of 700,000 residential fitted out with solar panels, which will include 50,000 rental properties (Maisch 2019). The South Australian Labor government proceeded with a project to install solar panels for an additional 1000 or more Housing SA homes after a successful trial had reduced the power bills of residents' renting Housing SA homes (Energy Matters 2018b). The Queensland Labor government is trialling a solar panel installation programme for public housing, the Affordable Energy Program, which offers low-interest loans to support panel installation and batteries (The Solar Man 2018; Matich 2019).

There has been increased interest in subsidising the costs of installing rooftop solar panels and battery storage systems for low-income households (Judson et al. 2019). The Australian Capital Territory has a programme that provides a 60 per cent subsidy to meet the costs of installing rooftop solar panels and updating connections and meters, and the scheme will provide an interest-free loan to assist the householder to meet the upfront cost of getting installation underway (Bondio et al. 2018). The ACT government is also proposing to provide a medium-sized battery to residents of a new suburb to enhance supply reliability (Mannheim 2020). The Victorian Labor government introduced a scheme in 2018 that commits the government to provide a 50 per cent subsidy for low-income residential homes and access to an interest-free loan to meet the balance of the cost, although the requirement that the loan be paid off within four years could discourage interest, while the income threshold is sufficiently high as to be open to 90 per cent of households (Stevens 2019;

Parkinson 2018b). The New South Wales government is trialling a solar installation support scheme for 3000 low-income households funded as an alternative to a cost of living rebate (New South Wales Department of Planning and Environment 2018; Vorrath 2018). Tasmanian government is offering interest-free loans to low-income households to install solar panels (Bondio et al. 2018).

Governments have also expanded their solar panel assistance programmes to include rental properties. The Victorian government is proceeding with one such programme (Stevens 2019). South Australia and Queensland have also committed to a trial of subsidising installations for rental properties, extending the Affordable Energy Program to landlords on the condition that benefits are passed onto tenants, although not all States will extend subsidies to anyone apart from homeowners (Queensland Government 2018; The Solar Man 2018; Matich 2019; Vorrath 2019b). After winning the election, South Australia's Liberal government Premier Marshall announced that the programme would be wound back, means testing introduced and that the government focus would be shifting the policy emphasis on energy security through renewable energy programmes to constructing an interconnector with New South Wales (Williams 2018). Complementing these developments are some community initiatives that encourage landlords to invest in solar panels, such as Z-Net Uralla and CORENA, which is offering discounted loans to landlords (Don 2018; Salleh 2018). Businesses are also setting up similar systems to install panels on investment properties that are mutually beneficial for property owners and renters (Vorrath 2019b).

Notwithstanding these developments, the promise that the liberalisation project would prove an empowering process for consumers generally has to be qualified. Not all households have the capacity, financial and social, to extend their role in the energy system to encompass that of consumers also engaging in generation. Understandably, those renting accommodation have limited opportunity to become household generators. The proportion of rental properties fitted with solar panels is low. The absence of title to property precludes eligibility for the RET scheme. Generally, there are very few apartments that have panels because of the limited roof capacity and because residents generally cannot claim exclusive access to a roof to install rooftop solar panels.

The Rationale for Household Investment in Solar Panels and Battery Systems

Several studies that have sought to explain the reasons motivating household investment in rooftop solar panels or those which have indicated their intention to (ACIL Allen 2013; Macintosh and Wilson 2011; Best et al. 2019; Bondio et al. 2018; Chapman et al. 2016; Maisch 2019; Sommerfeld et al. 2017a, b). Drawing on surveys of Australian households each of the studies reports that the principal motivation is financial: the desire to limit the ever-increasing cost burden of electricity and to 'take control' of their energy bills to minimise the call that power bills make on their income (Australian Energy Regulator n.d.-b; Brakels 2019). The benefit-cost calculations are also made in the context of the opportunity to install panels at a reduced cost because of the rebates on offer and the promise of feed-in tariffs. There is a temporal dimension to this calculation because the capacity to generate their own electricity needs is also a hedge against future price increases. Environmental considerations are important but, among the households surveyed, the environmental consideration is not the overwhelming motivating factor that climate concerned advocates would wish for. However, households that report their intention to install solar panels place relatively greater weight on the environmental benefits (Best et al. 2019).

The studies also point to socio-economic and demographic characteristics that are common features distinguishing the solar-charged households. Residences that have installed solar panels are invariably owner-occupied and free-standing. Larger residences are disproportionately more likely to have installed panels, and there is a disproportionately higher rate of middle-income households that have drawn on the rebate. As Best and others have noted, there is a decided wealth effect (Best et al. 2019). It is generally agreed that the system of incentives on offer is socially regressive in that the chief beneficiaries who have the financial resources to meet the upfront costs of installing panels, and which place low-income households at a disadvantage, compound the inequities in access to energy (an issue that is discussed in the following chapter), while high-income households have demonstrated comparatively little interest in investing in panels to reduce their carbon footprint (Nelson et al. 2011; Sommerfeld et al. 2017b).[14]

The wealth effect also comes into play with respect to different age cohorts. The length of residence in a given home is a factor framing the uptake of solar panels, and this can also be a proxy for identifying the age grouping of residents. Younger adult cohorts are generally less likely to invest in renewable energy technologies, and this can be attributed to them lacking the necessary capital resources. On the other hand, two studies identify post-war 'baby boomers' and imminent retirees as figuring prominently among those taking advantage of the opportunity to invest in subsidised solar panels. The behaviour is explained in terms of the welfare maximising egocentrism of the subjects. The quaint subject of conventional economic theory, *homo economicus*, economic man is reckoned to have calculated the flow of benefits through time that will be generated from investing in solar panels and concluded that enhanced energy self-reliance will shield them from the possibilities of future power price increases when their retirement income could be flatlining (ACIL Allen 2013; Bondio et al. 2018).

This attribution of a calculative logic that informs individual household decisions to invest in solar panels being made independently of others warrants critical reflection. The framing of the surveys tends to invite responses that are measurable; the cost of installation, the rebate and the feed-in tariffs are known and measurable in monetary terms and this privileges the weighting of these more emphatic responses. Survey respondents responding to questions inquiring about whether the household believed that, upon reflection, it had made the right decision generally gives priority to the measurable achievements, such as the savings on the energy bill, pointing towards the decision to install panels as being commendable, whereas reflections on the worthiness of the investment in advancing emissions mitigation goals are given less weight. This is not unexpected, since, *ex poste*, it is logical to rationalise the priorities for installing panels when the benefits are material and the investment is paying off in terms that can be calculated, whereas the environmental gains are more opaque and elusive, reliant on policy goals and the related achievements, especially in the face

[14] This wealth effect should not be overstated. Some lower-income cohorts figure prominently among households that have financed the installation of panels with the assistance of the rebate, and there are other social and environmental benefits that must be taken into account (Nelson et al. 2011).

of a federal government resistant to giving ground on strengthening its emissions reduction policy target. The gains are generally weighted in relative terms, not commensurate with the monetised calculations, and less tangible and able to be linked directly to individual household decisions. This is more especially so given that environmental gains.

This narrow, if not economistic, framing also tends to be reflected in the analytical focus of the evaluations of survey responses. The evaluations tend to elevate the importance of the measurable, foregrounding the confidence that households display in their responses when calculating the benefits *vis-à-vis* costs that are quantifiable, valued in monetary terms and referenced in terms of transactions wholly framed within the functioning and decisions of the household. Whereas perceived benefits and costs associated with social and environmental factors that cannot be directly attributed to the household's decisions are not commensurate and are invariably downplayed. This, I suggest, provides an explanation as to why those households that have installed panels when asked what motivated their decision to invest in panels, *ex poste*, place less significance on environmental factors, whereas those households that indicated an intention to install rated environmental factors more highly among their reasons for considering installing panels (Bondio et al. 2018).

This tendency to explain the exercise of household preferences in economistic terms also appears to shape the interpretation of the older cohort's decisions to invest in solar panels. The interpretation of the decisions is viewed through the lens of a self-interested, egotistical party concerned only with maximising the economic or more correctly the financial returns from the investment. The method is in line with the rational decision-making of the quintessential economic subject, ***homo economicus***, the egotistic welfare maximising figure. Yet, the possibility that this cohort might be motivated by public good, concerned with generating benefits enjoyed by the broader community, is overlooked.

Beyond the Market Logic—A Different Take on Household Climate Agency

The research method adopted in the studies that seek to explain the rationale for households investing in rooftop solar systems is based on the same logic that informed the decisions to corporatise and privatise the state-owned utilities and to establish the National Electricity Market. The method rests on the promise of enhancing competition to give free rein to the ability of household decision-making in terms of the calculative logic of the rational, economically egocentric individual agent. The household is conceived in the same terms as the autonomous subject of the conventional economic paradigm, *homo economicus*, as an introspective, self-contained utility maximising entity.[15] The limitations of this method are evident in research findings that interpret the investment decisions of the older-aged household residents as preoccupied with securing their economic future by investing in solar panels. This embodies a significant flaw both in the framing of survey questions, including those conducted by Energy Consumers Australia, and the interpretation of responses given anecdotal evidence of the cohort's interest extending beyond hedging potential future price inflation. The demonstrated conviction in participating in public events, in climate change rallies and protests, calling for emissions mitigation and membership of community climate action groups is evidence of support for household interest in clean energy being more than based on egotism. This reflects engagement in climate politics, in the embrace of consciousness to act on climate change and mark out the terrain of a progressive energy democracy.

Households are more than atomistic economic islands in a neighbourhood. In the preoccupation with viewing decisions as the result of a monetised calculative logic, conventional analyses overlook the development of a sense of communal spirit that seeks to contribute to efforts to decarbonise the economy. One physical manifestation of this is identified in studies that capture different ways in which community sentiment results in geographic concentrations of solar panel installation. American research argues that upswings can be interpreted as the product of contagion, catalysed by neighbours discussing the import of the technology (Graziano and Gillingham 2014; Plumer 2016; Bollinger and Gillingham 2012).[16] Monitoring of solar panel installations by locality in Australia paints a similar picture, with some localities having a very high proportion of solar-powered residences according to the Photovoltaic Institute (Australian PV Institute 2020). Membership of the national climate advocacy group Solar Citizens displays a similar pattern (Solar Citizens 2018). What we are witnessing is a process of the climate challenge taking the democratisation of energy to another level in what can be described as a covert form of col-

[15] The sub-discipline of Modern Home Economics

[16] For a similar argument with respect to small businesses, see Baranzini et al. (2017).

lective action to do what the state is not doing, and that is to lead the transition to renewable sources of energy.

Viewed through the lens of households being more than atomised constructs provides a richer appreciation of empowerment, as something more than an auditing exercise. This is apparent in the formation of local neighbourhood climate action groups across the country (Rosewarne et al. 2014). The groups became an important political force, fuelling debate and arousing interest in local communities and in what households, and the communities more generally, could personally do to arrest the growth in greenhouse gas emissions associated with coal-based electricity generation. This is redefining the force of energy democracy. Empowerment has been manifest in how the community-based enterprise Solar Citizens emerged out of an inner-city climate action group, providing a rich library of resources, publicising campaigns to end reliance on coal and advertising practical initiatives that could be pursued such as supporting solar panel installation. This focus also figured in climate activist events, such as Climate Camps, which campaigned against fossil fuel 'hotspots' and were organised around occupation protests that sought to block coal mining, the transport of coal and the combustion of coal in power plants, but which also conducted workshops that discussed the promise of and the merits in renewable energy technologies. These events welcomed debate on the actions households and communities could pursue to contribute to emissions mitigation. They also provided a springboard for solar voltaic panel installation enterprises to demonstrate their wares.

This marriage of the venting of climate politics in the public sphere of the political economy with how the private sphere of the household could become part of the solution captures a richer tapestry of empowerment, underlying the synergy between the public and private spheres of the energy system. This is evident in the higher profile of women and mothers participating in the climate political process, and most visibly in young women's leadership in the School Climate strikes. This broadening of the social complexion of climate politics was evident in other climate events, diverse communities, young and old, coming together in events such as 'Walk against Global Warming'. This demonstrates that a critical understanding of what has been motivating households to invest in solar panels and become more energy self-reliant demands we reach beyond the conventional framing of the household as an individualised, insulated site of decision-making.

The estimate that some 735,000 households buying Green Power, paying extra to energy retailers for the privilege of reducing their carbon footprint, speaks volumes on the limits of the conventional welfare maximisation model (CORENA 2002; Vorrath 2019a).

The import of climate change politics in fostering the sense of community is evident in the instances of citizens promoting the transition to solar power and lobbying local governments to launch programmes to assist low-income families install panels, introducing an energy equity dimension to energy democracy. This is not motivated by self-interest, but out of a concern with the common good and emission abatement. Melbourne's Darebin Council, for example, dedicated funds to assist low-income households to invest in rooftop solar panel installation, empowering them to exercise some 'power over [their] bills' (Sansom 2014). In a similar vein, as we have noted, State and Territory governments have also been experimenting with incentive schemes to encourage owners of rental properties to install rooftop panels, benefiting both owners and renters who gain access to household-generated, and often less-costly, energy.[17]

Moreover, the conventional economic method privileges a focus on exchange value: empowerment is borne of households' ability to avail themselves of the opportunities presented by the *laissez faire* market by weighing up the pros and cons of investment in solar panels to generate electricity *vis-à-vis* purchasing power from the grid. Empowerment dressed up as 'taking control of the bill' ignores the different forms that underpin the use-value in electricity. Purchasing power from the grid presents electricity as a mere commodity. Taking control of generation is a testament to the realisation that the calorific force in the coal whose combustion is necessary to drive the turbines is embodied carbon. Investing in solar panels breaks with this fetishisation. The residential buy-in to solar panels is as much about contesting the construction of the liberalisation project that sanctions the freedom to choose whatever source of energy is least cost, irrespective of the non-priced consequences, as it is about redefining electricity as something more than a commodity. Electricity becomes a social and political construct shorn of its fetishised form, and this invokes an entirely different meaning to empowerment.

The scale of household investment in the installation of rooftop solar voltaic panels has wrought a significant shift in energy power relations. Energy market liberalisation has engendered a re-engineering of the household, and this has forced a transformation in the structure and organisation of the energy supply chain. The household, recast in the role of *prosumer*, has taken on the mantle for progressing the generation of

[17] If taken up on a similar scale to the owner-occupied free-standing and semi-detached housing stock, this could double the proportion of residences with rooftop solar panels (ABS 2017).

clean energy. In the process, the value of the commodity electricity has become inextricably linked to its status as being carbon-free, of the household reducing its carbon footprint, and this becomes the bases of empowerment: the household generating electricity free of emissions and embracing an ecological sensibility that is bound up with climate justice. The scope of empowerment is underscored by the choices that power-generating capacity invites with respect to how clean electricity is fed into the broader energy value chain. This becomes a choice between asserting control over the respective stages of the value chain, of generation, of storage and distribution, and all of the costs associated with this, or choosing to contract with one or more of the retailers, or other service enterprises, competing to secure their place in the chain of managing and distribution. This is the context in which the rhetoric of the efficacy of the liberalised electricity market has some merit. The confidence in the empowering force of the liberalisation project has become inextricably linked to the emergence of the *prosumer*, in the new institutional forms of organising and governing electricity generation, storage and distribution and to decentralisation of the energy system and breaking down the power relations that had come to define it. But to reduce empowerment to the prosumer as an economic subject concerned only with taking control of the bill downplays the import of investing in rooftop solar voltaic panels as a radical reconstituting of socio-ecological relations.

References

ABS [Australian bureau of statistics] (2017) 2071.0—Census of Population and Housing: Reflecting Australia—Stories from the Census, 2016: Apartment living.

ACCC [Australian Competition and Consumer Commission] (2018) 'Restoring electricity affordability and Australia's competitive advantage'. *Retail Electricity Pricing Inquiry* – Final Report June.

ACIL Allen Consulting (2013) *Drivers of domestic PV uptake: Characteristics of households with solar voltaic systems* Brisbane.

Australian Energy Regulator [AER] (n.d.-a) 'Switching retailers'.

Australian Energy Regulator [AER] (n.d.-b) 'Smart Meters'.

Australian Energy Regulator (2020) State of the market report 2020.

ARENA [Australian Renewable Energy Agency] (2015b) 'Toolkit for community renewable energy funding and financing', June.

Australian Made (2020) 'Red Energy is true blue Aussie outfit' *The Sun-Herald* 26 January, 7.

Australia PV Institute (2020) 'Australian PV installation since 2017' 27 January.

Australia PV Institute (2018) 'Mapping Australian Voltaic Installations'.

Australian Photovoltaic Institute Ltd (n.d.) https://pv-map.apvi.org.au/historical.

Baranzini, A., S. Carattini and M. Peclat (2017) 'What drives social contagion in the adoption of solar photovoltaic technology' *GRI Working Papers* 270, Grantham Research Institute on Climate Change and the Environment.

Best, R., P.J. Burke and S. Nishitateno (2019) 'Understanding the determinants of rooftop solar installation: evidence from household surveys in Australia' Centre for Climate Energy and Policy Working Paper 1902, Crawford School of Public Policy, Canberra: A.N.U., April.

Bloch, M (2018) 'Interest free solar power system loans for Tasmanians' Solarquotes blog 2 May.

Bollinger, B., and K. Gillingham (2012) 'Peer effects in the diffusion of solar photovoltaic panels' *Marketing Science* 31(6), 900–912.

Bondio, S., M. Shahnazari and A. McHugh, A. (2018) 'The technology of the middle class: Understanding the fulfilment of adoption intentions in Queensland's rapid uptake residential solar photovoltaics market' Renewable and Sustainable Energy Reviews 93(May), 642–651.

Brakels, R. (2019) 'Simple payback time for 6.5KW of rooftop solar for Australian capital' *Solar Quotes* 13 June.

Byrnes, L., C. Brown, J. Foster, L. Wagner (2013) 'Australian renewable energy policy: Barriers and challenges' Renewable Energy 60: 711–721.

Chandler, D.L. (2018) 'Explaining the plummeting cost of solar power' MIT News 20 November, http://news.mit.edu/2018/explaining-dropping-solar-cost-1120.

Chapman, A.J., B. McLellan and T. Tezuka (2016) 'Proposing an evaluation framework for energy policy making incorporating equity: Applications in Australia' *Energy Research and Social Science* 21, 54–69.

Clean Energy Council (2021a) 'Everyday Australians doing net-zero heavy lifting with three million solar systems now installed' 9 November.

Clean Energy Council (2021b) *Clean Energy Council Australia 2020.*

Clean Energy Council (2020) Clean Energy Council Report 2020.

Clean Energy Council (2019a) 'Solar energy facts', 31 December.

Clean Energy Council (2019b) 'Clean Energy Council submission to the NSW Empowering Homes Market Sounding Paper' CEC 9 August.

Clean Energy Finance Corporation (2018a) 'Mirvac communities with built-in clean energy initiatives to make big savings'.

Climate Change Authority (2012) *Renewable Energy Target Review: Final Report* December.

Collier, K. (2013) 'Victoria's electricity prices rise more than New South Wales and South Australia' *news.com.au* 14 December.

Community Own Renewable Energy Renewable Energy Network Australia (2002) 'What we do'.

Community Power Hub (2018a) 'Community Energy'.

Department of Industry, Science, Resources and Energy (2020) 'Backing next generation technologies' 11 November.

Digital Eagles (2020) 'Australia's renewable energy developments see the first drop in carbon emissions' *Energy Matters* 20 January.
Don (2018) 'Renters can reap the benefits of solar power too – here's how' *Energy Matters* 18 March.
Egan, R. (2019) *National Survey Report of PV power applications in Australia 2018* Australian Photovoltaic Institute
EnergyAustralia (n.d.) 'PowerResponse—Virtual Power Plant.
Energy Matters (2018a) 'Solar battery systems in 100 South Australian homes form a virtual power 16 July.
Energy Matters (2018b) 'Victorian government funds solar micro-grids for multi-tenant buildings' 28 August.
Energy Security Board (2019) *The Health of the National Electricity Market*
Engineers Australia (2017) *The Future of Australia's Electricity Generation* Canberra: Institution of Australian Engineers.
Graziano, M. and K. Gillingham (2014) 'Spatial patterns of solar voltaic system adoption: The influence of neighbors and the built environment' *Journal of Economic Geography* 15(4), 815–839.
Hannam, P. (2020) 'Renewables energy advance drives Australia's carbon emissions lower' *The Sydney Morning Herald* 16 January.
Hendy, N. (2020) 'Saving on power is in our hands' *The Sydney Morning Herald* 22 January, 1.
Hicks, J., N. Ison, J. Gilding and F. Mey (2014) *Community Power Agency* Sydney.
IRENA [International Renewable Energy Agency] (2018), Renewable Power Generation Costs in 2017, International Renewable Energy Agency, Abu Dhabi.
Judson. E., F. Zirakbash, A. Nygaard and A. Spinney (2019) *Renewable Energy Retrofitting and Energy Poverty in Low-income Households Research report for United Housing Cooperative Final Report* Swinburn Centre for Urban Transitions, August.
Lucas, C. (2019) 'Tenants raise roof over solar power, demand fair share of panel grants' *The Age* 17 May.
Macintosh, A., and D. Wilson (2011) 'Searching for public benefits in solar subsidies: A case study on the Australian government's residential photovoltaic rebate program' *Energy Policy* 39(6), 3199–3209.
Maisch, M. (2019) 'Low, middle income households install more than wealthy' *PV Magazine* 25 February.
Mannheim, M. (2020) 'Canberra will be home to one of Australia's biggest batteries as ACT weans off fossil fuels' *ABC News* 12 September.
Matich, B. (2019) 'ARENA funds WATTwatchers toward more consumer empowerment' *PV Magazine* 19 December.
Mihm, U. (2018) 'Green electricity review: Green ratings for 21 energy retailers, including AGL, Energy Australia and Origin Energy' Choice.
Mountain, B. (2018) 'If you need a PhD to read your power bill, buying wisely is almost impossible' *Renew Economy* 25 June.

Murphy, K. (2017) 'Australian household power prices rose 63% in last decade, says watchdog' *The Guardian* 15 October.

Nelson, T., P. Simshauser, S. Kelley (2011) 'Australian Residential Solar Feed-in Tariffs: Industry Stimulus or Regressive form of Taxation?' *Economic Analysis and Policy* 41(2) 133–129.

NSW Department of Industry, Planning and the Environment (2018) 'Trial to deliver power bill relief for low-income households' 28 September.

Ovo Energy (n.d.) 'Average electricity prices around the world: $/kWh'.

Parkinson, G. (2018b) 'South Australia urged to re-think household battery storage' *Renew Economy* 19 April.

Plumer, B. (2016) 'Solar power is contagious. These maps show how it spreads' *Vox* 4 May.

Queensland Government (2018) 'Connecting your solar system to the grid' 23 February.

Roberts, D. (2020) 'California now requires solar panels on all new homes. That's not necessarily a good thing' *Vox* 2 January.

Rosewarne, S., J. Goodman and R. Pearse (2014) *Climate Action Upsurge: The Ethnography of Climate Movement Politics* London: Routledge.

Salleh, A. (2018) 'Solar boom: new schemes may help renters get solar on their roof' *ABC News* 20 February.

Sansom, M. (2014) Solar saver program gives pensioners gives power over bills' *Government News* 18 September.

Sheth, K. (2017) 'Countries That Use The Most Solar Power' *WorldAtlas* 25 April.

Solar Citizens (2018) 'Renters and Low-income Households Suffer Worse Through Weekend Heatwave', 20 March.

The Solar Man (2018) 'Queensland opens rooftop solar market to low-income and rental households' 18 July.

Sommerfeld, J., I. Buys and L. Vine (2017a) 'Residential consumers' experience in the adoption and use of solar PV' *Energy Policy* 105, 10–16.

Sommerfeld, J., L. Buys, K. Mengersen and D. Vine (2017b) *Renewable and Sustainable Energy Reviews* 67, 315–323.

Solar Victoria (2019) 'Solar panel rebate'.

Stevens, M. (2019) 'ERM becomes part of Shell's energy power play' *The Australian Financial Review* 23 August.

Toscano, N. (2020) 'Energy reforms give customers power to find better rates' *The Sydney Morning Herald* 28 June.

Vorrath, S. (2019a) 'Coalition gives $1b to CEFC for 24/7 reliable renewable power' *Renew Economy* 30 October.

Vorrath, S (2019b) 'City of Melbourne leads second "bulk-buy" renewables auction' *Renew Economy* 22 November.

Vorrath, S. (2018) 'Origin to build 5MW virtual power plant in Victoria, using rooftop solar and storage' *Renew Economy* 5 September.

Williams, S. (2018) 'Sydney development the Orchards to have one of the world's biggest rooftop solar systems' *Domain* 29 January.

CHAPTER 6

The Metamorphosis of the Household: Household Electricity Generation the *Sine Qua Non* of Empowerment

The electricity market liberalisation project, somewhat paradoxically, has wrought an institutional transformation that is radically changing the character of the energy system beyond what was originally envisaged. The extraordinary magnitude of investment in residential rooftop solar systems has been driven by the failure of the privatisation of state utilities and the liberalisation project to live up to expectations. The ease with which the energy sector so readily succumbed to the ascendancy of large global corporations and translated into an unprecedented period of sustained price inflation has caught many observers by surprise. This occurred with the endorsement of the state, effecting a significant shift in market power in favour of capital and at the expense of residential electricity consumers. The paradoxical upshot has been that the liberalisation project had also heralded a loosening of regulations and the design of support that has permitted households to draw on the technological developments with photovoltaic solar panels to invest in solar panels and to free themselves, to varying degrees, from relying on energy retailers to meet their energy needs.

The proliferation of households generating electricity has created a new cohort of actors who combine electricity generation with consumption: the ***prosumers***. As a new institutional phenomenon, the prosumer now contributes to the remaking of the energy value chain in which power generation has become more fragmented and decentralised. The pace and

S. Rosewarne, *Contested Energy Futures*,
https://doi.org/10.1007/978-981-19-0224-6_6

ease with which this has occurred is represented as one of the achievements of the liberalisation project and is regarded as one of the strengths of the *laissez faire* market. Viewed through the lens of a conventional economic method that informs the case for economic liberalisation, according to state governance agencies and private enterprise service providers, opening up the market has enabled the household to invest in rooftop solar panels and to increasingly link this to battery storage systems and become more actively engaged in the energy system.

Climate change activists and advocates of renewable energy regard the more integral role of the household in the organisation of the energy value chain as marking a process of empowerment. Possessed of the ability to generate and distribute energy, the household is in a stronger position to exercise the choices as to whether to continue to contract for the supply of electricity to supplement the power they generate, with whom to contract, the terms of the contract and from what source electricity is generated, as well as to choose between the various options for trading the power they generate that is in excess of their needs, with whom and when and at what price. This enhanced freedom to choose is being underwritten by the formation of an assortment of enterprises offering a range of services to support investment in solar panels and battery storage systems, the management of energy use and the distribution of electricity. This has expanded the options for how households participate in the market and consolidates the competitive nature of the energy market. Herein lie the foundations of consumer empowerment, as household prosumers are in a stronger position to assert their will in the energy market.

This expansion of options in the energy market is all the more significant because of the distinctive status of electricity. Electricity is a necessity good, an essential energetic source for which there is effectively no alternative or immediate substitute. From the most basic of activities, such as turning on the lights, to cooking, cleaning and communicating with family, friends and the world of work beyond the household, electricity has become the lifeblood of the household. It is difficult to imagine households meeting their day-to-day needs in the absence of electricity. This reliance, indeed, dependence on purchasing electricity severely constrains consumer choice, and it reduces electricity to the status of a thing, a commodity. In this commodity form, as we have seen, the supply of electricity has proven to be an extremely profitable business for the energy corporations to be engaged in. On the other hand, it has proven to be a costly one for the household to be so reliant upon. Investing in the means to draw directly on the sun's radiant energy to generate electricity enables the household to free themselves from some of these constraints and to assume responsibility for meeting its energy needs.

Not only is this considered to empower households to escape reliance on accessing the resource through the market, it also allows the household to rise above regarding the energy resource as simply a commodity to be bought and sold, the terms of which have been set by energy retailers. In escaping this construction, the emergence of the prosumer reflects an empowerment that is coincident with challenging the fetishised form of this necessity good. While this might not always have been openly articulated, it is the effect and the consequence manifest in prosumer households contributing to the reconstruction of socio-ecological relations. There has been a flourishing of a diverse group of energy service suppliers, formed expressly to coordinate or support households' engagement in the multiple facets of this transition to clean energy. This has provided institutional sanctioning that is reinforcing this reconceptualising of energy and, with it, the merits in decarbonising. Established energy corporations now confront a more robust competitive impulse, competing against the wave of new enterprises endeavouring to capture a share of the value being generated as well as the shift in consumer preferences regarding the source of their energy.

The previous chapter identified the key factors that have shaped the transition of households from being passive consumers of electricity to becoming prosumers actively engaged in generation and storage. This signalled a measure of empowerment and reorients the platform on which energy democracy unfolds. The organisational transformation consequent upon household investment in generating and storage capacity provides another measure for garnering some sense of the force of household agency. This has been revealed in the efforts of established corporate energy retailers to partner with households in the formation of virtual power plants. A further manifestation of empowerment arises with the initiatives of a range of the new enterprises working to build a customer base by offering a range of services that enhance household agency.

The Foundations of Consumer Empowerment

The empowerment narrative is premised on an inversion of the contractual relations between the household and the energy retailer as oligopolistic retailers face the prospect of households becoming more energy self-reliant and the retailers responding to the challenges presented by the formation of independent retailers or those offering a variety of services for household prosumers, and which has amplified the level of competition in the energy market. As the ACCC and the Australian Energy Regulator have observed, one of the ways in which the energy corporations have sought to insulate their businesses from this competition is to manipulate the allocation of costs across the different tiers of the vertically integrated enterprise to maintain profitability and to promote less-than-transparent contract offers that are designed to retain customers, but which obscure the real price of power. The introduction of default reference prices is designed to bring an end to the practice, and the ACCC and the AER continue to persevere with efforts to regulate the industry to strike a 'better balance between industry and customers' (Macdonald-Smith 2020b).

However, the energy corporations' high rates of returns as well as the development of a range of technologies attracted an array of new enterprises into the energy market, and this has resulted in more competition being felt across the spectrum of the energy value chain (ARENA 2019; Matich 2019).[1] In the first instance, the changing dynamic of competition pivots around the retail corporations that draw their power from the large, industrial-scale coal-fired power stations, and in the case of AGL, EnergyAustralia and Origin Energy this is drawing from within the respective corporation-controlled vertically integrated energy supply chain. Because economies of scale are crucial to establishing the competitive advantage that these particular enterprises have in generating and supplying power, maintaining the size of the customer base is critical to the viability of their business model. The threat of a growing number of households setting aside complacency and customer loyalty to contract with other competitors or to invest in solar panels to become more self-reliant threatens to jeopardise the corporate business model. State moves to encourage switching reinforce this threat.

[1] The intensity of competition is evident in the fall in the cost of solar panel installations and the effect this is having on the viability of installation companies. One of the leading long-established installers is facing such intense competitive pressure that it has been placed in a quite financially precarious position (Fowler 2020). The COVID-19 virus has highlighted how significantly low are small retailers' margins (Williams 2020). The Energy Coordinating Mechanism was set up to bring the AEMC, AER and AEMO together to develop responses to the pressures that many small retailers are facing with the collapse in the economy with the COVID-19 lockdown.

The industry also faces a second challenge that arises with the increasing proportion of households that have installed rooftop photovoltaic solar panels utilising the power that they are generating, which, because generating capacity relies on the radiant energy of the sun that varies across the day, there is a surfeit of electricity being generated during the mid-afternoon and market, or spot prices, fall. Because the marginal cost of household generated power is negligible, households enjoy a distinct competitive advantage, and this places the energy corporations in a somewhat weaker position, disrupting their market dominance. The disadvantage impels them to intensify their efforts to recapture market share. In South Australia, for instance, the huge take-up of solar panels has led to a substantial increase in power generated during the middle of the day and early afternoon, and the abundance of supply has resulted in the collapse of demand for power generated by the energy corporations. This has placed in question the viability of those companies that lack the flexibility to adjust generation. Some are considering heavily discounting power prices, by up to 75 per cent in some cases, in order to try to boost demand and keep their turbines operating (Macdonald-Smith 2020c).

Uncertainty has become a more pronounced feature of the sector with the decentralisation of electricity generation, and the energy corporations have been negotiating partnerships with households in order to reposition themselves, to concentrate generating capacity under their direction to re-introduce some sense of order into supply. In the process, the energy corporations have modified their business model, morphing into *gentailers*. The corporations continue to lock in custom by forming partnerships with households to form virtual power plants.

These developments provide a measure of consumer empowerment, but it is the investments in residential rooftop solar panels and battery storage systems, which are changing the physical landscape of the energy system, that are really impacting the balance of power relations in the energy market. One way in which the force of household empowerment is shifting the balance of relations in the energy system is by considering how the growing importance of value creation in the household sector is engendering new sources of competitive pressure as enterprises are formed or shift their business focus to nurture and help organise the household engagement in generation and storage as well as on-selling. These enterprises are being established with the objective of capturing a share of the additional value being generated by households. The new businesses are seeking to profit from advancing funds to finance household installation of solar panels and battery storage systems, setting up leasing arrangements to back installations, assisting households to manage the utilisation of the power being generated through the introduction of smart technologies, negotiating feed-in tariffs that are being linked to demand response agreements, and providing resources that endow households with the capacity to initiate trading with, or joining others in aggregating their generating capacity, to set up virtual power plants. A clear measure as to how these ventures are changing the energy bargaining landscape can be gleaned by analysing the efforts of established energy corporations to pursue partnership agreements with prosumers.

Whereas the liberalisation of the electricity market initially presented households with a very limited range of retailers with whom to contract electricity supply, in a market that was completely dominated by major energy corporations, or monopoly State enterprises in the case of Queensland and Tasmania, investment in rooftop solar panels has broken this mould and sparked a new impetus for the development of a competitive energy market. This has introduced a degree of uncertainty for the established retailers who are under greater pressure to hold market share as the Queensland energy retailer has observed (AER 2015). The prospect of losing custom affects business viability, and this has become a more pronounced threat because households now have more options to contract with competitors, are becoming more informed about the range of terms and conditions on offer and are seemingly more inclined to discuss options with neighbours if the evidence of contagion is any indication (CleanTechnica 2012). The changes that are animating the agency of households in the liberalised energy market can have a cumulative effect, as households develop more confidence in navigating their place in the energy market, and, as Giles Parkinson of *Renew Economy* has observed, this can make good sense even to the point of withdrawing from the market by defecting from the grid (Parkinson 2016a).

One transformational development that has considerably strengthened the bargaining position of energy consumers, and commercial and other enterprises and which is discussed in Chap. 11, has been the formation of energy-user associations that negotiate the supply and price of power on behalf of the combined members delivered to members of the association. This is of particular significance in Australia because the ability to draw on household electricity reserves is playing an increasingly important role in firming energy supply in the context of the energy corporations investing in wind and solar farms. It is also critical in the retail market dominated by energy corporations because any loss of retail market share can have deleterious consequences across the corporation's vertically integrated energy supply chain, disrupting the generating operations, affecting the integrity and viability of the corporations' business in its entirety.

The Coalition's policy measures that require the energy corporations to meet affordability and reliability standards highlight the importance of their role in contributing to system stability more generally. Unless household energy is effectively managed, the supply uncertainties resulting from the intermittency of electricity being generated by the increasing number of households can have a disruptive impact that affects the operation of the network. Irregular flows of distributed energy introduce frequency variability and variability in network voltage, which can disrupt the orderly transmission of electricity in the network that was designed for the predictable and regular flow of dispatched energy. This is not necessarily an insurmountable problem, but it does present challenges across the electricity network, adding to uncertainty. The sector's regulators recognise that because of their dominant position in the market, the energy corporations can play a part in contributing to stabilising the transmission and distribution by maintaining their share of the household market.[2]

A Janus-Faced Metamorphosis: Household Empowerment and the Making of the Prosumer and Gentailer

In confronting the possibility that energy market prices would continue to increase and with the cost of solar photovoltaic panels dropping and a range of incentives making installation of panels an increasingly economic proposition, a growing proportion of households have been transitioning from being consumers to producers of electricity. The market dominance of energy corporations over the generation and distribution of electricity is gradually being whittled away, and the bargaining position they have enjoyed in the market over the past score years is being eroded. This shift in power relations is accelerating with the increased take-up of solar

[2] A project funded by ARENA demonstrated the ability of smart inverters to overcome this problem and for distributed energy to be integrated into and contribute to the stability of the national grid, but this is contingent on a degree of predictability of flows (ARENA 2016; Macintosh and Alexander 2017). AEMO has been dedicating considerable resources to addressing the management of distributed energy.

panels, the expanding generating capacity of rooftop panels alongside increased investment in battery storage system, and retailers have had little choice but to devote resources to exploring measures to hold onto their customers. State governments had set one precedent for engaging households in the form of feed-in tariffs, and this set a model for the energy retailers. But as the scale of the household take-up of solar panels grew, a more strategic approach became imperative if retailers were to avoid being marginalised in the sector. One solution that the corporate energy retailers hit upon was to try to negotiate enduring partnerships with households based on households committing to providing energy retailers with power that was excess to their needs or, better still, to agree to permit retailers to exercise some control over the distribution of the electricity being generated. The arrangement would see the formation of virtual power plants. As households transformed from being simply consumers to also generating power that would enter the market, so they morphed into the role of prosumer. The corollary has been that corporate retailers, on the other hand, would combine their principal interest in generating power with capturing and managing the electricity being generated by households and to incorporate this into the virtual power plant under their management. As gentailers, they would be coordinating and aggregating power being generated by households, in effect, re-centralising energy generation capacity. This has established a model that other energy enterprises have begun to emulate, ensuring that competition continues to ensure that households have continuing scope to negotiate their place in the energy system.

In negotiating electricity feed-in arrangements, energy corporations have restored some of their presence in the market. Given that coal-fired power stations account for two-thirds of the electricity being generated, and that these generators lack the flexibility required to adapt to changing demand and increased intermittent supply from renewable sources, the effect of households accounting for a larger share of the nation's generating capacity assumes a disproportionate significance because in meeting their own needs from their rooftops they are contributing to the volatility in demand as well as supply, and because they possess the potential for feeding energy into the market by contracting with a gentailer, they can engender some order in supply.

Retaining customers becomes a crucial quest in ensuring the reliable and orderly supply of electricity. One of the strategic responses to this has been for retailers to form long-term contractual partnerships with households based on them exporting surplus electricity to the retailer via the grid in return for the payment of feed-in tariffs. The retailers have sought to lock in arrangements that result in the reliable and orderly flow of power. The three dominant retailers have not retreated from large-scale industrial-size energy generation, but they have adapted by incorporating household generating capacity as an integral contribution to their overall generating capacity. The retailers, and particularly those that own and operate generators, have been transformed into the role of gentailer as they complement industrial-scale generating with managing the flow of household energy in consort with the formation of a virtual power station. The reliability and orderly supply of electricity becomes contingent on this partnership, the revival of some sense of loyalty and this becomes a means of dissuading households from defecting from the grid. (Such partnering is also seen as a counter to the prospect of the reestablishment of municipal utilities and community-owned and operated renewable energy enterprises, as we discuss in Chaps. 7 and 8.)

The twinning of the metamorphoses is regarded as the most efficacious means for invigorating a competitive spirit as the energy sector transitions towards a more flexible and cleaner energy generation system. Flexibility and household participation in the energy system are being further accommodated by the creation of new business opportunities. This is driving a remaking of the sector with a range of new enterprises entering the market to offer an array of tailored services to households that promise greater efficiencies and enhanced control. This is injecting greater diversity in the structure and organisation of the energy supply chain as new competitors explore ways of capturing some of the revenue that is being generated in the process of the reconstruction of the energy value chain. The consequence is that there is a new impetus to the competitiveness across the sector. Not only are established retailers having to focus their energies on competing with their established peers, they are also facing government policies that are mandating energy retailers to provide affordable and reliable supplies of power which is underscoring the deepening of the competitive milieu.

The viability of the sector's transformation rests on the gentailers' ability to harness a sufficiently sizeable cluster or community of households generating and storing energy and which are committed to longer-term contractual arrangements.[3] In the absence of recruiting a sufficient number of households, the gentailers face uncertainty in the capacity of the virtual power plant. A reasonably sized capacity is necessary to safeguard the plant's ability to meet supply demands while accommodating marginal changes in its customer base and the variability in the flows of energy that could result from households switching suppliers or defecting from the grid. In this rhizome-like organisation, the viability of the virtual power plant depends on the disparate collection of household generators, the nodules, being marshalled into service on a regular basis and, more crucially, being able to be called upon to feed more power into power plant during periods of peak demand. Households are in the position of being able to assert the terms and conditions of their participation in a virtual power plant in a way they cannot when they are totally reliant on drawing their power needs from the grid.

The prosumer's empowerment in the market is underscored by the relative cost advantage of generating electricity based on the sun's radiant energy, which is negligible compared to the marginal cost of generating dispatched and firming electricity. This advantage is being multiplied by the increasing pressures that are working their way into the market as fossil fuel generators are being pressurised to acknowledge and to report on how they are planning on ameliorating the risks associated with burning coal. Partnering with the household clean energy generators has become one way in which retailers have responded to this pressure, although this is not the principal motivation as we have noted. Nevertheless, insofar as greater recourse to harnessing distributed energy from a large number of prosumers affects the energy retailers' capacity to secure their status as a reliable supplier, this is 'putting pressure on their (the energy corporation's) traditional business model' (Solar Choice Staff 2013).

Stabilising the Energy Market: Corporations Fettering Empowerment

One of the more interesting developments with the emergence of the gentailer as an institutional force in the energy system is that it is not only newly formed enterprises that are transforming the way the system is

[3] AGL set a target to recruit 1000 households for its Adelaide-based virtual power station. Enel envisages that it would have to marshal one-third of the residential market to satisfy its ambitions.

generating and distributing energy and competing and courting prosumers to garner a share of the market The organisational transformations are attracting energy corporations from across the world. Three of the largest international energy corporations, BP, Shell and Total, envisage their future as being bound with extending their business beyond the production and sale of the commodity energy. Shell's takeovers of Germany's battery manufacturer Sonnen and software company ERM mark significant shifts in the organisation and operational priorities of global energy corporations as they envisage a future in which emissions-intensive electricity generation is supplanted by solar and wind generation and in which profitability is increasingly defined in terms of service provision (Macdonald-Smith 2019; Stevens 2019).

The effects of these are beginning to be felt in Australia, with Sonnen for instance establishing a manufacturing presence, which it is tying in with service providers. The more expansive operations, however, involve the for-profit energy corporations that have continued to project their dominance in the industry as it transforms. AGL, EnergyAustralia and Origin Energy have each launched virtual power plant schemes. Some of these projects are being undertaken in conjunction with state backing. This is, needless to say, a highly competitive phenomenon that is regarded as underwriting household empowerment, but it is also a contested process as will be discussed in the following chapter.

For the energy corporations, the shift in bargaining relations has affected their market dominance, which is being challenged by the need to modify their business model to capture and incorporate household generating capacity into their energy value chain. Fairly typical of this metamorphosis has been AGL's transformation into gentailer. The transformation is intriguing because of the cloak of the colourful rhetoric in which it has been dressed. AGL—a gentailer that owns major power plants and is the second biggest energy retailer in the country and the number one greenhouse gas emitter in Australia—launched the idea of a virtual power plant in Adelaide early in 2017. It was part of a suite of initiatives to tap the growing enthusiasm among households and businesses for renewable energy sources to counter soaring prices in the existing network (White 2017).

In 2017, AGL's head of the then-recently established service department, New Energy, Elisabeth Brinton, celebrated the idea of household empowerment in inviting households to participate in the virtual power plant. AGL supported the benefits that would accrue to households partnering in the virtual plant by providing households with smart technologies. The technologies would facilitate households' more effective management of their use of electricity and this would increase the available power that could be fed into the power plant. The smart technology would also provide AGL with the ability to monitor 'behind-the-meter' energy use and storage and, crucially, the capacity to access household energy reserves. AGL announced plans to recruit 1000 partners, signalling that there would be 'power in numbers' in the making of what would be the largest virtual power plant in the world (AGL 2016a, 2017). Brinton is quoted as stating that the prosumer-gentailer relationship marks a new era in 'optimization [and] information for the consumers'. This was a new business model based on partnerships being forged with their, hitherto, customers consuming electricity and later represented as members of the AGL community (cited in Macdonald-Smith 2017; Parkinson 2016a, b, c, 2017, 2018; AGL 2018). Households become better equipped to manage their energy use, and as customers they were empowered according to Brinton:

> It's a modernisation that enables the customer to have a really short and sharp and meaningful and relevant experience, whatever their question might be, she explains.
>
> That's actually the investment in technology, knitting together all the back-end systems so there's a seamless, easy view for the customer.

Brinton observed that the company, AGL, was well-placed to make this virtual power plant the well-head of its energy system because one in four households in Australia already have rooftop solar panels, 'providing a vast potential base for new energy networks that include storage' (Macdonald-Smith 2017). The system would help lower household energy costs by combining the technology that manages distribution into the grid with the provision of 'behind the meter' technology that determines the optimal times to draw on energy. This, it was contended, would be empowering for households because deployment of smart meters would provide the technical means for increasing efficient use of power.

Given the instability and unreliability of energy supply, Brinton stressed the urgency in getting the *virtual power plant* in place as soon as possible, and that AGL would have to move quickly in doing so because there were other enterprises that were also moving to establish a presence in this burgeoning market. AGL launched the virtual power plant project in mid-March 2016 (AGL 2016b). Clearly, the haste with which the project proceeded was about market capture. Brinton presented this new energy regime as a strategic initiative that was forging a new future, one that would move Australia 'away from scarcity into abundance', a future based on having 'to earn the respect of our customers on a daily basis, so it's really real', [and] 'an incentive to innovate … to try new things, and that lines up well with consumers'.

Underpinning the transformation of the energy giant into a *gentailer* was a conviction in the positive environment created by the revolution in Australia's economic system, the energy liberalisation project which heralded the construction of a national *laissez faire* energy market. Brinton, who was recruited from the liberalised American energy market, applauded this because: 'all this is made possible … [because:] Here's what's exciting about Australia—it's because it's deregulated, it's because it's competitive.' Yet, Brinton neglected to reconcile this acclamation of the *laissez faire* model with the fact that the venture was underwritten financially by the state, by the Australian Renewable Energy Agency.

Similar schemes have been launched by the other dominant energy corporations, most with some encouragement of support from State governments. EnergyAustralia has expressed comparable but more instrumental sentiments in launching its *virtual power plant*, which it has named Power Response. It too represents the project as a partnership with the EnergyAustralia community, albeit a little less lyrical than Brinton's description of the AGL project:

> With PowerResponse, we're mobilising the EnergyAustralia community to help learn whether a community effort to shift or reduce electricity use can help maintain grid reliability when demand is high.
>
> If we all make a small change to how we use electricity during peak times, we could make a collective difference to reducing overall demand on the grid.

Origin Energy also began exploring the establishment of a virtual power plant in Victoria. It too needed State government assistance before it was prepared to commit resources to the venture (Origin Energy 2018). The company aimed to recruit 600 partners for the project.

Such arrangements are transforming the corporate retailers. They are investing resources in financing the household take-up of panels and battery storage systems, in setting up behind-the-meter energy flow management systems designed to secure energy supply reliability, and in providing a range of other services. The for-profit energy corporations have joined the State and Territory governments in financing household investment in generating and storage capacity, although in at least one State financing is undertaken in partnership with the government. This transformative development of energy retailers morphing into the role of gentailer has been one response to the increasingly competitive environment. In taking on the mantle of prosumer and supporting household's active participation in the market, the retailers have reinforced the significant investment that households are undertaking to assume a leading role in the remaking of the energy market (Fowler and Macdonald-Smith 2019).

There is nothing novel in this. State and Territory governments continue to play an influential role in the elevation of household electricity in the energy system, including supporting the installation of solar panels and battery storage systems in low-income households. They are also encouraging the uptake of 'smart technology' to enhance the efficient utilisation of electricity, which should boost the quantum of power that can be fed into the grid. The ACT government has launched a project to equip every household with battery storage facilities in what will be among the most ambitious, if not, the largest scheme of its kind in the world (Smart Energy Council 2019a, b). Similar projects are underway in New South Wales, South Australia and Victoria. In each instance, support is contingent on households contracting with an energy retailer that is being reconstituted as an energy hub, coordinating the aggregation of generating capacity, re-centralising the management of power, with households being recast as the essential cogs in this virtual power plant. While they have the right to choose with whom they will contract, the energy retailer will have the right to draw on their stored energy to meet any shortfall in network supply. The NSW Empowering Homes Market Sounding Paper proposes a no-interest loan scheme for batteries which would be provided on the condition that stored energy would be able to be drawn upon to meet the demand in the network, that is, that households become members of a virtual power station community (Smart Energy Council 2019a).[4]

[4] Interestingly, the Clean Energy Council recommends against mandating participation in a virtual power plant (Clean Energy Council 2019). Consistent with the dominant liberal philosophy, it argues that: 'Forcing customers into VPPs will not deliver the outcomes the NSW Government is looking for. Instead, the Empowering Homes program should encourage the development of VPP markets by ensuring that the scheme favours VPP-ready devices. This is a different and preferable approach to mandating VPP participation. Customers will

The interest in virtual power plants has been bolstered by the potential multiplier effects that households' involvement in electricity generation could have in underwriting the viability of local manufacturing industry. There are examples in each State where the nexus of an energy-related manufacturing enterprise and the rush to generate emissions-free electricity has accelerated the engagement of households in the energy system. Among the more interesting endeavours to provide a fillip for local manufacturing has been the South Australian manufacturer of solar panels, Tindo Solar, which received a $30 million loan from the Clean Energy Finance Corporation to finance the installation of panels in 5000 homes and businesses (CEFC 2014). The funding would enable panels to be installed free of charge, and those taking up the offer would pay for the power generated, paying up to 20–30 per cent less than the market rate. Panels would eventually become the property of the household after 15 years (Parkinson 2014). The support included the South Australian government contracting with Tindo Solar to fit out 200 public housing homes with panels. The company continues to offer fully financed installation agreements that households can pay off by either buying power generated from their roof panels and at a rate comparable to the rate at which they would be remunerated for feeding into the grid, as well as rental agreements. Tindo is now partnering with Tesla and Sonnen to include batteries as part of the package (Brakels 2018).

A similar partnering involves battery producer Energy storage pioneer AlphaESS with ShineHub, and energy retailer. Powershop has signed up 3000 households that will be paid a fee every time that Shine Hub draws on power from the residence's battery (Smart Energy Council 2019b). The South Australia State's Home Battery Scheme will support the scheme to establish a virtual power plant by subsidising householders' battery installation, while AlphaESS will provide a financial advance for households installing solar panels and battery systems that can be paid off over time. The producer is expanding capacity to increase outputs by 1000 units per month.

Some State governments that retain an interest in energy retailing are exploring or have established virtual power plants. Queensland's Energy

enter into VPP arrangements where they see the economic benefit.' The South Australian government's Home Battery Scheme is likewise designed to support development of local battery production alongside increasing virtual power plant capacity (Smart Energy Council 2019a).

Services is partnering with GreenSync to establish a virtual power plant in northern Queensland that promises to reduce price volatility and affordability (GreenSync 2017). Western Australia State-owned Synergy in Western Australia is exploring setting up a virtual power plant with the objective of creating a market that would allow households to export power when prices are high or store power when prices were low, thereby acting as a price leveller and stabiliser (Synergy 2019). The system would strengthen the sense of community because it would include provisions to facilitate the donation of power to charities.

Re-empowering the Prosumer?

The interest in establishing virtual partnerships with prosumers has become an important strategic lever for newcomers to establish a presence in the energy market, and this has intensified the competition to recruit households to join the energy corporations in partnerships. The retailers' motivations may vary from simply gaining a foothold in the provision of power in conjunction with managing energy distribution to capturing a quantum of generating capacity that can be traded as a hedge in the grid stability market, as appears to have been the primary motivation for the Italian energy giant, Enel, decision to establish a virtual power plant. It is now the largest frequency control and ancillary services supplier in the Australian market (Fowler 2020).

With retailers, and now manufacturers competing for their custom, the power relations in the electricity market appear to have been flipped. From the situation when households had very little choice, having to go into the market to meet their energy needs and contract with one of the three dominant retailers, with all but no freedom to choose, they are now presented with a situation where their capacity to generate electricity is being hotly pursued. The evolution of the National Electricity Market in a surprisingly short period of time appears to have injected real momentum into consumer empowerment transforming the thrust of energy democracy.

The promise of empowering households has been explicitly articulated by one new entrepreneurial venture that has been positioning to capture a share of the prosumer market. The business possibilities in organising the aggregation of households' generating capacity and servicing decision-making by the prosumer has concentrated the efforts of the energy intermediary, IT start-up Power Ledger. Power Ledger has been presenting its service model as a more profitable opportunity than what gentailers offer.

The Power Ledger model is quite explicit in promising to shift the balance of bargaining power more systematically in favour of the household by restoring the household's negotiating power in the market while extinguishing that of the energy retailers. The co-founder and chair of Power Ledger, Jemma Green, characterises the inhabitants of electricity-generating households as 'citizen utilities' whose contribution to increasing the community's energy generating capacity should be coincident with the 'citizen utilities' being able to determine how that extra capacity should be utilised, including trading and with whom if so desired (Gifford 2016).

Power Ledger is among the more interesting IT start-up service providers, given its claim to be offering solutions to two problems. The first arises with the need to provide the means of strengthening the ability of households to maintain records of the information associated with the energy generation, storage and the flows and distribution transactions systems. Power Ledger's software program utilises blockchain technology to provide a technical solution to this challenge, and because the blockchain data is publicly available, or at least available to those who subscribe to the particular data set, it can instil confidence in the reliability of the information that informs decisions. Reinforcing this sense of confidence is that the model denominates the value of energy transactions in cryptocurrency and this provides a standard, independent and stable reference, which unlike currencies is not subject to the whim of the state and other speculative activity.[5]

The second design feature of the model goes to the question of enhancing the ability of the household to process all the data that is generated, and to place it on a comparable footing with other much better-resourced participants in the energy market so that they are in a position to make optimising decisions. This is not unlike the challenge confronting the Australian Energy Regulator's endeavours to counter the effects of the asymmetrical access to market information arising from the oligopolistic dominance of the energy corporations. But the problem that is encountered is somewhat different insofar as Power Ledger identifies the information shortcomings as being intrinsic to the market; there is a problem of data overload given the volume of data that has to be processed in order

[5] Power Ledger selected Ethereum as the preferred cryptocurrency, over bitcoin, because it was regarded as being more stable than bitcoin. Clearly, the speculative trade in cryptocurrencies makes nonsense of this contention.

to give rise to the fully informed individual. The Power Ledger model provides the technical means for households to process the vast amount of data, drawing on the bank of data recorded in the blockchain system and calling on algorithmic procedures to process the data. Subjecting the data to the calculative force of algorithms results in pinpointing the optimal options, such as when it is best to trade and with whom. This opens up the opportunity to see the potential benefits of not getting locked into contractual arrangements with the dominant oligopolistic gentailers and, for instance, to steer households to engage in 'peer to peer trading'. This promises to put 'the power to manage the energy economy into the hands of consumers, while maintaining the value of existing distribution networks' (Power Ledger n.d.).

The Power Ledger model proposes a somewhat different construction of the virtual power plant model on the grounds that the conventional construction of a virtual power station draws the prosumer into an arrangement that simply accommodates the market power of the energy retailers. The Power Ledger model is organised around households exercising greater autonomy in how they trade the electricity surplus to their needs. To draw on the rhizome-nodule analogy, the power of coordinating energy flows is located within the nodules, and the rhizome hub is reduced to that of a mere repository and data processor. The prosumer is more actively engaged in trading, exercising decisions through a forum or a network of buyers and sellers, the collection of nodules, a community in which 'peer-to-peer' trading is conducted. The model shifts the focus to individuals who decide on the trades. This is represented as having a liberating potential because responsibility rests with the individuals, and the individuals are able to exercise agency because they avoid being subsumed in a virtual power plant where the plant, *viz.*, the gentailer's, priorities dictate the terms of household engagement. The Power Ledger model is presented as a counter to the market dominance of the energy corporations, and the opacity of market transactions that have characterised the corporations' approach to its customers is overcome by the watchful eye of the prosumer with the aid of blockchain (don 2017, 2018). One convert to Power Ledger, New Zealand' energy company Vector's chief executive, Simon Mackenzie, has argued that: 'This arrangement empowers consumers to better manage and profit from their energy supply and demand' (Gifford 2016). Green, herself, has celebrated the Power Ledger project as being about the 'democratisation of power' (Power Ledger n.d.).

Power Ledger is not alone in developing systems that provide prosumers and others with the information resources that can assist them in deciding on the nature of their engagement in the energy market. In 2018 AEMO also launched the Open Energy Networks (OpEN) project in partnership with Energy Networks Australia to explore how distributed energy resources could be integrated into the future energy system. Under the project, different models for integrating distributed energy resources would be investigated with an emphasis on customers and other participants buying and selling 'energy and related services at a local level in a more dynamic way, responding to price signals and customer preferences' (Holmes À Court and McConnell 2018; Energy Networks Australia 2020). This would entail building a database that was readily accessible to allow informed decision-making.

Nor is Power Ledger the only enterprise that is specifically focused on developing business opportunities with the expansion of household investments in renewable energy technology by providing coordinating or innovative ways of managing energy storage and flows. The increasing digitalisation of energy management is proving a springboard for software companies to break into the energy industry. GreenSync's Simple Energy Virtual Power Plant, which ARENA is supporting financially, is a similar enterprise being developed in South Australia (GreenSync 2017; ARENA 2018, 2019). The project is being promoted as a positive for households because it would link solar panels to battery storage systems to enable households to sell stored energy when prices are high and thereby reduce households' electricity bills by as much as 40 per cent. GreenSync has developed the technology to form a decentralised electricity market exchange (GreenSync 2019). The GreenSync deX technology allows open exchange between utilities, households and businesses. The AMP Capital-backed software company Evergen is partnering with the solar panel and battery installation company Solar Service Group to recruit the Group's customers as members of the Evergen virtual power plant, which is envisaged to become Australia's largest (Fowler 2019; Mazengarb 2000). The venture would utilise software that would allow customers to use their batteries to store energy and to sell it into the wholesale market when prices are high. The system promises to halve the payback on the household investment, making it more economically attractive than other virtual power plants.

The opportunities for contracting households to manage their generating and storage capacity have attracted a number of new participants in the energy market that enhances the bargaining position of households because they have so many more options to choose from. International aggregators and other service providers are partnering with local enterprises to capture some of the benefits that flow from the supply chain. Global energy corporation Enel X has partnered with the South Australian government to develop the latent power in storage batteries (Enel X 2020). This is just one example of energy corporations adapting their business model to capture some of the revenue generated in the renewable energy value chain. There are a number of virtual power plant models that are more narrowly concerned with dealing with peak-load demand challenges in instances when demand exceeds system supply. This has become a particular concern for AEMO and the ESB in the management of emissions-intensive industry, but it has also attracted attention with demand management schemes designed specifically for the household sector. The adoption of behind-the-meter smart technology provides the means to increase efficient utilisation of electricity at the same time that generating and storage capacity can be managed and coordinated to reconcile and ensure system supply stability. EnergyAustralia, for instance, introduced a less-domineering arrangement recently, offering $20 reparation for each time it drew on a household's power (EnergyAustralia 2019; Macdonald-Smith 2020d).

One offshoot of this model can be observed in some new housing developments that have built solar panel generating and centralised battery storage capacity into the development. This provides scope for the housing community to be more energy self-reliant as well as export any surplus into the grid.[6] Based on minigrids, there is still potential for feeding excess power into the grid to the advantage of residents (Naturalsolar 2019).[7] Other similar models include major developers. Brookfield and Mirvac, for instance, have received backing from ARENA and the CEFC, respectively, to build similar housing complexes (ARENA 2015; CEFC 2018a, b).

[6] Such estate developments are contributing to more diverse forms of distributed energy resources being integrated into the network. Based on nodules in a broader rhizome-like arrangement, they also significantly reduce the cost of connecting to the grid because there is only one connection in place of each residence being connected.

[7] Kurmelovs, (2021) 'Community batteries: what are they, and how could they help Australian energy consumers?' *The Guardian* 5 April.

Conclusion

The liberalisation of the energy market in Australia has freed households from the constraints of an energy regime dominated by three energy corporations, at least at the retail end of the energy supply chain. With the falling cost of renewable energy technologies and the increasing cost of energy, households can be seen to have become active agents in setting their own energy destiny. This investment is not simply the response to an economic calculus based on costs. It is also a reflection of the community's conviction in renewable energy and the efforts being pursued to confront the challenge of climate change, to drive the decarbonisation of the economy. As world leaders in investment in rooftop solar voltaic panels and battery storage, Residential Australia is dramatically transforming the physical architecture of the energy system and along with socio-ecological relations.

The household investments are changing the scale of energy generation and distribution, decentralising the energy system, and this has created opportunities for entrepreneurial service providers to engage in the energy market to advise households on how they can make the most out of their investments. This has heightened competition in ways that could not have been foreseen by the architects of energy market liberalisation, and this is being viewed as an empowering process for households. The corollary of household venturing into energy generation and storage has been a remodelling in the make-up and modus operandi of the corporations. In order to secure their future, the corporations are having to reposition themselves in the energy market. This repositioning goes well beyond the simple harnessing of energy being generated by households. To secure their place in the energy system and, of course, their market hegemony, the corporations are pursuing a more engaged interaction with households, providing the technology that gives the household greater capacity to manage the use of the electricity they are generating and storing, at the same time as aligning this with the operational needs of the corporation as a reliable supplier of affordable power. This new business model now includes shepherding, a counter to the service provisions being offered by a raft of new competitors, and as a means of capturing some of the value that is being generated across the supply chain by introducing new distributional arrangements. In the process, they also seek to impose some limits on household empowerment, to limit energy autonomy and to retrieve their standing in the energy regime.

Against this market dynamic, the energy corporations must confront the rhetoric that energy market liberalisation has heralded a transformation in market relations that has excited the agency of households. Green, for instance, has celebrated the Power Ledger project as an innovation that is underwriting the 'democratisation of power' (Power Ledger 2017). There is, though, cause for caution in taking this too literally because 'democratisation' can be a Janus-faced metaphor for empowerment, as Ransan-Cooper et al. (2018) have observed. When viewed through the lens of the workings of the *laissez faire* market, democratisation, like empowerment only more so, is a narrowly framed construct. It is premised on gaining and exercising a voice through the market, which in the context of the partnering of the prosumer and gentailer invokes a fundamental tension in the definition of electricity. In inserting their place in the energy supply value chain, households contest the dominant conception of electricity whose value, for the most part, is based solely on the embodied carbon calorific content that is released with the combustion of coal. The households' investments in solar panels are, in effect, as much about asserting their authority in the market to review the place of energy in the carbon cycle and to substitute energy generated from the solar radiance to displace the privileged status of carbon-embodied energy, as it is about inserting their place in the energy value chain to 'take control of their bills'. The fetishisation of the commodity electricity is challenged and this is a process of transformation of the socio-ecological relations upon which the energy regime is formed.

This challenge comes up against the *raison d'être* of the energy corporations and, for the most part, the host of energy retailers and other service providers whose principal motivation lies in the capture and distribution of the value embodied in the energy value chain, irrespective of the source of value. For them, their oeuvre is an instrumentalist one, viewed through the lens of energy production system that is removed from its ecological significance. The strategies deployed to reassert their place in the energy value chain are the subject of the following chapter.

References

AGL (2018) 'What is a virtual power plant' 27 July.

AGL (2017) 'AGL virtual power plant goes live' 16 March.

AGL (2016b) 'AGL launches world's largest solar virtual power plant battery demonstration to benefit customers' 5 August.

AGL (2016a) https://www.agl.com.au/solar-renewables/projects/power-in-numbers.

AER [Australian Energy Regulator] (2015) 'Statement of the Energy Market 2015'.

ARENA [Australian Renewable Energy Agency] (2019) 'Innovative distributed energy platform to streamline Australia's future decentralised energy system' 5 April.

ARENA [Australian Renewable Energy Agency] (2018) 'Simple Energy Virtual Power Plant' 28 March.

ARENA [Australian Renewable Energy Agency] (2016) 'Networks renewed', Media Release 22 November.

ARENA [Australian Renewable Energy Agency] (2015) 'AGL Solar project'.

Brakels, R. (2018) 'Sonnen battery update for Australia: New prices, features & specs revealed' *Solarquotes Blog* 26 September.

Clean Energy Council (2019) 2019 'Clean Energy Council submission to the NSW Empowering Homes Market Sounding Paper' CEC 9 August.

CEFC [Clean Energy Finance Corporation] (2018a) 'Mirvac communities with built-in clean energy initiatives to make big savings'.

CEFC [Clean Energy Finance Corporation] (2014) 'CEFC investments lower costs for business and improve energy productivity' *Media* 22 July.

CEFC (2018b) 'CEFC and Mirvac to help homebuyers reduce energy costs at new masterplanned communities' 15 March.

CleanTechnica (2012) 'Rooftop solar: it's contagious' *Renew Economy* 22 October.

Don (2018) 'Power Ledger partnership brings P2P solar energy trading to Melbourne' *Energy Matters* 4 April.

Don (2017) 'Blockchain boosts peer-to-peer green energy trading' *Energy Matters* 29 August.

Enel X (2020) 'South Australian back-up boost program' 30 April.

EnergyAustralia (2019) 'Power response'.

Energy Networks Australia (2020) 'Open Energy Networks: An Energy Networks Australia and AEMO joint project' 13 May.

Fowler, E. (2019) 'Evergen touts virtual power plant to help make solar panels pay' *The Australian Financial Review* 19 December: 20.

Fowler, E. (2020) 'Fortune 100 company cashes in when power grid wobbles' *The Australian Financial Review* 13 July.

Fowler, E., and A. Macdonald-Smith (2019) 'Playing catch-up on green energy *The Australian Financial Review* 5 November: 36.

Gifford, J., (2016) 'Power Ledge expands trials of blockchain electricity trading' *Renew Economy* 28 October.

GreenSync (2019) 'GreenSync secures $10 million to rollout Australia's first Decentralised Energy Exchange' 5 April.

GreenSync (2017) 'ARENA and GreenSync launch deX: a new renewable energy digital marketplace that will transform Australia's energy industry' 23 February.

Holmes À Court, S., and D. McConnell (2018) 'OpenNEM: An open platform for National Energy Market data' *Renew Economy* 8 February.
Kurmelovs (2021) 'Community batteries: what are they, and how could they help Australian energy consumers?' *The Guardian* 5 April.
Macdonald-Smith, A. (2020d) 'EnergyAustralia offers $20 to cut power as grid suffers'. *The Australian Financial Review* 18 December.
Macdonald-Smith, A., (2020b) 'Energy Regulator steps up the pressure for price cuts' *The Australian Financial Review* 1 July.
Macdonald-Smith, A. (2019) 'Shell enters home battery market with Sonnen acquisition' *The Australian Financial Review* 15 February.
Macdonald-Smith, A. (2017) 'AGL Energy's tech high-flyer to redefine energy for customers' *The Australian Financial Review* 21 April.
Macintosh, L., and D. Alexander (2017) 'Crisis, what crisis? How smart solar can protect our vulnerable grids' *Renew Economy* 8 February.
Macdonald-Smith, A. (2020c) 'Grids struggle with solar splurge as records tumble' *The Australian Financial Review* 7 October.
Matich, B. (2019) 'ARENA funds WATTwatchers toward more consumer empowerment' *PV Magazine* 19 December.
Mazengarb, M. (2000) 'CSIRO spin-off raises $6m to accelerate deployment of battery optimisation software' *Renew Economy* 1 June.
Naturalsolar (2019) 'Construction of first new age residential townhomes in Australia's most exciting smart city underway'.
Origin Energy (2018) 'Origin to fire up first power plant in Victoria' 5 September.
Parkinson, G. (2018) 'South Australia urged to re-think household battery storage' *Renew Economy* 19 April.
Parkinson, G. (2017) 'S.A. Govt fast-tracks switch to renewables with ZEN contract' *Renew Economy* 17 December.
Parkinson, G. (2016c) 'AGL picks new head of solar, battery storage and EV strategy' *Renew Economy* 10 August.
Parkinson, G. (2016b) 'AGL eyes virtual power plants' *Renew Economy* 10 February.
Parkinson, G. (2016a) 'Why you are not selfish if you disconnect from the grid' *Renew Economy* 21 January.
Parkinson, G. (2014) 'CEFC to provide $120m to unlock Australia rooftop solar finance' *Renew Economy* 22 July.
Power Ledger (n.d.) 'All power to producers and consumers'.
Power Ledger (2017) 'The democratisation of power' *You Tube* 24 October.
Ransan-Cooper, H., A. Chapman, P. Scott, and V. Hann (2018) 'Tesla's 'virtual power plant' might be second-best to real people power' *The Conversation* 19 February.
Smart Energy Council (2019b) 'Advancing state and territory renewables' Media Release 16 August.

Smart Energy Council (2019a) 'Empowering homes in NSW' Media Release 15 February.

Solar Choice Staff (2013) 'A distributed energy market: Customer and utility attitudes towards PV' The Australian Voltaic Institute, 9 July.

Stevens, M. (2019) 'ERM becomes part of Shell's energy power play' *The Australian Financial Review* 23 August.

Synergy (2019) 'Find the right energy plan—electricity plans'.

White, A. (2017) 'Powering up, one home at a time' *The Australian* 10 June, 25.

Williams, P. (2020) 'Power bill debts to hit retailers' *The Australian* 28 July, 17.

CHAPTER 7

Subordinating the Household to the Rhythms of Energy Capital

The project of liberalising the Australian energy market has had quite contradictory outcomes. On the one hand, the processes of corporatising and privatising State-owned utilities have not delivered lower prices, although there are signs that prices are about to fall greater reliability of supply. On the other hand, the inflation in power prices, alongside the escalation in energy corporation profits, has impelled households to reconsider their reliance on the market to meet their power needs, and advances in and the decreasing cost of renewable energy technologies have prompted growing numbers of households to invest heavily in the installation of rooftop solar voltaic panels and battery systems to become more energy independent. One consequence has been that the structure of the electricity system has been radically transformed, and this has been a fascinating development because the transition to renewable energy has been inspired by grassroots investments. This has been a transition spearheaded 'from below'. Established energy corporations have sought to accommodate residential power generation and a range of energy service retail enterprises have invested in the sector facilitating this transition. These developments have engendered a capacity for households to negotiate their place in the energy market that is unprecedented. According to advocates of transitioning to a zero-emissions energy future, energy market liberalisation has provided the opportunity for household empowerment. Households 'taking control of their bill' is one aspect of this, although the more significant consequence is the shift in reliance on electricity generated from the combustion

S. Rosewarne, *Contested Energy Futures*,
https://doi.org/10.1007/978-981-19-0224-6_7

of fossil fuel to households generating more of their energy from renewable sources and radically transforming socio-ecological relations.

The scale of this uptake in household electricity generation has set Australia on a quite distinctive path in a world in which energy market liberalisation became the norm. The trailblazing role of households in investing in renewable energy makes the idea of empowerment an even more powerful contention. However, there is cause for re-evaluating the efficacy of the market liberalisation-empowerment thesis given that residential Australia appears to have been paying a high price for its access to energy. The empowerment narrative is rooted in the actions of those households that have made the leap into rooftop solar generation, yet the majority of households has not taken that step. Buying power from retailers supplying power through the grid is an expensive proposition and the investment in residential solar has been one means of limiting this. For many households, though, this is not an option and those residences that have not invested in the installation of rooftop solar panels face the full force of the vagaries of the market and the inflation in power prices set by the corporate energy conglomerates. This has engendered increasing attention to the prevalence of energy poverty, of households unable to afford the cost of energy in the context of energy market liberalisation, and governments have been moving to rein in energy price inflation to alleviate the extent of energy poverty. These moves have become more voluble in the context of the COVID-19 pandemic, but they are tentative because of the contradictory developments at play. The pandemic lock-down has reduced the business sector's demand for power, eroding energy retailers' profitability, while residential demand has increased because of the lock-down with residents spending more time at-home practising social isolation. The combination of rising unemployment and the increase in household power demand appears to have exacerbated the problem of energy poverty, placing mounting financial pressure on a growing number of households. Those experiencing financial stress are now reliant on the goodwill of retailers to grant some leeway in the settlement of their accounts. Not only does their precarious situation suggest anything other than empowerment, but the reliance on corporate goodwill also means that there is less scope for them to search for competing power supply offers and switch suppliers.

The contrasting experiences of households in the energy market are underscored by the system of incentives established by governments to encourage investment in solar panels and battery storage systems and how this is being funded. The federal government determined that the small-scale renewable energy rebate should be budget neutral and that the cost of the rebate should be incorporated into the market price for electricity. Incorporating this 'environmental cost' charge in the retail price has meant that those households drawing their energy needs from the grid have been forced to pay an impost that bankrolls the rebate (Faulkner 2021). Two million-plus households have benefitted from the rebate, whereas the three-fifths of all Australian households that have not installed panels and are entirely reliant on power from the grid are, in effect, subsidising those households that are saving on their electricity bills by generating power on-site (ABS 2019). Recognising that not all households are in the position to invest in rooftop solar panels suggests the need for more circumspection in celebrating consumer empowerment.[1]

Opening up the question of the extent of household empowerment and the inequities this engenders is not necessarily an endorsement of the argument that those households that invest thousands of dollars in the installation of solar panels and, increasingly, in battery storage systems, enjoy unfettered power in the energy system. Certainly, they are insulated from rising power prices to varying degrees, yet there is no certainty that the returns on their investment, especially with respect to the tariffs they are paid for feeding power into grid will be maintained. In fact, wholesale power prices have begun to fall, the consequence of declining industry demand and the growing volume of renewable energy, and feed-in tariffs are being reduced. Not only does this raise the prospect of households losing out in the distribution of the value generated in the energy value chain. Energy retailers have been setting up partnerships with households, in the form of virtual power plants, to aggregate energy that is being generated across the residential landscape. The virtual power plant enables gentailers to exercise authority over the distribution of the power generated and/or stored by the prosumer and to reassert the primacy of the market and exchange relations as the institutional frame in which electricity is defined. With a quarter of solar households engaging in an energy system that continues to be dominated by the accumulation imperative of capital, organised by energy corporations and an array of energy service providers, and three-quarters whose energy needs are met through the

[1] The opportunity for households that are unable to invest in the installation of panels on their residence are increasingly turning to investing in solar gardens that enable them to contribute to reducing their net carbon footprint.

market, we cannot be assured that transitioning 'from below' will necessarily prevail to transform socio-ecological relations to engender a radically different energy dynamic. Despite the heroic conjectures of the promise of empowerment, the transformation of the energy system remains very much a contested process. This chapter explores the limited reach of empowerment and the distributional inequities that arise with the terms of state support for solar households before considering the efforts of the energy corporations and other energy service providers to assert their dominance in shaping the role of solar households in the energy system.

The Structural Asymmetries Framing the Energy Market in Transition: Energy Poverty

The political forces that successfully lobbied for the liberalisation of the energy market looked to abandoning the State-owned and controlled energy system to engender efficiency dividends, lower prices and attract new sources of capital to revitalise the system. They also looked to capitalising on the sale of generators, transmission networks and retail distribution as one way of ameliorating fiscal constraints, which tended to translate into concessions to prospective private buyers to maximise the sale value of the different components of the system. But the advocates for liberalisation were either dishonest and ideologically driven, or at best naïve, in pressing the case for corporatisation and privatisation of state-owned utilities because there were obvious limits to the extent to which the utilities could be unbundled into multiple enterprises and incorporated into a competitive market. The not-unexpected consequence was the replacement of state-owned utilities by international oligopolies that, given the natural monopoly and public good nature of transmission networks and to some extent generation capacity, clearly did not fit the mould of the perfectly competitive industry structure idealised in the *laissez faire* model. The corporatisation and then privatisation of state-owned energy systems established a high profit-driven energy system based on sustained inflation of electricity prices. As we have observed, price inflation, and profitability, was underwritten by the Energy Regulator which set the rates that the owners could charge for transmitting electricity to incentivise investment to increase the capacity of the networks (ABC 2016; Hetherington 2017).

As the ACCC (2018) concluded in its investigation into pricing in the sector, there is 'insufficient competition in the generation and retail markets'. Competition has not been a defining feature of the National Electricity Market and the result was that an 'increasing number of consumers are reporting having trouble meeting their electricity costs' (ACCC 2018; Blowers 2018). The ACCC recommended that more effort be put into encouraging consumers to shop around the various energy retailers for lower prices and, in lieu of competition, it also suggested that a default standard offer price be introduced, which the federal and some State governments have done, although with not a hint of irony the Energy Regulator expressed concerns to avoid setting the competitive default price too low for fear that this could discourage consumers from shopping around for cheaper prices (Australian Energy Regulator 2019: 13). There is also a raft of other measures to enhance competition being considered by governments, the Regulator and the Australian Energy Market Operator. Alongside the pervasive role of administered pricing and the moves to strengthen regulatory frameworks to mitigate the excesses of oligopolistic pricing, the structure of the energy system and state oversight is clearly at odds with the *laissez faire* construct upon which the liberalised National Electricity Market was ostensibly modelled.

Notwithstanding the contradictions in the embrace of energy market liberalisation, the ascendancy of household investment in rooftop solar panels and battery storage systems is represented as a development of some considerable substance in the liberalisation thesis. Indeed, State policies introduced to support households investing in the installation of generating capacity are quite explicitly expressed as being possible because of the liberalised market, and these policies are promoted on the basis that they are empowering households (Department of Industry, Science, Energy and Resources n.d.; Department of Planning and Environment n.d.; Mazengarb 2019b; South Australia Power Networks 2020). The ACCC and Energy Regulator advocate enhancing the information at the disposal of the consumer and making it easier to switch suppliers. This does not acknowledge that choice is constrained by the nature of electricity as an essential good and the resources required to navigate the various supply options that are available in the market. As we have observed, over three million households have sought to escape the reliance on purchasing their power needs through the market. The majority of households choose not to, while an ever-growing number are not in the position financially to contemplate the option. This is not the consequence of market failure—information asymmetry—as supporters of the *laissez faire* market contend, but a reflection of the structure of the energy market and how governments have managed support for solar uptake. Systemic inequities frame households' places in the energy system.

It is estimated that almost a third of families are experiencing difficulties in meeting the cost of electricity and that this number is growing, especially in the context of COVID-19 pandemic (Australian Energy Foundation 2018; Hogan and Salt 2017; Taylor 2020). Between July 2015 and June 2019, one retailer, AGL, disconnected some 395,000 households in South Australia, Victoria, New South Wales and South East Queensland from the network, equivalent to 1.5 per cent of the households connected to the grid (Labaste 2019). Energy poverty has become an entrenched feature of the National Electricity Market, and low-income households are having to dedicate an ever-increasing proportion of their income to paying their energy bills. The asymmetry in power relations is more pronounced for low-income households, and a confluence of factors exaggerate this asymmetry.[2] Sole-parent families and single-person households spend more of their income proportionately on electricity than households headed by a couple (Phillips 2018; ABS 2019). The majority of sole-parent households are headed by women, two-thirds of whom reside in rental accommodation, are less likely to be employed and thus more reliant on social security benefits. Meeting the costs of energy and housing absorbs some two-thirds of their income, even though low-income households, by stint of circumstances, tend to economise on their energy, using less power than their wealthier counterparts (ACOSS et al. 2017; Tasmanian Council of Social Services 2018). Yet, at the same time, low-income households spend a disproportionately greater share of their disposable income on energy compared with all other income cohorts (Bankwest Curtin Economics Centre 2016; ACOSS et al. 2017; Salt and Hogan 2017). The situation of single-person households is similarly gendered, with women, many in receipt of aged pensions, accounting for a disproportionate number of residences.

The systemic nature of energy poverty is the unspoken contradiction in the empowerment narrative. The difficulty that some have in meeting the cost of their power bills highlights that the freedom to choose is not an endemic feature of the energy market. Forget about the ability to 'take

[2] The ACCC also questioned the ethics of retailers targeting different household cohorts that discriminated against low-income households, setting terms that encouraged them to consider shifting to other suppliers, while retaining more valuable customers by offering a range of incentives (ACCC 2018: 143–151).

control of their bills', low-income households are denied that opportunity, forced into negotiating household hardship programmes in order to work out ways to pay outstanding debts and maintain access to electricity. Others are being excluded from the energy system because they do not have the capacity to meet the cost of their electricity, fail to settle accounts, and are disconnected, and the rising number of disconnections indicates that the problem is getting worse (ACOSS et al. 2017; Salt and Hogan 2017). For a growing number of households, the liberalisation project has been a decidedly disempowering process, and this is highly gendered.

The embedded nature of the lack of equity in the energy market is one in which the state has been integral in shaping (Australian Energy Regulator 2019: 14). Given the continuing policy uncertainty in the federal political arena, and the claims that this uncertainty is contributing to price inflation, it is difficult to imagine how the concerns with respect to energy equity can be addressed. There are, however, some initiatives that do aim to attend to the concerns. In September 2018, the Energy Regulator began publishing details of the energy companies with the objective of providing more transparency in the companies' market operations (Australian Energy Regulator 2019: 26). At much the same time, the industry established the Energy Charter that is intended to follow this initiative (Energy Charter 2019a).

The Structural Asymmetries Framing the Energy Market in Transition: Who's Paying?

Market failure is also not just the result of the oligopolistic character of the National Electricity Market. The inequities arising from market failure are being compounded by the barriers that prevent universal backing for the installation of solar panels and battery storage systems. Home ownership is, in practice, a condition for securing a rebate on the cost of installing a solar system under the terms of the small-scale renewable energy certificate scheme. It does not make sense for residents renting accommodation to invest in solar panel installation because they do not have security of tenure and because panels are not portable it is not a straightforward task to take the panels and inverters with them if their lease expires. Few apartment dwellers install panels because the buildings cannot accommodate panels and the prospective investing residents' exclusive rights to the power generated cannot necessarily be secured,[3] and there is no point in installing panels that cannot capture a reasonable amount of solar radiation. In effect, it is the owner-occupiers of free-standing residences, the quintessential Aussie homeowner, who is the privileged beneficiary of the rebate. At the very least, the one-third of the residential population who are not owner-occupiers are excluded from the subsidy (Australian Institute for Health and Welfare 2019).

This privileging of the owner-occupiers of free-standing residences is further compounded by the benefits that accrue from the investment in panels. Generating electricity reduces or does away with the call to contract with energy retailers to supply electricity and feeding power into the grid provides a stream of revenue. These options become available because of the support of the federal and State schemes to support household installation of solar panels in the form of the rebates as well as the payment of feed-in tariffs, but the distinguishing feature of these schemes was that they were established on the condition that they would be budget neutral. This condition was designed to allay potential criticism that government was singling out the renewables energy sector for special treatment to the detriment of other sectors of the energy sector. The cost of the respective market incentives was incorporated into the market price of electricity, the electricity bill, which is paid by all households that are supplied energy from the grid. Represented as an 'environmental cost', this accounted for some 5–15 per cent of the total retail price charged to households (ACCC 2017). All households that rely on electricity supplied through the network are paying this premium. Those households that do not have solar panels pay the premium for all of their electricity, whereas residences with solar panels pay for only that proportion of their energy use that is supplied via the network.

The financial viability of the two schemes has become dependent on the financial contributions of communities that are not eligible to reap the benefit of the schemes. Those residences that cannot afford to meet the cost of installing solar panels and, more recently, the cost of battery storage systems, or whose residential circumstances preclude them from installing panels, have been carrying the burden of subsidising those households that have installed panels. This amounts to a threefold advantage for the prosumer residence: households' investment in panel

[3]The technology that enables the power-generated panels on multi-occupancy builders to be sequestered for use of the investing resident is now available.

installation under the terms of the small-scale renewable energy target scheme is subsidised; households installing solar panels are able to avoid altogether, or reduce, their consumption of high-priced electricity (including that resulting from 'gold plating' returns on network investment), and avoid meeting some of the 'environment cost'; and are rewarded for feeding surplus power into the grid.

There is one further disadvantage experienced by those households that are not in the position to install panels. More recently, as the Australian Energy Market Operator reins in the cost of 'gold plating',[4] it is concentrating resources on investing in measures to stabilise the flow of electricity from distributed energy sources to better manage the disruptive effects of more distributed energy being fed into the network by wind farms, solar farms and households. This presents another irony because the cost of managing an orderly transition to more intermittent sources of power, including that being generated by households, is carried by all who draw electricity from the network, the low-income and rental households, while the households that are the source of the distributed energy are being paid a feed-in tariff (7 News 2019). There is now some pressure for this cost to be borne by solar households (Harmsen 2020c).

It is also apparent that disadvantage begets additional disadvantage. Those not in the position to generate power from their rooftops, and this is particularly apposite for low-income households, tend to economise on their energy use and use less power than their wealthier counterparts. Yet they spend a disproportionately greater share of their disposable income on energy compared with all other income cohorts; their power usage has higher opportunity costs (Bankwest Curtin Economics Centre 2016; ACOSS et al. 2017; Salt and Hogan 2017; Tasmanian Council of Social Services 2018). This disadvantage is compounded by the likelihood that they are less likely to utilise the technology that is being deployed alongside solar panel systems, such as smart, behind-the-meter technologies, which enable more efficient and cost-saving energy usage, and which can assist in maximising the power fed into the grid to increase tariff earnings (ACOSS et al. 2017; Chester and Morris 2012; Chester et al. 2018).

Energy poverty also acts as a deterrent to the exercise of choice of suppliers, which frames another dimension of the asymmetries in the bargaining power of low-income households vis-à-vis the retail energy providers

[4] It is accepted that the rate of return was set too high (Rate of Return Consumer Reference Group (CRG) 2018).

(Australian Energy Regulator 2019: 50–51; ACCC 2018). They are more or less locked into trying to negotiate a payment plan with their supplier and are most unlikely to explore the option of switching suppliers and, in fact, are advised not to do so (Department of Industry, Science, Energy and Resources 2020). Unless some compensatory measures are introduced, and there are some initiatives being pursued, there is not much evidence that the precarious position of these households will be alleviated. The promise of energy market liberalisation empowering households through investment in solar panels has had a distinct effect in exaggerating the gulf between the haves and the have-nots. The inequities manifest in a fourfold distributional disadvantage for non-owner-occupied households, apartment dwellers and low-income households more particularly and even more so sole-parent families and especially women-headed households.

The ACCC and State pricing regulators have acknowledged the distributional inequities resulting from the small renewable energy technology rebate and feed-in tariffs. In 2018, the New South Wales' independent pricing authority slashed feed-in tariffs to reduce the price of power that all, but especially low-income, households were required to pay to meet the costs of the tariff (Hannam 2018), and the ACCC's inquiry into energy price inflation recommended that the end date for closing the rebate be brought forward (ACCC 2018).

Institutionalising Prosumer Privileges

The ACCC recommendations to address energy inequities have not been adopted, although exposés that have drawn attention to the mounting evidence of energy poverty as a demonstrable consequence of market failure is at last prompting some interventions in the liberalised energy market to address the situation. Expanding access to solar panel and behind-the-meter technologies is being promoted as a means of alleviating energy poverty. Equipping households with the means to generate some of their energy needs would reduce dependence on commercial retailers. Towards this end, State and Territory governments have launched some initiatives, installing panels on public housing (Lucas 2019) and subsidising the costs of installing rooftop solar panels and battery storage systems for low-income households (Judson et al. 2019). The Australian Capital Territory has a programme that provides a 60 per cent subsidy to meet the costs of installing rooftop solar panels and updating connections and meters, and the scheme will provide an interest-free loan to assist the householder meet the upfront cost of getting installation underway (Bloch 2017). The Victorian Labor government introduced a scheme in 2018 that commits the government to providing a 50 per cent subsidy for low-income residential homes and access to an interest-free loan to meet the balance of the cost and is proposing to extend this to meet the full cost of installation (Solar Victoria 2019; Tchetvertakov 2018; Estcourt 2020).[5] The government is extending the programme to include rental properties (Solar Victoria 2019; Matich 2019). The Queensland Labor government is trialling a solar panel installation programme for public housing, the Affordable Energy Program, and is extending the programme to offer low interest loans to landlords to support panel installation and batteries, and the assistance will be provided on the condition that benefits are passed onto tenants (The Solar Man 2018; Vorrath 2018; Matich 2019). The South Australian Labor government proceeded with a project to install solar panels for an additional 1000 or more Housing SA homes after a successful trial had reduced the power bills of residents' renting in Housing SA homes (Energy Matters 2018). The New South Wales government is trialling a solar installation support scheme for 3000 low-income households funded as an alternative to a cost-of-living rebate (New South Wales Department of Planning and Environment 2018; Vorrath 2018). Tasmanian government is offering interest-free loans to low-income households to install solar panels (Bloch 2017). South Australia and Queensland have also committed to a trial of subsidising installations for rental properties (Queensland Government 2018; Vorrath 2018).

These are positive initiatives, but the schemes are not nearly as focused as they could be as governments display a reluctance to be seen to be providing too much support for low-income households to the exclusion of other constituents. For example, the Victorian government scheme sets an income threshold that is sufficiently high as to be open to 90 per cent of households, and it requires the loan be paid off within four years.

Another example of the endeavour of State and Territory governments to expand generating capacity by establishing an all-inclusive system of residential engagement in the energy system has involved partnerships with building companies, especially based on housing estate

[5] While the scheme would benefit tenants, providing them with access to renewable energy and decreasing their power costs, it also subsidises the installation of panels on rental properties to the advantage of the landlord.

developments, such as that initiated by the South Australian government, for instance. The government is partnering with a project developer to mandate the installation of solar panels and battery storage systems. The ACT has committed to providing a storage battery for a new housing estate (Mannheim 2020).

Addressing energy poverty has also prompted some local government councils to launch programmes that offer financial assistance to remove the barriers faced by low-income pensioner households, and particularly aged and disability pensioners wanting to install solar panels (Sansom 2014; Solar Savers 2017). Complementing these developments are some community-based initiatives involving social enterprises that offer discounted loans to landlords to encourage them to invest in solar panels on rental properties, such as Z-Net Uralla and CORENA (Mike 2018; Salleh 2018). There are also some commercial enterprises that are setting up similar systems to install panels on investment properties that are mutually beneficial for property owners and renters (Vorrath 2018).

The federal government remains the laggard in developing measures to address energy poverty. It has, however, sought to allay criticism of the absence of a defined policy in arguing that a principal objective of the energy and emissions policy being designed will be affordability. It has introduced a default price, but this is directed as much at addressing concerns expressed by businesses as much as it is households. It does not resolve the problem of the growing number of households that are in financial distress and finding it difficult to pay their electricity. It had also developed into a classic and crude case of class politics, an apolitical wedging exercise with households that had installed panels and battery storage systems being criticised for supposedly profiteering at the expense of low-income households (Faulkner 2021; Lucas 2019). Directives to retailers to negotiate household payment plans and to not call in debt collectors adding another layer to this politicking and simply shifting added another layer who to this politicking and simply shifted responsibility to the independent energy retailers who are also experiencing financial difficulties.[6]

[6]The independent energy retailers are being pressured by the Australian Energy Regulator not to disconnect customers who are behind in their payments by negotiating a payment plan. The independent retailers are arguing that relief plans be extended to include their dealing with network operators who are slow in, or resisting, adjusting their charges in line with falling wholesale prices. The large retailers, the three energy corporations, maintain that they are carrying a disproportionate share of the burden (Foley and Toscano 2020; Williams 2020). The Australian Energy Regulator has lobbied the Australian Energy Market Commission to change the rules to permit retailers six months' grace on meeting their payments to the transmission network companies (Foley 2020).

Acknowledging that energy poverty is the hidden face of household solar panel installation does not also necessarily translate into those that do escaping impoverishment in the energy system. The groundswell of interest in solar panel installation has created opportunities for suppliers to reap the rewards of engaging the interest of households, and in a highly competitive market, this can high pressure selling, mostly involving a chain of sub-contractors beginning with sales representatives who sell households the installation package, possibly advising on or actually negotiating finance, and then the installation and connection to the grid and feed-in contracts. There are thus several points at which there is scope for profiteering. One of the most deceptive is the provision of 'buy now, pay later' scams, which have come to the attention of the ACCC. The ACCC has recommended the introduction of responsibility requirements that mandate those companies negotiating solar panel installations to establish that the credit arrangements are not beyond the financial capacity of borrowers and result in hardship. The solar panel suppliers are opposing such requirements (Eyes 2020).

Virtual Power Stations and Other Servicing Enterprises: Capturing Energy Revenue

The ever-increasing number of Australian households installing rooftop panels, and the installation of systems with greater generation capacity and storage systems, highlights the fact that household empowerment is far from being a general, let alone universal, phenomenon in the energy landscape. The capacity of households to exercise choice in power supply is quite uneven across that landscape. However, State and Territory government initiatives are addressing this inequity in a reasonably systematic manner, albeit somewhat cautiously and notwithstanding that some remain resistant to the idea of extending assistance beyond homeowners (Hiley 2019).

The equally salient aspect of this is that the financial assistance is provided on the assumption, or condition, that the households will remain connected to the network and on the understanding that they will feed into the grid power surplus to their needs, and that this will be mostly marshalled through the auspices of energy-retailer-controlled virtual power stations. The South Australian government has been quite explicit in declaring that its support is contingent on all new households becoming a small cog in a larger electricity-generating machine (South Australian Government 2019; Russell 2020). Most state assistance programmes are now being framed on the basis of households agreeing to be a party to virtual power stations. Initially considered to be a supplementary source of electricity, households are now regarded as an integral component of the energy supply system and are predicted to become an even more significant energy supplier and justify more systematic efforts to aggregate the decentralised generated energy.

The significance of this organisational transformation is being reinforced by technological advances that provide households with the resources to manage their use of energy more efficiently. The 'smart technology' that has been installed in most residential properties as part of the solar panels and battery storage systems package enhances the tools available to households to optimise the benefits derived from becoming prosumers (Australian Energy Market Commission 2019). The real-time data that is generated by the technology means that households are much better informed and equipped to improve efficient utilisation of domestic technologies, thereby reducing electricity costs and increasing the power that is surplus to their needs and which can be fed into the network to maximise the benefits (Stagnaro and Benedettini 2020; Sioshansi 2020a, b). In this narrative, 'behind the meter' and other smart technologies are represented as the handmaidens of household empowerment.

However, the technologies also permit the gentailer partners to coordinate and manage household energy flows. This provides the means to establish some order in the market, which, because of the dispersed and decentralised nature of household electricity generation, would otherwise be unpredictable and uncertain. The ability of the virtual power plant operator introduces some control over what AEMO characterises as 'uncontrolled energy'. The ability of the plant operator to utilise 'behind the meter' technology to coordinate the harnessing of the household's energy enables a more intrusive form of energy flow management which can override household energy management plans and compromise empowerment.

Virtual power stations are regarded as a crucial organisational vehicle for remedying the unpredictability and uncertainty in the supply of residential distributed energy. Indeed, energy retailers were granted sole rights to establish virtual power plants with operators of transmission networks being blocked from registering virtual power plants and denied the opportunity to establish the means for managing household distributed energy (Vorrath 2016). Presumably, limiting the number of parties that could recruit households to aggregate generating capacity was deemed to provide a stronger foundation for concentrating the marshalling of the dispersed households' generating capacity and make for a more orderly distributed energy market (AEMO 2020, 24).[7] The ruling also restricts the number of energy enterprises that are in a position to capture a share of the revenue that is generated by the household.

The energy retailers' moves to take on the role of coordinating and managing household generating and storage capacity set a different agenda to that that influenced householders' decisions to invest in generating and storing electricity. Whereas households are motivated by the ambition to 'take control of the bill' and to decarbonise, and reframe socio-ecological relations, energy retailers are first and foremost moved by the objective of reasserting their market dominance to shore up the returns on their investments. They are more instrumentally driven by profit-making, to pursuing financial returns than households. Profit maximisation and maintaining or expanding market share are always at the fore. The rationale for gentailers setting up virtual power stations is to capture a share of the value being generated in the supplementary household energy value chain. Partnering with households in virtual power plants extends the reach of their garnering beyond the legal bounds of the retail enterprise to capture a share of the value from household solar power. In the process, the morphing of the household into a prosumer transforms the *raison d'être* of the solar-powered households, and virtual power plants become the means for forcing and generalising this transformation. As a result, households are pressured into becoming more entrepreneurial in their focus as they seek to grapple with ensuring their power needs are met in the context of gentailers having the ability, and the authority as managers of the virtual power plant, to draw down household energy reserves to meet their market supply demands.[8] The motivation for investing in solar panels as a means for decarbonising becomes marginalised.

This organisational transformation subordinates the solar residential community to a different accumulation dynamic. The gentailer's value capture amounts to a form of rentier capitalism as household investment in electricity generation for domestic use becomes a source of revenue when exported into the grid. The gentailers actively cultivate this commodification process to their material advantage. One strategic manoeuvre

[7] No doubt, the concern to contain the network-dominated vertically integrated energy corporations strengthening their position in the market also influenced this decision. But this also limited households' partnering options and with whom they could contract with to sell their surplus energy. Enhanced control over 'uncontrolled energy' comes at the expense of households.

[8] The AEMC applauds the rise of 'a new type of energy entrepreneur' (AMEC 2018).

designed to give effect to this and expand the scale and reach of their surplus harvesting has been to provide financial advances to support an ever-growing number of households to install panels and storage capacity. The financialisation of the partnership that fosters household investment in generating energy establishes a stream of revenue for the gentailer. This is not necessarily to the immediate disadvantage of the household. Such financial arrangements can be advantageous for households. The gentailers' contracts with prosumers are generally based on the household drawing on the rooftop power they are generating, paying the gentailer a price that is comparable to the feed-in tariff rate or above but less than the market rate for electricity drawn from the grid. Any energy surplus to their needs can then be fed into the grid for which they are credited, typically an amount equivalent to the feed-in tariff rate. The payments for electricity consumed and credits received are debited as instalments against the financial advance, and once this has been settled, the panels and battery storage system and the returns from energy generated will normally become the property of the household. Such arrangements inject an entrepreneurial demeanour into the functioning of the solar household.

This shift in orientation has its counterpart in the characterisation of virtual power plants as a form of rentier capitalism, as retailers turn to extracting the value generated in the residential sector by exploiting the disparity between payments for feed-in tariffs and the rate charged for power imported from the grid. The markedly lower rate of remuneration for power fed into the grid compared with the cost of power exported from the grid provides a material measure of the power imbalance that defines the virtual power plant partnership. The disparity becomes the principal means for gentailers to capitalise on the value being generated by households. Moreover, it is a gap that has widened as more and more households generate electricity and partner with gentailers to distribute the energy surplus to their needs. The gap is becoming ever more substantial as the energy retailers lower the rate of the feed-in tariff ostensibly in response to the increased flows of distributed energy and the decline in the wholesale price of electricity and especially following the end of most State-funded schemes in 2017 (Arup 2011; Jewell 2015; Parkinson 2015; Energy Matters 2017b).

The rate of feed-in tariffs is now set by the energy retailers, and this varies considerably from one jurisdiction to another and varying according to when feed-in contracts were made, but the tariffs are now substantially less than in the decade or so following their introduction in 2008 when the rates were set to encourage interest in investing in solar panels. Some states have published reference rates to signal a reasonable rate of remuneration but are reluctant to regulate these, justifying the decline in tariffs in terms of the fall in spot prices even though retail prices have not declined to anywhere near the same extent (Jewell 2015; Energy Matters 2017a; New South Wales Department of Environment and Planning 2020; Solar Markets 2017; Victoria, Environment, Land, Water and Planning 2020). Feed-in tariff rates are now substantially below, by up to a quarter, the rate that customers are charged for power supplied from the grid (Parkinson 2015; Energy Matters 2017b; Solar Markets 2017; Baran 2020). State pricing regulators are generally supportive of not intervening in markets to arrest the decline in tariffs (Vorrath 2019). One leading industry commentator has opined that the feed-in tariff will become a thing of the past before too long (Macdonald-Smith 2019).

Paradoxically, the calls for ending the material advantage of solar-powered households that comes with state support have recently been complemented by calls for these households to be charged for the right to access the grid, given that additional investments have to be made to accommodate the feed-in of decentralised power. Welfare groups and the transmission company SA Power Networks (SAPN) have highlighted the inequitable nature of the rebate and lobbied the Australian Energy Markets Commission to change market rules to allow charging of household exporters (Harmsen 2020c; ACCC 2018; Eacott 2021). While there is substance to this concern with inequities in the energy market, the principal motivation for the rule change is not equity but stability, with rationing access to the network to minimise congestion (Foley 2021b; Roberts 2021; Williams 2020; Markham 2019).[9] In fact, there is a lack of balance in the rule changes regarding access to the distribution network. AEMC's rules prohibit large generators from having to pay for their use of the network. Indeed, new rules that foster demand management are based on large energy users being compensated for foregoing energy use at times of excess demand. The measure to regulate one source of distributed energy resources will penalise households. Indeed, the changes could well exacerbate inequities because households will no longer be able to capitalise on their investments in panels and batteries; retailers will profit by having

[9] At the same time AEMO has successfully sought a rule change that gives it and retailers the right to block households exporting power to the grid (Harmsen 2020a, b; Foley 2021a, b).

access to household-generated power when it suits, paying a negligible sum for this power, charging a premium price for supplying power to households not in a position to become more energy independent and continuing to ensure that generators and distributors receive an above-norm return on their investments. It is clear in whose interests AEMC is operating, and this exposes a material measure of the extent to which the liberalised market provides an arena that is eroding the substance of the empowerment contention.

Energy Market Liberalisation: Affordability and Reliability Mandates and Who Carries the Risk?

While federal and state governments have been reluctant to address some of the inequities in the energy market, the federal government has been quite forthright in mandating energy retailers to assume responsibility for delivering affordable electricity and ensuring the reliability of supply. Both demands have had implications on household experiences in the energy market. Whereas the federal government has introduced a default price and threatened energy corporations with penalties for not reducing prices, including the extraordinary warning to divest energy companies of assets if they do not meet these obligations, there have been no comparable initiatives to protect solar households from retailer profiteering. Threats have been made, repeatedly by the Minister for Energy, but this seems little more than hyperbole. Some pressure has been brought to bear on retailers to provide more concessions for low-income households experiencing financial stress and having difficulties finding the resources to pay their electricity bills and to the negotiate payment plans, but this has mostly taken the form of moral suasion. The reduction in the rate of feed-in tariffs suggests that one of the key ways in which energy retailers will respond to the affordability direction is to cut tariffs still further.

The advocacy of enhanced supply reliability has had more traction with significant consequences for household rights in the market. The concerns with the impact of the increased proportion of energy from distributed sources have incited a quite different order of interest and action by state authorities. The Energy Regulator and the Energy Market Operator have focused considerable resources on formulating measures to ensure reliability of supply. Foremost among the initiatives has been the pressure imposed on energy network operators and retailers to meet reliability requirements as a condition of their operating licences. The promotion of demand response management has been a critical aspect of this, shifting the focus from supply to managing demand to overcome shortfalls in supply. There is reasonably widespread support among the regulators, energy generators and energy-intensive industry for the management scheme to be regulated and extended to the distribution/supply sector, including energy consumer advocates (Memery and Livsey 2019).[10] The regulations governing demand response management are being put in place for the supply of energy to large-scale, energy-intensive users and do not yet extend to the retail sector although plans for such are being formalised with the introduction of the Retailer Reliability Obligation in July 2019 (Australian Energy Market Commission 2020; Australian Energy Market Operator 2019; Energy Security Board 2020; Parkinson 2019). State governments have been working with retailers to trial demand response techniques for some time now, principally with the big three energy corporations and often in conjunction with IT companies, and ARENA has funded some projects (Origin Energy 2018; Energy Matters 2018; Vorrath 2018; Chang 2017; Energy Australia 2017; ARENA 2019; Energex 2019; Ergon Energy n.d.).

The federal government's reliability supply policy, with the government threatening to divest energy companies of assets if they do not meet reliability and affordable supply obligations, is also pressuring the corporations to lock in demand response management arrangements. These would provide them with the ability to requisition power based on household's generating capacity and stored energy reserves to be used as a buffer for meeting supply benchmarks. However, the Commission and the Regulator are holding back regulating and extending demand response management to involve prosumers and gentailers while rules governing practices in the distribution/supply sector are formulated.[11]

[10] Support by Consumer Energy Australia seems to reflect a conviction in the ability to rectify market failure through more liberal reforms.

[11] The network operators, including the vertical integrated energy corporations that have an interest in the distribution/supply sector and who are already engaged in demand management projects, are comfortable with this delay because it is holding back some service providers from competing with each of the three majors (Hobday and Goswell 2019; Memery and Livsey 2019; Long 2020). Consistent with privileging the market as a mechanism for promoting efficiency and containing costs, the Australian Energy Regulator has canvassed industry and consumers on the value they place on reliability (AER 2019).

The introduction of demand response management that would involve households is being supported by governments and energy corporations alike. The rhetoric of empowerment is now linked to the deployment of smart technologies: 'valuing customers in the energy system and supporting them in smarter ways that benefit them, the energy system and Victoria' (Victoria Environment, Land, Water and Planning 2019; New South Wales. Parliament. Legislative Council 2018). According to EnergyAustralia's Managing Director, demand response management is 'an approach that puts customers in control and keeps costs down' (EnergyAustralia 2017). EnergyAustralia promises to reward customers by providing 'a financial incentive in exchange for reducing their demand or making excess energy available'.

The reality is in fact otherwise. The Australian Energy Market Commission's decision to restrict rules governing demand management to the wholesale market and energy-intensive industrial and commercial users has placed electricity retailers in a strategically powerful position. And if the retailers are not prepared to lead the way in sequestering household energy reserves to overcome supply shortages, AEMO has indicated it would have no qualms in acting, announcing that it has the authority to shutter South Australian households' energy use to protect the State from blackouts (Harmsen 2020b). AEMO CEO Zibelman has argued that clawing back energy from households, including shutting down household use of energy, would occur only in emergencies, but some residents have expressed their concern that they would lose all control over the energy that their rooftop panels are generating (Mercer 2019; Donnellan 2020; Harmsen 2020b).[12] While the precise nature of the Retailer Reliability Obligation is still to be determined, the Obligation has significant implications for solar households. The execution of demand response management will effectively place control over household energy with the energy retailer, which is why State and Territory government initiatives to

[12] This is also likely to be the case with the South Australian government's pioneering the installation of Tesla batteries for public housing for free, even in those residences that cannot or do not have solar panels. The batteries would be fed any power exceeding market requirements in a virtual power plant arrangement. The batteries would act as an energy sponge, storing power when generation levels exceed demand, from which households could draw on the stored lower-cost power when needed or the market operator could release power from the batteries to ensure orderly supply at peak-load times. These battery households would be acting like subcontractors who the market operator had outsourced storage (Bakhtar News Agency 2021).

promote residential solar generation on the condition that households partner in virtual power plants are so significant (Government of South Australia 2020).[13] Demand response management at the level of the retail market will extinguish household control over the electricity that they are generating. There are some quite striking indications of just how much this might play out to the disadvantage of households. AEMO's preparedness to step in and take unilateral action requiring retailers draw down energy to stabilise frequency or meet supply shortfalls is one. Another is the level of recompense that is paid, while another arises with the lack of certainty with respect to the variability in and volume of energy extracted EnergyAustralia's parsimonious plan to offer to pay a household either $10 or $20 for each time a request is made for the residence to cut its power use, the compensation varying depending on the magnitude of the energy requested be cut, highlights the limited benefits of households participating in the market (EnergyAustralia 2019).

Given the institutional make-up of the energy system, the AEMO's observation that there is an urgent need to bring 'uncontrolled energy' under control if an orderly market is to be secured subordinates solar households to the dictates of the energy retailers. With most virtual power plants organised under the governance of energy corporations, control is imparted to the most powerful enterprises in the market. This is obviously at odds with the *laissez faire* construct on which the National Electricity Market is ostensibly modelled.[14] Technological advances have exacerbated the shift in market power relations.

More generally, this also signals a dramatic shift in the management of risk in the face of uncertainty. Energy retailers can turn to solar households to make good any compromise in the supply uncertainty. The emergence of prosumers as an institutional force in the energy system has also freed the energy corporations from having to assume the role of the principal

[13] The South Australian government is among the strongest proponents of virtual power plants. The Homes Battery Subsidy Scheme offers financial assistance to households to install battery storage systems, promoting virtual power plants as a win for households and retailers. There are several schemes being led by Simply Energy Virtual Power Plant, AGL Virtual Power Plant, ShineHub Community, Tesla Energy Plan and Sonnen Flat.

[14] This was a concern raised by Andrew Dillon of Energy Networks Australia: 'One of the key concerns I would have, as a consumer, we may soon get to the stage when networks will be OK to connect new solar customers but not let them export because we can't be sure we can safely allow more to connect partly because we're not sure of the capability' (Macdonald-Smith 2019a).

investors in the future of the energy generation. Households have taken on that role, initially on the basis of 'taking control of the bill' and then on the promise of a return on their investment, but neither of these can be assured just as their ability to exercise some autonomy and energy self-reliance is compromised by the mandating of demand response management. In the process, some portion of the risks thrown up by a market and an industry in transition are now going to be borne by households and the savings captured by retailers in the form of profits (Lucas 2021). The actual marked reductions in feed-in tariffs being paid and the anticipated deflation in market prices, which can only be reinforced by the continued increase in residential solar generating capacity as well as in the increased capacity from industrial-scale wind and solar farms, point to the likelihood that returns of investing in panels and battery systems is no longer the economic proposition that was envisaged. Indeed, it is to be expected that declining feed-in tariffs will erode any prospect of households' investments offsetting the savings from reducing their purchases of power from the grid. What control they have had over the power generated from their rooftop solar will largely evaporate as retailers are delegated the role of managing households' call on household energy generation capacity.

There are still other costs faced by households that are generated by the loss of autonomy. In considering the promise of empowerment it is also essential to reflect on how optimising electricity utilisation and distribution affects the organisation of activities within the household. Certainly, the wealth of data collected on energy generated, used and distributed places the household in the advantageous position of being able to make more informed decisions about how to optimise the benefits that can be derived from their investments in solar panels, battery systems, behind-the-meter and other smart technologies. The digitalisation of information collected has made this evaluation process so much easier, but there are some qualifications that must be rehearsed with this positive scenario. Firstly, while maximising energy efficiency is the goal, which can be facilitated with recourse to the new technologies, most of these household technologies are powered by electricity and generally require they remain connected to the power source, and this be turned on. Some data processing systems and most notably blockchain draw down an inordinate amount of energy. Secondly, weighing up the benefits and costs of the various priorities that frame the utilisation of energy *vis-à-vis* feeding energy into the grid and which generate a financial return can be problematic because the household is required to compare the value of various options that are incommensurable, and this problem is compounded by electricity being a necessity good. For example, what factors figure in estimating the cost of postponing or abandoning a domestic chore, as varied as preparing a meal, vacuuming or sitting in front of a screen, each of which draws on energy, in order to have a sufficient supply of power to meet the virtual power plant's planned requisition, and in what currency are the costs and rewards denominated? Is it in terms of kilowatts or dollars, and what is omitted in the process? Valuations will presumably vary according to whether power is fed into the grid voluntarily or requisitioned as a demand response management decision. Invariably, the process of commensuration of the various options will result in the reification of some activities as the most likely currency for valuation will be in dollars.

The Metamorphosis of the Household: The Prosumer and Reification

The contractual relations that bring the household, or prosumer, together with the energy retailer, or gentailer, are represented as being a relatively straightforward process, the prosumer agreeing to a fixed reimbursement for forgoing its unfettered control over the energy that is being generated and stored. However, without the clarity and definition of a regulatory framework that would set some benchmarks and standards, retailers will be free to define the terms of those households being drawn into participating in demand management schemes under the guise of feed-in tariff agreements (Memery and Livsey 2019; Parkinson 2012; Long 2020). Indeed, the Australian Energy Market Commission has warned that once households become embedded in such contractual arrangements their consumer protections become compromised (AEMC 2019: 63). Notwithstanding this reservation, retailers are arguing that the existing regulations that require households to sign onto demand response management schemes under the direction of an energy retailer should be retained (Hobday and Goswell 2019; Memery and Livsey 2019). The retention of the existing arrangements will be to the financial advantage of the retailers because, through the deployment of 'behind-the-meter' and other smart technologies, they are able to determine when it is to their financial advantage to quarantine household power and sell this into the market. The application of the reliability and affordability of supply requirements that energy retailers are to be subjected to makes this capacity to control household energy flows all the more crucial.

One emblematic illustration of this is Power Ledger's energy trading system. More than any other system designed to incorporate solar households in market deliberations, Power Ledger promises to ensure that the exercise of household agency lies at the heart of the model trading system it is implementing. The model is based on garnering all relevant data related to the parties who are members of Ecochain and generating, distributing or buying energy and storing this in a proof-of-stake blockchain system, which has the capacity to install an unlimited amount of data. Access to the data resource invites the members of Ecochain to draw on the record of all transactions related to the generation and distribution of energy. This information resource is argued to enable households to make the most of their investment in energy generation and trading of this energy. Unlike the virtual power model, where decisions on energy flows are unilaterally determined, the Power Ledger system is based on transparency of information related to transactions which establish an equality or symmetry in bargaining for all Ecochain members, and this is said to instil trust in the trading relations. Access to the database establishes a confidence in the capacity to gain immediate insight into the state of the market to inform educated responses about the changing conditions and the effect of the prosumer responding to these. For example, what is the consequence of increasing the export of energy in response to the incentives that present with price inflation or with choosing to trade with particular entities? In this model, rather than the operator of the virtual power plant exercising unilateral decisions, the household continues to have a voice in trading to exercise agency.

Yet, as much as the system promises to maintain the sociality of transactions through peer-to-peer trading, optimisation depends on the processing of a vast amount of information.[15] This necessarily presents the same problems that present with virtual power plants. Calculating requires

[15] The Power Ledger model proposes to record all transactions in cryptocurrency to avoid the effects of political interventions in the market which would make for an invariable measure of value. This, of course, is misleading because cryptocurrencies are traded which mean that the relative value of the reference value, the currency, is not invariable. Turbulence in the markets for bitcoin and other cryptocurrencies has demonstrated that it is not immune from volatility. This exposes another level of risk that confronts the solar household, and, as the Australian Energy Market Commission has observed, once households become embedded in such contractual arrangements their consumer protections are compromised (AEMC 2019: 63).

establishing some commensuration of the various variables that are included in the database. Algorithmic sequencing provides a solution for processing the wealth of data, and for positing optimising trades, but the calculative logic is only as good as its ability to capture and value in a meaningful way all the variables that must be accommodated in optimal decision. As is the case with the conceptual foundations of the liberalised energy market, recourse to algorithmic sequencing is based on a process of reification, presenting the illusion that it captures all the variables that determine an optimal outcome. It also misrepresents the capacity for the household to exercise agency when in fact the household plays no role in setting the specifications that frame the sequencing and thus the calculation of the optimal options.[16] Optimisation is in fact based on a calculative logic and household agency is sacrificed at the altar of the calculative power of the blockchain technology. Moreover, this power is framed by what happens within the grid, or the micro-grid, and value of trades in other markets, including energy derivatives, such that determinations made in the individual households in which human agency has a say are going to all but be overshadowed by the infinitesimal machinations that define and frame that calculative power.

To reiterate, the optimal options in the Power Ledger calculative model are necessarily as good as in inclusive for all the transactions that influence decision-making, but it is apparent that there are limits to this because of the difficulties of commensuration. The lack of a common valuation standard that enables the capture of all transactions will necessitate some filtering out of some variables, and this can result in the algorithm sequencing generating optimal options that do not reflect the household's priorities. Optimising the generation, storage, utilisation and distribution of electricity, for instance, could be made using kilowatt-hours as the standard of value, but this cannot capture the energy expended, or saved, involved in the exercise of domestic labour, which cannot be designated or valued in kilowatt-hours. The same problem presents if the measure of value is in

[16] Power Ledger is also advertised as a means of complementing the decentralised nature of solar household generating distributed energy because the Ecochain system is decentred, based on blockchain. With all parties having access to the database, the tendency for decisions to be made by oligopolistic energy corporations is avoided. However, the calculations are made utilising a model in which sequencing is set by the Power Ledger operators and which households are not party to so that the model limits the exercise of human agency and in this respect is only marginally more democratic than the virtual power plant.

dollar terms, and calculations of optimal options are likely to be generated by filtering out the labour variable. Optimisation becomes a mechanistic and reified process devoid of its human content.

This is being played out concretely with the energy sector signalling the most 'efficient' times to undertake certain activities. It amounts to the programming of household labour power and energy use to align with the daily fluctuations in market demand and supply in order to be regarded as maximising the return on the household investment in generating and storage capacity. With the value of human labour overlooked and who undertakes this labour ignored, the organisation and timing of the use of various electric appliances are set to minimise the use of electricity and to maximise the electricity being fed into the network when market demand and electricity prices are high, and holding off feeding electricity into the network so that appliances are used when market demand is low as are market prices. This optimising exercise can, for instance, inform when is the most 'efficient' time for the washing machine to be operating and switched on. As a relatively energy-intensive device, energy efficiency would likely be optimised during the off-peak hours of the middle of the day when the sun is shining and spot prices for electricity are at their lowest. The vacuuming might most 'efficiently' be done in the wee hours of the morning of a weekend when the demand for electricity in the wider world is weak. Moreover, given the capacity of such technologies to communicate with one another (Sioshansi 2020a), these operational decisions can be programmed in line with the digitalisation process. Of course, what is missing from such optimisation processes is an appreciation of the functional consequences on the social division of labour within the household.

The marketisation of electricity generation and use across the public and private spheres of the energy political economy as the means of promoting energy efficiency displaces the social division of labour as the foundation on which the energy system is organised. Optimisation presumes a genderless household, as if the household functions according to the logic of *homo economicus* completely neglecting the sexual division of labour in the structure of work within the household and the gendered differences in the expenditure of human energy that is required to oversee the various apparatuses that make the household the social and reproductive realm that it is. Gender is critical in defining the division of labour and who carries responsibility for performing these domestic functions. There is a reasonably rich literature analysing the gendered implications of energy poverty in developing countries, where differential access to energy results in women's labours being more intensively called upon, but the different implications of the electrification of the household and the optimisation of electricity use in OECD households on the sexual division of labour have attracted comparatively little attention (Dutta et al. 2017; Winther et al. 2017). Dictating when household labour is most energy efficiently employed turns the household workers, and particularly women who continue to assume a disproportionately large share of the responsibility for household work, into prisoners of the prosumer energy cycle. The rhetoric of the liberating force of technology has a long history. The advent of the electric washing machine and the vacuum cleaner are classic illustrations of this, both holding out the promise of freeing up women's labour. In practice, these technologies wrought changes in women's work practices resulting in an intensification of women's work. While the electrification of the prosumer residence and digitalisation may reveal the pathway to optimising household wellbeing and the financial rewards from the electrification of the household, the choices about the organisation of household affairs are subordinated to the calculative logic of electricity market forces and can be disempowering to all or some of the occupants.

The point about commensuration is laboured here because similar criticisms can be made with respect to electricity. Irrespective of whether the value of power is measured in terms of kilowatt-hours or, more likely, in money terms, the distinctive qualities of renewable energy and the value that households place on their generation of renewable energy cannot be reflected in the value that is determined in the process of trading. The transformation in the socio-ecological relations that is effected in the residential endeavour to present electricity as more than a thing is lost when it is commodified and subsumed within the market. The import of the motivation to reduce the household's carbon footprint is overshadowed as renewable energy is marketised. What appears as a process of fetishisation contributes to an undermining of household empowerment.

Conclusion

The value of the energy that is being generated by household solar panels has contributed to the transformation in socio-ecological relations. However, scale of this decarbonising transition has also become a magnet of opportunity for a range of energy service providers eager to explore opportunities for capturing a share of the value. It is not just the energy retailers that have transformed their *raison d'être*, reconstituting as gentailers, to reassert their dominance in the National Energy Market. Other intermediaries have been established offering services to assist households to navigate the range of options for managing distributed electricity. With the smart technology creating a wealth of data on electricity generation and use, the processing and management of data is also attracting IT enterprises and start-ups that look to energy as a growing source of profit-making. Digitalisation has opened up the energy system to the involvement of new ventures that have had no previous acquaintance with the sector. These ventures assure a prospective clientele that they have the technical capability of being able to monitor and manage trade in markets as effectively as, if not more so, than the established energy corporations and deliver relatively higher returns to that client base. Power Ledger's model of a decentralised grouping of prosumers engaging in peer-to-peer trading is illustrative of such an arrangement, an arrangement that is based on Power Ledger capturing a share of the value generated by charging a commission on electricity that is traded. The promise of empowerment is clearly being qualified by the extent to which some proportion of the efficiency gains and the value generated in the residential energy value chain are being captured by parties beyond the household. This is all the more prescient given the evidence that the returns on household investment in generation and storage are being eroded and the control that households had over the energy value chain is being rapidly extinguished.

Certainly, the liberalisation of the electricity market has awakened the household agency in shaping the transition to a renewable energy future. This has contributed to transforming the energy generation and transmission system with more residential generation and distribution being decentralised challenging the dominance of the corporate capital accumulation model. Increasingly, however, the extent to which solar residences have been able to assert their agency and energy self-sufficiency has come under pressure to be integrated into the energy mix. Virtual power stations have become a vehicle for harnessing and aggregating the renewable energy generated by and stored in the household. In the process, households are being exhorted to assume a more entrepreneurial spirit as they, in effect, take on the role of sub-contractors to the gentailers (Parkinson 2012; AEMC 2018). Like many other examples of petty commodity producers, they bear the cost of investing in generating and storage capacity and are subjected to the intrusion of the gentailers sequestering power to meet their power needs. They carry most of the risks and uncertainties associated with the vagaries of the market. With feed-in tariffs falling quite markedly the partnerships in virtual power plants become a source of primitive accumulation for the gentailers, power requisitioned at a price anywhere between 10 and 40 per cent of the price of power supplied from the grid with the prospect of more price decreases in the pipeline ostensibly to regulate frequency. While this is underscored by the state's demand response management policy, the state is also envisaging extracting some of the value generated in the household and pushing down tariffs in the process, creating a 'solar sponge' to reduce the magnitude of the power fed into the grid during the day (Australian Energy Regulator 2020: 129; Parkinson 2020; Russell 2020). The chief executive of the Energy Security Board has also raised the notion of a feed-in charge as a means of sending appropriate pricing signals to ration the rate at which households feed in power to reduce variation in voltage flows (Parkinson 2020). This turn around transformation has happened in an extraordinary short time frame.[17]

The different claims on the household point to processes of contestation that cannot be reduced to the orderly and optimising exchanges that govern that autonomous subject, the economic island on *homo economicus*. The liberalisation of the electricity market has spawned winners and losers. This asymmetry in outcomes is mostly pushed aside by the state apparatuses that are formulating the regulatory structure of the NEM. Their preoccupation with the objectives of ensuring an orderly transition to the NEM, a transition that is hoped will deliver greater efficiencies and temper price inflation and possibly one that mitigates emissions in the energy system, is based on a vision that is having to be reconciled with a Coalition government's political agenda that remains wedded to a moribund fossil fuel-based industrial-scale power system. This is simply providing more opportunities for those energy corporations that superseded the privatised State-owned utilities to assert their dominance in the market. This hardly amounts to a story of empowerment 'from below'.

The freedom to choose that is ostensibly ignited by this *laissez faire* market is circumscribed and the urgency to transition to a clean energy future becomes an option subordinated to the rhetoric of affordable and reliable flows of electricity irrespective of the source of that energy. It is the limitations of this free market to accelerate the transition that has impelled the initiatives to forge alternative structures for driving change. The positive sense of purpose that has inspired so many households in Australia to invest in rooftop solar panels and battery systems is to be applauded. But the structural and political obstacles that have held back what could have

[17] AS AGL's Elisabeth Brinton observed: 'It is really about security of supply and it needed to be done quickly because a large plant can't be done quickly' (Macdonald-Smith 2017a).

been achieved have impelled other initiatives that contest established structural and political arrangements that define the market. This is evident with the re-municipalisation movement or the endeavours to abandon engagement with the national electricity market as we can see with the community-owned renewable energy cooperatives.

References

7 News (2019) 'Electricity prices to rise in Victoria with households to pay up to $50 more' 12 November.

ABC (2016) 'Fact check: Does privatisation increase electricity bills?' 3 March.

ABS [Australian Bureau of Statistics] (2019) *Energy Accounts 2018–19* Canberra.

Arup, T. (2011) 'Tariff cuts eclipse VIC household solar panel benefits' *Domain* 17 October.

Australian Competition and Consumer Commission (2018) *Retail Electricity Pricing Inquiry: Restoring electricity affordability and Australia's competitive advantage—Final Report* June.

Australian Competition and Consumer Commission (2017) *Retail Electricity Pricing Inquiry: preliminary report*, September.

ACOSS [Australian Council of Social Services], Climate Institute and Brotherhood of St Laurence (2017) *Empowering disadvantaged households to access affordable, clean energy*, Sydney.

Australian Energy Foundation (2018) *Energy Justice—We need to look after our neighbours* Melbourne, September.

AEMC [Australian Energy Market Commission] (2020) Annual Report.

AEMO [Australian Energy Market Operator] (2020) *Demand Side Participation forecast methodology consultation.*

AEMO [Australian Energy Market Operator] (2019) *Interim Reliability Forecast Guidelines* 2019.

AER [Australian Energy Regulator] (2020) *State of the Energy Market 2020.*

AER [Australian Energy Regulator] (2019) *State of the Energy Market Report 2019*: 14

AEMC [Australian Energy Market Commission] (2019) *Reflections from behind the meter symposium* 5 September.

AEMC [Australian Energy Market Commission] (2018) *Retail Energy Competition Review* June.

ARENA [Australian Renewable Energy Agency] (2019) 'Demand Response Project Performance Report' January.

Australian Institute for Health and Welfare (2019) *Australia's Welfare: Data insights*, Canberra.

Bakhtar News Agency (2021) 'South Australia to install free Tesla batteries in homes without rooftop solar' 2 June.

Bankwest Curtin Economics Centre (2016) *Energy Poverty in Western Australia: A comparative analysis of drivers and effects* BCEC Research Report 1/2016 June.

Baran, S. (2020) 'EnergyAustralia cuts feed-in tariff rates' *Finder* 20 February.

Bloch, M (2017) 'Interest free solar power system loans for Tasmanians' *Solarquotes blog* 2 May.

Blowers, D. (2018) 'Energy policy failure behind 10-year story of spiralling electricity bills' *The Conversation* 4 January.

Chang, C. (2017) 'Households will be paid to reduce power consumption during peak periods' *News.com.au* 11 October.

Chester, L., A. Elliot, and P. Crossley (2018) 'Improving energy affordability for Australian low-income renter household', 29–32 *IAEE Energy Forum*.

Chester, L., and A. Morris (2012) 'A new form of energy poverty is the hallmark of liberalised electricity sectors' *Australian Journal of Social Issues* 46(4), 435–458.

Department of Industry, Science, Energy and Resources (n.d.) 'Energy markets'.

Department of Industry, Science, Energy and Resources (2020) 'Support for Australian households' 6 July.

Donnellan, A. (2020) 'Concerns over plan to switch off household solar panels when grid is unstable' *Weatherzone* 20 May.

Dutta, S., A. Kooijman, E. Ceceliski (2017) 'Energy access and gender: getting the balance right' *SEAR* International Network on gender and sustainable energy, New York: World Bank.

Eacott, A. (2021) 'Proposed solar export charges leave owners feeling like they've been 'penalised'' *ABC News* 26 March.

Energex (2019) 'Demand management plan & initiatives'.

EnergyAustralia (2019) 'Households could see their power bills drop by helping to take the pressure down on hot days' *News* 5 December.

EnergyAustralia (2017) News—Energy Australia to deliver 50MW in demand response trial, 11 October.

Energy Matters (2017b) 'Is it time for an energy health check?' 15 December.

Energy Matters (2017a) 'Solar feed-in tariffs in Australia: what's the latest?' 15 December.

Energy Matters (2018) 'Solar battery systems in 100 South Australian homes form a virtual power plant' 16 July.

Energy Security Board (2020) *Interim Reliability Measures—Reliability Reserve Consultation on Draft Rules*, May.

Ergon Energy (n.d.) 'Demand management'.

Estcourt, D., (2020) 'Victorian government to introduce interest free loans for new rooftop solar panels' *The Age* 27 July.

Eyes, J. 2020 'Solar installers fight ACCC on buy now pay later; *The Australian Financial Review* 18 February.

Faulkner, D. (2021) 'Low-income families back proposed export fees in hope of reducing their power bills. *ABC News* 26 March.

Foley, M. (2021b) 'Energy experts slam plan to charge households supplying solar to the grid' *The Sydney Morning Herald* 26 March 3.

Foley, M. (2021a) 'Solar homes face charges for powering energy grid' *The Sydney Morning Herald* 25 March.

Foley, M. (2020) "Financial contagion' from coronavirus may infect electricity market' *The Sydney Morning Herald* 28 May.

Foley, M., and N. Toscano (2020) 'Struggling consumers 'must retain connection' *The Sydney Morning Herald* 28 July, 11.

Fowler, E. (2019a) 'Energy Charter needs to be energised' The Australian Financial Review 5 December, 18.

Government of South Australia (2020) *Home Battery Scheme: Join a VPP.*

Hannam, P. (2018) "Sabotage': Regulator slammed for plans to slash feed-in tariffs' *The Sydney Morning Herald* 3 July.

Harmsen, N. (2020c) 'Rooftop solar panel owners could be getting charged fees to sell back to the grid' *ABC News* 16 July.

Harmsen, N. (2020b) 'Power granted to switch off household solar in SA to prevent statewide blackout' *MSN* 19 June.

Harmsen, N. (2020a) 'Energy operator wants to remotely switch of rooftop solar systems amid 'uncontrolled growth" *ABC News* 2 April.

Hetherington, D. (2017) 'From public good to profit: how privatisation is failing our communities' *The Guardian* 6 March.

Hiley, T. (2019) 'Huge new NSW solar scheme on the way' *Wattever* 10 February.

Hobday, L., and G. Goswell (2019) 'Households cut power bills with 'demand response', but big energy retailers want to keep grip on market' *ABC News* 24 June.

Hogan, C., and B. Salt (2017) 'Census insights: the rise of energy poverty in Australia' KPMG 11 September.

Jewell, C. (2015) 'Solar industry fumes as feed-in tariffs slashed' *The Fifth Estate* 1 September.

Judson, E., F. Zirakbash, A. Nygaard, and A. Spinney (2019) *Renewable Energy Retrofitting and Energy Poverty in Low-income Households Research report for United Housing Cooperative Final Report* Swinburn Centre for Urban Transitions, August.

Labaste, S. (2019) *Households in the Dark II* Analysis and Policy Observatory 1 October.

Long, S. (2020) 'Electricity users will get paid to cut energy use under historic new market reform' *ABC News* 11 June.

Lucas, C. (2021) 'Power failure: Home hit by solar limits, as distributors protect network, and profits' *The Age* 14 March.

Lucas, C. (2019) 'Tenants raise roof over solar power, demand fair share of panel grants' *The Age* 17 May.

Macdonald-Smith, A., (2019a) 'Plan targets saving as solar strains grid' *The Australian Financial Review* 22 July, 19.

Macdonald-Smith (2019) 'Solar households to face hit from zero prices' *The Australian Financial Review* 22 October.

Macdonald-Smith, A. (2017a) 'AGL Energy's tech high-flyer to redefine energy for customers' *The Australian Financial Review* 21 April.

Mannheim, M. (2020) 'Canberra will be home to one of Australia's biggest batteries as ACT weans off fossil fuels' *ABC News* 12 September.

Matich, B. (2019) 'ARENA funds WATTwatchers toward more consumer empowerment' *PV Magazine* 19 December.

Markham, D. (2019) 'Suns of anarchy' Australian Energy Council, 20 June.

Mazengarb, M. (2019b) 'City of Sydney strikes landmark deal to go 100% renewal' *Renew Economy* 22 October.

Memery, C., and A. Livsey (2019) 'Households get less in landmark demand response decision that will cut power bills' *ABC News* 19 July.

Mercer, D. (2019) 'Authorities look to control household rooftop power systems to stabilise the grid' *ABC News* 8 December.

Mike (2018) 'Solar power for rental properties' *savingwithsolar* 19 February.

New South Wales. Parliament. Legislative Council (2018) Select Committee on Electricity Supply, Demand and Prices in New South Wales—*Final Report.*

NSW Energy, Department of Planning and Environment (n.d.) 'Renewable Energy Action Plan'.

NSW Department of Industry, Planning and the Environment (2018) 'Trial to deliver power bill relief'

Origin Energy (2018) 'Origin to fire up first power plant in Victoria' *Press Release* September.

Parkinson, G. (2020) 'Solar households may soon be charged network fee for exports to grid' *Renew Economy* 2 July.

Parkinson, G. (2019) 'Blaming renewables for reliability issues is "wrong and dangerous" *Renew Economy* 11 February.

Parkinson, G. (2015) 'Networks propose compulsory fees for all—to stop grid defections' *Renew Economy* 5 August.

Parkinson, G. (2012) 'Energy retailers: do we really need them' *Renew Economy* 5 July.

Phillips, B. (2018) *Energy stressed* ACOSS, Brotherhood of St Laurence and Centre for Social Research and Method.

Queensland Government (2018) 'Connecting your solar system to the grid' 23 February.

Rate of Return Consumer Reference Group (CRG) (2018) Submission to the Australian Energy Regulator Rate of Return Guideline Review, May.

Roberts, E. (2021) 'Making solar pay is coal comfort' *The Sydney Morning Herald* 26 March, 23.

Russell, C. (2020)'Free solar for new homes' The Advertiser 30 January: 13.

Salleh, A. (2018) 'Solar boom: new schemes may help renters get solar on their roof' *ABC News* 20 February.

Salt, B., and C. Hogan (2017) 'The rise of energy poverty in Australia' Census insight series: KPMG.

Sansom, M. (2014) Solar saver program gives pensioners gives power over bills' *Government News* 18 September.

South Australia Power Networks (2020) 'Empowering South Australia' 3 February.

Sioshansi, F.P. (2020b) 'Creating value behind-the-meter: Digitalization, aggregation and optimization of behind-the-meter assets', in F.P. Sioshansi (ed) *Behind and beyond the meter: Digitalization, Aggregation, Optimalization, Monetization* London: Academic Press: 47–82.

Sioshansi, F.P. (2020a) 'What lies behind the meter' in F.P. Sioshansi (ed) *Behind and beyond the meter: Digitalization, Aggregation, Optimalization, Monetization* London: Academic Press: 3–29.

Solar Market (2017) 'To pay households fairly, we need to understand the true value of solar' 7 February.

Solar Savers (2017) 'Removing the barriers to solar' 31 October.

Solar Victoria (2019) 'Solar panel rebate'.

South Australian Government (2019) 'Solar voltaic systems and battery storage' 30 September.

Stagnaro, C., and S. Benedettini (2020) 'Smart meters: the gate to behind-the-meter?' in F.P.

Tasmanian Council of Social Services (2018) '120,000 Tasmanians living in poverty', Hobart.

Taylor, Andrew (2020) "Billing anxiety": How COVID-19 is driving up the cost of staying warm this winter' *The Sydney Morning Herald* 19 July.

Tchetvertakov, G. (2018) 'Victoria approves further offshore gas exploration and sails into early criticism' *Small Caps* 18 May.

The Solar Man (2018) 'Queensland opens rooftop solar market to low-income and rental households' 18 July.

Victoria, Environment, Land, Water and Planning (2020) 'Minimum Feed-in Tariff' 25 March.

Victoria, Environment, Land, Water and Planning (2019) 'Demand management achievements' 1 July.

Vorrath, S. (2019) 'Queensland cuts regional solar feed-in tariffs as renewables push prices down' Renew Economy 4 June.

Vorrath, S. (2018) 'Rental solar scheme targets 15,000 Queensland investment homes' *Renew Economy* 20 April.

Vorrath, S. (2016) 'Retailers score major win in battle for behind-the-meter market' *Renew Economy* 30 November.

Williams, P. (2020) Power bill debts to hit retailers' *The Australian* 28 July, 17.

Winther, T., M.N. Matinga, K. Ulsrud, and K. Standal (2017) 'Women's empowerment through electricity access: scoping study and proposal for a framework of analysis' *Journal of Development Effectiveness* 9(3): 389–417.

CHAPTER 8

Climate Citizens Embracing Renewable Energy: Re-municipalisation

In the context of national governments resisting the transformation of the political economy by underwriting the transition from coal-fired electricity generation to renewable sources of energy households have been exploring a range of initiatives to personally lead the take-up of renewable energy technologies. A particularly striking manifestation of this in Australia has been the investment in rooftop solar voltaic panels. This has been represented as the archetypal rational individual economic agent maximising the opportunities afforded by the free market. But this obsession with the market-framed behaviour of the rational economic agent, the prosumer or 'utility citizen', rests on an atrophied construct that overlooks the place of the citizen in the broader political economy. It ignores the other attributes that go into the making of the citizen in civil society and the polity and its institutional context. The citizen's exercise of choice in the quest of shaping a future is not simply limited to the economic domain. The full force of human agency operates in these other realms, as social and political beings, and citizens' concerns about the challenges presented by climate change can manifest in them contesting other dominant structures that contribute to the constitution of the emissions-intensive fossil fuel-based industrial complex.

One arena in which the energy order is being contested is through the institution of the state and particularly the municipal state. This has proved to be a markedly different route for pursuing the quest to break the reliance on fossil fuels as the primary source of energy. This has been made

S. Rosewarne, *Contested Energy Futures*,
https://doi.org/10.1007/978-981-19-0224-6_8

possible with the animation of citizens pressuring local government municipalities to take the lead in transitioning from fossil fuel-based energy. This has been taking the form of pressuring municipalities to reduce their carbon footprint. It has also focused citizen energy on pressing those local governments, councils or municipalities that operate power utilities to seize the initiative and invest in clean energy or lobbying those municipalities that had abandoned a role in the generation and/or supply of electricity during the successive waves of privatisations of public utilities in the mid-1970s and then the 1990s to resume the role of energy provider as a clean energy supplier. The focus of such citizen pressure has been brought to bear overwhelmingly on municipalities to resume this role largely because it is, and historically was, predominantly this level of the state that had taken on the task of in investing in the technology to produce electricity and gas for the local community because of its public good nature and increasingly as a necessary good in the modern age.

Politicising the State: Drawing Municipal Governments to Climate Change Politics

Citizen lobbying of local governments to play a proactive role in reducing an area's carbon footprint, such as re-municipalisation initiatives that involve the local state to resume generating electricity to generate clean energy has turned the local state into a something more than an administrative apparatus and provider of services. The municipal state has emerged as a site of climate struggles, establishing another terrain on which energy democracy is played out, and recasting this tier of the state as a political subject. In fact, through their associations, local or municipal governments have become quite proactive in the efforts to forge the transition to renewable energy, and thereby demonstrating institutional agency. The political advances that have been forged and how these have informed other climate change initiatives are examined in this chapter, focusing mostly on developments in the United States and Germany before reflecting on the bearing of these on developments in Australia.

No doubt, the exercise of popular pressure has also had, and continues to display, more traction with municipal governance than it has had, and does, with the regional and national tiers of the state system. Equally, if not more so, it is evident that local governments have also responded to the opportunity to assert some influence in developing measures to mitigate emissions and support the transition from coal-fired to renewable sources of electricity. European and North American municipal council developments have provided model advances in the transformation of energy systems and the formation of international forums that bring municipal councils together with the specific objecting of sharing experiences highlights how climate politics has transcended the local and garnered further momentum. The municipality has thus become an important site for the exercise of citizen agency at the same time as it is giving vent to institutional agency. As a result of community campaigns, municipal governments are, in many instances, playing a critical part in the transition from fossil fuel-based electricity generation to renewable energy, and while the nature of the engagements at the behest of citizens varies considerably, the engagements have been highly contested processes and no more especially so than with the campaigns for re municipalisation. This has been the case in the United States and Germany where campaigns to win municipal support for taking the leap to produce clean energy have been confronted by stiff and extremely well-funded resistance from energy corporations. The role of municipalities in energy generation and distribution in Australia has been much less pronounced, but this has been changing as a result of community pressure as well as because municipalities are also seeing economic and other supply advantages in shifting to clean energy and becoming agents of change in the process. We can gain some critical insights into the significance of municipal governance in energy system transformation in Australia by first considering the developments of energy systems in the much-celebrated Denmark for its energy initiatives, and in the United States and Germany.

Before moving onto this analysis, it is sensible to note the distinction between the individual household's engagement with the energy system mediated through the market, and the citizen who lobbies the state for action on climate change as a political process. The household investing in solar panels, the prosumer or 'citizen utility' and the environmentally and politically engaged citizen, the 'climate citizen', operate in different social and mental realms to *homo economicus* and his pursuit of different goals. This makes for a different construct, introducing a fluidity in the multiple ways in which the response to the climate challenge transcends individualist responses to assume more social forms. The celebration of the individual energy consumer as an agent of change overlooks the fact that the representation of her/his aspirations is narrowly conceived. Viewed through the lens of the prosumer construct, there is no reason to consider that there is any rationale for contesting the fundamental structure of and the power relations that frame the market system in which decisions to invest in solar systems are being made. The structure of the energy-emissions system remains beyond scrutiny. Decision-making generally proceeds as an individualised phenomenon. As observed in the previous chapter, this construction can downplay the social contexts in which some of these household decisions are made to proceed with installing solar panels, such as a community's collective endeavour to take action to mitigate emissions or households that are influenced by and following the lead of neighbours to install panels. The prosumer is an attenuated construct; she/he belies the existence of civil society and its energy organising institutions, and the power relations that structure the market and delimit options. Acknowledging the citizen as a more broadly constituted social construct disrupts the idea of the renewable energy surge being defined by this one-dimensional oeuvre, the market. The climate citizen pursues her goals both within the bounds of the energy market system as it is constituted and also gives voice to her/his ambitions in realms beyond the market. In acknowledging the citizen, we can posit the emergence of a social and political actor whose animation ignites the potential for challenging the dominant institutional arrangements and the emissions-intensive modes of generating and utilising energy to drive energy democracy.

The designation of the *climate citizen* quite deliberately invokes a quite different social subject whose oeuvre is much broader, politically engaged, than that of the individual whose demeanour is cast as not looking beyond the household (Gløjmar Berthou and Vind Ebbsen 2016; Roberts et al. 2014; Slocum 2004).[1] In the place of individuals, the responses to the climate change challenge can be seen to be galvanised around community-defined social enterprise, a collective endeavour projected onto the political plane, as can be observed with the Solar Citizen's *A Plan to Repower Australia* (2016, 2018). Acknowledging the purchase of climate politics requires a shift in the field of focus from the private sphere of the household to the public sphere of the community and the political economy.

[1] An alternative nomenclature that has been voiced in Australia is *solar citizen*, which more realistically captures the essence of climate citizens who invest in solar panels (Solar Citizens 2020; Ison et al. 2018). Here we retain the construct climate citizen to convey an agent who looks beyond the installation of household solar panels to underwrite a broadly framed foundation for empowerment. See, for instance, Rubik and Kress (2014), Flemsæter et al. (2018), and the Citizens' Climate Lobby (https://citizensclimatelobby.org/). The idea of climate citizens also entertains the notion of climate democracy.

Local communities have become more organisationally and politically engaged to advocate a collective cause, to realign established institutional arrangements and to forge new ones. One of the ways in which this is being expressed is by projecting the call for action through the medium of the state, by elevating the role of local municipal government more particularly.

This provides a quite different construction for considering the constitution of the energy system and how the system is being challenged and transformed. One of the agendas that have framed the emergence of the climate citizen has been the call for the democratisation of energy. The democratisation project engages a different pathway to empowerment, the morphing of the consumer-come-prosumer into the climate citizen, shifting the bases of empowerment from the sphere of the market to the political realm, the arena which defines the make-up of the market. The exercise of energy democracy is aimed at disrupting this arrangement, to try to wrest control of energy generation and distribution from the oligopolistic energy corporations, and in some instances monopolistic state-controlled generators, and to transform existing institutional arrangements to forge new ones that foreground the decarbonisation of electricity generation and ensure community access to clean energy.

Identifying the climate citizen can also help us understand the predilection for focusing lobbying efforts on the municipality as a political institution instead of the other tiers of the state system. There is the obvious explanation that it is uncommon for the national state to own and operate electricity generation plants. Likewise, this is mostly the case with regional states, although State governments in Australia did for a time control power utilities before the utilities were privatised in the 1990s.[2] Notwithstanding important differences in the structure and organisation of state systems, the *raison d'être* of the national state can generally be characterised by it conveying the impression of it not being subject to the influence of particular stakeholders, that it represents all of the nation's constituents. Its legitimacy is contingent upon governance being represented as relatively autonomous from the different constituents that seek to assert their will upon the definition of state policy and state practices, however much the actuality tells a different story as we have observed in

[2] In fact, the smaller populated State of Tasmania and the Australian Capital Territory continue to own and control energy utilities, as does the State of Western Australia, which is not a party to the National Energy Market.

Chaps. 1 and 2 in discussing the fossil fuel industrial complex. The municipal state is more directly subject to the influence of the citizen residents, its oeuvre formed on a local stage. This makes for a more organic connection between the institutional apparatus and the local community. One possible manifestation of this is how successfully Greens Party members position themselves to be elected to municipal government. On the other hand, those actors with the political and economic power are more likely to direct resources to influence decision-making on the national stage, although as is argued below this is not always the case.

Here we focus on the re-municipalisation campaigns, particularly in German and the United States, to gain some insights into the place of municipal governance and how this articulates with developments in Australia and their role, or otherwise, in effecting the transition from a fossil fuel-based energy system to a clean energy system. Chapters 9 and 10 turn the focus to community renewable energy cooperatives.

The Birth of the Climate Citizen and Charting a New Climate Politics Terrain

The latter years of the 2000s' decade marked a shift in the tenor of the discourses of empowerment. From the idealism that carried the climate movements onto the global stage in Copenhagen, there emerged a range of parallel movements at the local level that were giving new meaning to the catchcry 'think globally, act locally'. A burgeoning climate politics movement campaigned in local political jurisdictions to transform the institutional and organisational bases for hastening the transition to a zero-emissions energy system. We can identify several strategic initiatives that are being advanced with the animation of those members of communities determined to pursue this transition. One focus has marshalled the power of the citizenry to contest the dominance of energy corporations in defining the scale of the fossil fuel-based energy-industrial complex by considering the potential of local government to revive its role to take the lead in energy generation and governance. This has been evident in those instances where municipal government utilities have maintained their place in the energy system, but, more strikingly, it has been evident in the re-municipalisation campaigns to re-establish local government involvement in public utilities generally, and, given our focus, energy generation and distribution in order to spearhead initiatives to provide clean energy.[3]

The investment in renewable energy technologies in Denmark, and especially by municipal governments and community cooperatives often in partnership with one another following the energy crisis of the mid-1970s, has been considered a watershed moment in the decentralisation of and reliance on distributed energy. While well-established, the dominance of the local energy systems model has been subjected to persistent challenges. This has played out with the liberalisation of the European energy market. Energy corporations and some business groups have been pressuring the Danish government to remove the protections and support for the municipal and cooperative ventures to align the energy market with the larger European market. Although this pressure has resulted in some aggregation of governance arrangements and the development of some larger publicly owned wind farms, the popular commitment to local government and community cooperative involvement in energy has proved enduring (Cumbers 2012).

The distinguishing feature of the Danish investment in wind power is that the nation's clean energy pathway predates the growing global concerns with the climate change impacts of reliance on fossil fuel as the principal source of energy. The commitment to wind power amounted to the defence of the status quo in Denmark, but in demonstrating the possibilities of mitigating energy-related emissions alongside energy self-sufficiency, the Danish model of municipal energy utilities and community-based energy cooperatives provided a model for progressive change (Hoff 2016; Søgaard Jørgenson and Rahbeck Pedersen 2016). Denmark is the world's leader in the proportion of electricity generated from variable sources. The institutional arrangements enabled citizens to assert their preference for a low- or zero-emissions future and challenge the hegemony of the industrial-scale, oil and coal-fired energy machines under the control of global energy corporations. It is this citizen-initiated challenge that is the subject of this chapter. We draw on the experiences of two country contexts—Germany and the United States because of the significance of the

[3] Re-municipalisation around energy can sensibly be considered as reflecting a more general shift in the political economy of public ownership of other public goods, and most particularly water (See, for instance, Bakker 2013; Cumbers and Becker 2018; McDonald and Swyngedouw 2019).

re-municipalisation movements, as well as municipal-supported energy cooperatives which we discuss in the subsequent chapter—before reflecting on the counterparts of these movements in Australia.

The State as a Site of Contestation and Agent for Change

Two aspects of these developments provide the backdrop for the analysis. The re-municipalisation and energy cooperative initiatives in Germany and the United States have emphasised two key ambitions. One is directed towards challenging the hegemony of the energy corporations by reframing the role of the state. This has been based on climate activists marshalling political support within the community and being strategic in pressuring municipal governments to promote clean energy. It has also been frequently formalised through the citizen-initiated referenda to direct municipalities to take control of energy utilities and to transition to renewable sources of energy. The probability that targeting the municipality will achieve more traction than lobbying the regional or national state has helped to define the projection of energy democracy because successful lobbying in one location can have a catalytic effect in strengthening the campaign resolve in another location. Still the political agenda is distinguished by the determination to recruit the municipal apparatus in the fight against the emissions-intensive generation technologies that are driving climate change. This process of democratisation to foster decarbonisation is about engaging with the state system in order to work with the state. Unlike the prosumer, the climate citizen is engaged in a distinctly political project that is consciously designed to displace the big polluters. Moreover, it holds out the prospect of the state exercising its authority to drive the decentralisation of the energy system to better advantage the local community and which can further introduce scope for ensuring equitable access to electricity becoming an integral feature of the energy transition.

The democratisation of energy is about the politics of the possible and the climate citizens' turn to the municipality, to what can be regarded as citizen democracy's natural home, has been a logical move. It is also strategic since, as recent history has demonstrated, this is the sphere of the state where the exercise of climate citizen's democratic can have most traction. It is the level of the nation-state system that is generally presumed to be of relative minor significance, especially with respect to the policy formulation that frames anything beyond the local. Yet, it is the level of the state system in which climate citizens are able to be readily heard and responded to for the purchase of democracy can be quite muted at the regional and national levels of the contemporary state system. Many elected governments, especially in national configurations, have been resistant to implementing substantive measures that the citizenry is calling for to curb the generation of emissions from fossil fuel-fired power stations. Australia and the United States are indicative of this resistance and the powerlessness of popular campaigns to be able to break through the impasse in calling for action, at least at the federal level. This can be attributed to several factors, although the stand-out has to be that fossil fuels have been so integral to the structure of the economy and such a defining feature of the geopolitics of the respective nations, that the national states have rejected out of hand any measures that could be considered to negatively impact on the economy or be to the detriment of dominant industries or disrupt the hegemony of the emissions-intensive fossil fuel-based industrial complex.

This transformation is being swept along by an ever-growing popular sentiment, by deeply held personal and cultural convictions among the world's populations of the need for urgent action to cut greenhouse gas emissions and to transition to a zero-emissions energy. The surge of household investment in renewable energy technologies in the latter part of the first decade of the new millennium was one way in which this manifested in Australia. The extent to which this can be regarded as more than simply a transactional act concerned with 'taking control of the bill' can be observed in 'utility citizens', in their demeanour as climate citizens, pressing governments to build on their efforts to transform the energy system. The personal and cultural conviction was also evident in the very dynamic climate politics that were forged in the lead up to the 2009 Copenhagen Climate Change Conference, in the rallies that brought thousands of climate citizens together demanding government action. The vibrancy of this climate change politics, which had sprung to life in the grassroots, community-oriented movements advocating locally controlled decarbonisation initiatives, and which had worked its way to Copenhagen, drew municipal governments into this political dynamic. Municipalities became a focus for the possibilities for effecting progress, especially following the failure to progress agreement at Copenhagen. This prompted some climate citizens to concentrate their resources on organising around revitalising existing municipal utilities; others looked to resurrecting municipal utilities that had been privatised, while the political momentum also impelled municipal councils to join others in collectively endorsing the goals of the climate movement and to promote the climate cause (Hoff and Gaussett 2016; Hoff 2016; Søgaard Jørgenson and Rahbeck Pedersen 2016; Møllenbach and Hornbæk 2016). Climate citizens turned to the municipality as an arena that afforded the opportunity for giving effect to the climate cause. The municipality was tasked with the responsibility for charting a path to a clean energy future.

Climate Citizens and Engaging the State: *(Re-)municipalisation*

The state has been the dominant institutional and organisational force in the development and definition of energy systems, although how this has played out has varied across time just as the relative significance of the different tiers of the state system—national or federal, State and local government—has varied across time and from one political jurisdiction to another. But there is a common thread running through the history of the evolution and role of the state in the development of the power sector. More recently, a critical feature of this thread has been how calls for action to mitigate emissions and confront the challenge of climate change have ignited political focus on the ability of the state system to be the conduit for progressing these calls.

Interestingly, it is the third tier of the state, local government or the municipality, that has figured as the institution that communities have moved to engage with to progress action on climate change. This partly draws on the historical role of the municipality as a pioneering force in the electrification of communities. In many instances, nascent industrial and urban development was contingent on municipal utilities establishing the means to produce and deliver gas and electricity to power these developments. Because of the risks involved in establishing this novel technology and the public good nature of the enterprise, with power distributed by means of a transmission network, an infrastructure of poles and wires that was capital-intensive and costly to install, state authorities generally sanctioned a monopoly market to support the viability of the generating enterprise and to ensure reliability of supply. And this feature of a regulated market monopoly has characterised the generating sector until relative recently and continues to typify transmission infrastructure arrangements in which the returns on investment are commonly set by statutory authorities. Placing the municipal utility at the helm of this system of generating and supplying electricity has, over the course of the last decade, become a most politically charged milestone in contesting the established corporate-dominated energy regime, and has become all the more charged as energy corporations appreciate the directions in which the winds of change are blowing and manoeuvre to secure their future in the industry.

Historically, municipal authorities played a leading role in the energy system, very much because of the monopoly market in which they operated, but this role began to be questioned in the mid-1970s. Fiscal constraints and the oil crisis of 1973–1974, and an increasingly persuasive ideological milieu that the state should not be involved in ventures that private enterprises could manage more efficiently, so it was argued, prompted some municipal governments to consider capitalising on the value embodied in the enterprises by selling the utilities. Around the world, there was a flurry of privatisations and regulatory changes engineered to partially liberalise electricity markets. The historical standing of municipal utilities came under considerable pressure as energy generation became an attractive site for private capital accumulation. The continued presence of municipal utilities in energy markets came under fierce challenge. An agenda to liberalise markets, particularly by breaking down state as well as national borders, was reignited in the 1990s. The reach of this liberalisation project varied across different political jurisdictions. A number of the remaining restrictions that had shielded municipal utilities from competition with private companies and energy corporations have been gradually lifted. Cross-border investment and takeover restrictions have been relaxed, as were regulations that restricted new entrants gaining access to protected retail markets or barriers that prevented competition in wholesale markets, and cross-border investments in transmission infrastructure by large companies and energy corporations have been permitted to flourish. As the level of industry concentration has increased, so too municipal utilities have lost market share, and this is particularly pronounced in a large population and major industrial centres. The ever-expanding global reach of capital with the engagement of international energy corporations has diminished the position of municipal utilities.

Notwithstanding this displacement, not-for-profit municipal utilities have remained a strong physical presence in the energy system. Throughout much of the industrialised world, municipal utilities have remained the principal energy provider in smaller rural communities, and notably in Germany and the United States where they have been such an important and numerically significant feature of the energy landscape (Baach 2016). Across America, municipal utilities presently service some 2000 towns and cities, supplying power to more than 49 million people and employ 93,000 workers (The American Public Power Association 2020; Ciampoli 2018a, b; Ditto 2020).

In America, acknowledgement of the positive contributions that publicly owned utilities engender for communities is reigniting interest in their importance in underwriting wellbeing and sustainability. They supply power at competitive prices, provide employment and recycle their earnings back into the community and are the more likely means of addressing pollution and driving a turn to clean energy (Blaylock 2019).[4] This awareness has apparently been behind the renaissance of community initiatives to re-establish some utilities that had been privatised as well as behind a groundswell of enthusiasm for launching some new utilities.[5] In the face of considerable opposition from private energy operators and energy corporations, with some advice on 'how to' from the peak not-for-profit municipal utility industry association, the American Public Power Association (APPA), climate citizens have been organising to build popular support for municipal governments to conduct ballots to give citizens a say on municipalisation. This has seen some success over time with communities expressing their confidence in municipal utilities as their preferred energy supplier. Eighty-eight new utilities have been established since 1973, and 20 since 2000 (Schryver and Moline 2019).[6]

[4] Although there are instances of public utilities, such as at Long Island, proving costly failures (Dalzell 2020).

[5] This is not to argue that there has been some community opposition to municipalisation. The International Brotherhood of Electrical Workers, the union representing unionised workers at Pacific Gas & Electric, opposes a public takeover of PG&E arguing there are higher priorities where funds could be better spent and fearful of the utility being subject to being politicised (Morris 2020; San Francisco Priority Check 2019).

[6] The numerical significance of municipal utilities in the United States is partly attributable to the New Deal policies designed to boost the recovery of the economy from the 1930s. Depression included measures to promote the development of rural communities. The 1935 New Deal-era Rural Electrification Administration dedicated funds to electrifying rural industries, and this served to revive the fortunes of municipal-owned utilities, with over 1000 new utilities being formed, as well as, more crucially, the viability of community-operated cooperatives (Hess 2011; Smyth 2019).

The promise of municipal utilities assuming a lead role in the transition from emissions-intensive energy systems is more questionable. Most of the municipal utilities have historically been reliant on and, importantly with respect to how they envisage the future of electricity generation, wedded to coal as the principal energy source. Coal continues to be the main energy source for power supplied by the majority of utilities, with a considerable number of municipalities committed to coal into the future. Nuclear power has been the not-too-distant second main source of power. The deference to coal has been manifest in the stance of the peak industry association representing community-owned municipal utilities. The APPA has not been all that forceful in contributing to debate on energy and climate change. The Association obviously wants to represent the broad spectrum of its membership, and this helps explain the lack of resolve in the APPA leadership in declaring any policy initiatives that really speak to the challenge of climate change (The American Public Power Association 2019b, 2020; Kuckro 2019).[7]

Although there are a number of utilities, and the communities in which they are located, resistant to the pressure to get them to invest in renewable energy technologies, others are being held back by the lack of financial resources that would make such a transition possible (Bozuwa 2018; Partain 2020). But the political tide has been turning with so much attention being given to the need for action on climate change. The prominence of publicly owned municipal utilities as such a critical feature of the energy landscape in America has steeled climate citizens into directing political energy into pressuring utilities to invest in renewable clean sources. This is prompting some rethinking of the APPA's policy position on climate change. After over a decade of agnosticism, which has amounted to defending the status quo of relying on coal, the APPA is bowing to the pressure coming from customers and an increasing number of utilities. As well as confronting the reality of the shift in the relative costs of coal vis-à-vis renewables, the Association is in the throes of jettisoning its long-standing defence of coal in the course of developing an updated Climate Policy. This is reflected in its approval of proposed changes to the federal tax regime that would allow utilities to receive comparable investment write-offs to the corporate sector, which would allow shortening the

[7]The APPA opposed the Obama administration's *Clean Power Plan* as an overreach by EPA into the realm of electricity regulation and supported Trump's Affordable Clean Energy because it was deemed flexible.

amortisation time frame, and ramp up the sector's investment in 'modernisation' programmes—a metaphor designed to allay fears that APPA's increasing support for renewable energy technologies could accommodate those municipalities that retained reliance on coal (The American Public Power Association 2019a, b; Zummo 2020).

An interesting element in this shift to supporting renewable energy is that it draws on a long lineage of critical reflection on the benefits of abandoning fossil fuel as the chief source of energy just as the oil crisis hit. The debate on the shape of the energy system in America was set in train with the publication of E.F. Schumacher's *Small Is Beautiful: A Study of Economics as if People Really Mattered* (1973). Schumacher delivered a telling blow to the myopic preoccupation with boundless economic growth and the ecological threats posed by the reliance on fossil fuel that fired this growth. Nor was the solution to be found in nuclear power, which was being held up as resolving resource depletion and pollution concerns that so occupied Schumacher. The sentiments of *Small is Beautiful* were echoed in research commissioned by the United Nations on energy options by Amory Lovins. The research was developed over the ensuing years, culminating in the seminal study *Soft Energy Paths: Toward a durable peace* (1976, 1977). This cemented the idea of renewable energy and energy efficiency firmly in the global imagination in positing an alternative energy vision to the industrial-scale fossil fuel complex that had come to define America's energy system. This canvassed the legitimacy of community-scale energy systems.

Soft Energy Paths became a reference point for second-wave environmentalism that had taken off in the 1970s and the philosophical springboard for explaining the merits in communities' confidence in municipal utilities. It also provided a bridge uniting environmental movements around the world, and the reason for discussing *Soft Energy Paths* is that its concerns and the alternative energy systems it advocated were explicitly foregrounded in the formidable campaigns launched by the anti-nuclear movement in Germany beginning in the 1980s (Roselund 2016; Morris and Jungjohann 2016; Wuestenhagen 2012; Lovins 2013). It was not long before the movement also turned to campaigning for action to halt the combustion of coal to mitigate the threat of climate change and this renewed interest in alternative energy systems and on challenging the dominance of the energy corporations and the industrial-scale energy-intensive system.

Although the Schumacher and Lovins' critiques had traction internationally, they were first and foremost concerned with America's development trajectory, and the visions of an alternative energy future entered the imagination and figured in several notable citizen-initiated moves to bring an end to the fossil fuel industrial complex. Municipal utilities were viewed as one alternative institutional arrangement that could provide the platform for servicing the energy needs of the local community. As was the case in Germany, coal was a culprit, but it was the growing disquiet with nuclear power plants that initially catalysed the focus on the benefits of renewable energy. Nuclear power plants were proving to be an extremely costly and at times dangerous experiment. The ominous threat of a meltdown at the Sancho Seco nuclear plant in 1989 led to its closure by the Sacramento Municipal Utility District (SMUD) in 1989 (Smeloff and Asmus 1997). This set in train passionate debate on a 'soft energy' path, and the City canvassed the preferences of citizens before investing in wind and solar farms. SMUD also began subsidising installation of household rooftop solar panel systems.

Sacramento was not alone in paying the cost for investing in nuclear power plants. Other cities have made similar mistakes and their utilities have then turned to investing in clean energy as a sustainable future. This has intensified criticisms of the energy market liberalisation projects that helped to foreground a place for nuclear power in the American energy system. This has also engendered a revival of reflection on how this energy model, based on the centralisation of energy generation and establishment of vast transmission networks, has compromised local participation in energy governance. Promoting the democratisation of energy is back on the political agenda of many communities across the American landscape, and most notably in the Pacific Northwest, which has a long history of citizen activism. The failings of the energy market liberalisation model, in privileging the rent-seeking energy corporations and in constructing networks that have failed to deliver a reliable energy supply, have renewed interest in the value of municipal utilities (Smeloff and Asmus 1997).

Irrespective of whether the American Public Power Association makes the leap to endorse the anti-coal sentiments, an increasing number of communities are voting in support of transitioning to clean energy and lobbying for municipalisation as the means for doing so. There have been several iconic campaigns, which have emerged in response to the disruptive effects of global warming, that are based on some radical rethinking in favour of municipal utilities taking on responsibility for energy supply and distribution and stepping up efforts to transition to clean energy. One interesting illustration of this has been the efforts of citizens in the small city of Boulder, in the state of Colorado, to establish a municipal energy utility—to municipalise—with the express purpose of meeting climate goals (Buxton 2016). Having secured the commitment of the city to explore the options to increase the supply of clean energy, which included pressuring the established energy corporate Excel Energy to increase supplies of energy from renewable sources, citizens have been actively involved in framing the objectives of a municipal power utility and, in a manner consistent with the *Soft Energy* philosophy, envisaging a completely different construction of the place of energy in the community. Responding to consumer pressure and determined to reduce emissions, individual municipal utilities in many rural areas, as well as in some politically progressive urban communities, have become the agents of the radical transformation in linking the turn to clean energy to a raft of other socially progressive changes (Richard 2016; Ciampoli 2018a, b; Motyka et al. 2018; Palazzi and Seif 2018; Page 2016; Smyth 2019; Trabish 2017; U.S. Energy Information Agency 2019). This includes measures that would advance energy equity, reduce energy poverty, provide local employment opportunities and, in reducing reliance on coal which is mined in areas that are predominantly low-income neighbourhoods, promote climate justice (Bozuwa 2018).

Local governments, as agents of change, have initiated developments designed to foster zero-emissions electricity and abandon power generated by fossil fuels in the context of a federal state that has resisted any moves that would compromise the pivotal role of coal in the American energy regime (Hoff 2016). In pushing ahead with measures to secure renewable energy, the municipalities have become the institutional vehicle for displacing corporate energy operators, and this is having a very real material effect on emissions mitigation (Roth 2019; Smyth 2019). One estimate calculates that municipal utilities have reduced emissions by 33 per cent over 2015–2017, and some municipal utilities have announced that they are on track to achieving net-zero emissions (Schryver and Moline 2019; Shemkus 2019).

The questions of the scale of energy enterprise and supply reliability have also been galvanised by community interest in how to respond to the links between the evidence of global warming and the apparent increase in extreme weather events. The succession of annual wildfires in California has sharpened reflection on the possibility that rising temperatures have caused the recent spate of wildfires that have resulted in the loss of hundreds of residents' and firefighters' lives, and razed large swathes of forests. This has stirred interest in the merits of municipal utilities taking the lead to establish more decentralised energy systems. Community anger has been magnified by the largest energy corporation in the United States and the principal energy supplier in California, Pacific Gas and Electric (PG&E), shutting down the transmission network to minimise the risk of its power lines and substations sparking further fires. In the wake of these events, there have been widespread calls for the re-assertion of public ownership of power utilities, and public control is being seriously debated. One widely supported solution that has been advocated is to remodel the energy system to foster local involvement, promote reliance on distributed energy and microgrids operated by municipal utilities (Liccardo 2020; Morris 2020; Roberts 2019).[8] The City of San Francisco, for example, looked to municipalisation as one solution to the energy debacle. The city was considering establishing a utility to take over and operate PG&E infrastructure, but the Governor Newsom recently agreed to a deal supporting PG&E's exit from bankruptcy in return for a change in the board structure and management and undertakings to ensure safety and reliability (Bozuwa 2019; Baker 2019; Witkowski 2020).[9]

Germany and Re-municipalisation

A similar and equally politically charged struggle focused on climate citizens' confidence in the potential of municipalities as conduits for transitioning to renewable sources of energy has been a defining feature of green politics in Germany. Re-municipalisation campaigns have occurred across much of Germany. These have been pivotal in hastening the transition to renewable sources of energy most notably in two of the nation's major cities, Hamburg and Berlin. As has been the case in the United States, re-municipalisation builds on a long history of the innovative role of municipal government in the establishment of the energy system. While

[8] Should the idea of municipal utilities not gain traction, one other suggestion to bring pressure to bear on energy corporations to invest in renewable energy technologies is for climate activists to facilitate aggregating consumers into one buying cohort to strengthen their bargaining power in the energy market and for them to indicate a preference for clean energy supplied by small distributors.

[9] The deal proceeded amidst allegations that the Governor has a conflict of interest because of his close links with PG&E.

this involvement was abandoned in many instances, and particularly in the larger, more populated cities and industrial centres, and taken over by expanding energy corporations, municipal utilities continued to dominate rural and smaller energy markets.

Re-municipalisation movements emerged in the mid-2000s as one aspect of the broad campaign being waged against nuclear power and coal-fired electricity and the energy corporations' energy market monopoly. The grassroots movements focused on re-municipalisation because this tier of the state system has primary constitutional authority for energy governance. To indicate the measure of support of citizens, the re-municipalisation campaigns pressured municipal governments into conducting referenda to determine whether a majority of the citizenry favoured the idea of a municipality assuming responsibility for the generation and distribution of energy and, in the process, for supplanting the energy corporations that dominated the market. Displacing the corporations has been a critical element of the re-municipalisation project because the energy corporations have been so wedded to the industrial-scale energy generation system and so resistant to abandoning the status quo.

The success of re-municipalisation political projects is evident in the resurrection of some 284 municipal utilities over the period prior to 2017 and in the formation of 72 new municipal utilities, or *stadwerke*, between 2005 and 2013 (Dubb 2018; Wagner and Berlo 2015; Hockenos 2013). The re-municipalisation campaigns proved timely, unfolding as the thermal coal generators were investing heavily in the extraction of lignite for power generation. The campaigns have proved crucial in challenging the hegemony of the coal-industrial complex and the state's endorsement of the four energy corporations that dominate the energy system forcing them to reconsider their resistance to transitioning from nuclear and coal to renewable energy. The impact could be felt even when a referendum did not deliver an affirmative vote as was the case with the Berlin municipalisation campaign. In this instance, the referendum on whether to end Vattenfall's contract to supply electricity to Berlin did not attract a sufficient call-out of voting citizens to endorse the move to displace Vattenfall (Rocholl and Bolton 2016; Kunze and Becker 2015). Despite the failure to carry the referendum, the political pressure that was brought to bear strengthened the sense of purpose and the bargaining position of the municipality in negotiations with Vattenfall, forcing the corporation to agree not to proceed with plans to construct a new coal-fired power station (Cumbers and Becker 2018; Blanchet 2017).

The re-municipalisation movements in Germany have highlighted the power of climate citizen politics in effecting change through the state. This has proved to be a classic case of the social movement politics 'from below' being re-energised. The lineage of this climate activism was acknowledged in 2016 when Lovins was awarded the German Order of Merit for his vision of the transition. But progressing the transition, while being guided by this vision, was also predicated on the strategic adroitness to exploit the social democratic foundations of the state by giving voice to political claims in the local political sphere where they were most likely to gain traction (Schroeter et al. 2016; Jessop 1982). The momentum of these campaigns was matched by a militant climate movement that directly challenged the resistance of the federal and State governments and the energy corporations to bring an end to the fossil fuel-driven energy system. It was also bolstered by the extent to which they drew on the strength and successes of other re-municipalisation movements across the nation. These movements shared a common motif, captured in the catchcry 'think globally, act locally', and neatly encapsulated in the struggle for a referendum to reinstate the municipal energy utility in Hamburg where the campaign was flagged by the slogan: 'Our Hamburg—Our Grid' (World Future Council 2016, 2017; Cumbers and Becker 2018; Moss et al. 2015).

Much of the progress in the campaigns to democratise energy through reinstating, or elevating, the role of municipalities in the energy-state complex was made possible by the distinctive cultural and organisational structure of the German state. Governance in the German state system is characterised by an integrative philosophy which frames how lobbying and negotiating practices by different stakeholders are channelled towards an agreed resolution (Krick 2018; Morris and Jungjohann 2016; Jessop 1982). This integrative approach has been the basis of the corporatist system of federal governance, which has historically rested on engaging key stakeholders, principally representatives of capital and labour alongside state officials and bureaucrats in policy planning deliberations.[10] It is underscored by planning processes serviced by advisory committees,

[10] Another institutional dimension of the corporatist system was organised at the industry sectoral level in tripartite arrangements. Tripartitism brought the state together with management and worker representatives to iron out disputes before they escalate and is carried through to individual enterprises involving co-determination forums that are organised to bring management and labour representatives together to try resolve tensions. The system provided another level at which industry and organised labour could be engaged in the industrial transformation engineered under the *Energiewende* umbrella.

composed of academics, bureaucrats and citizens representing societal groups, that inject a 'technical rationality' to the mapping of proposals. This input diminishes the potential for stakeholders getting locked into intractable and competing positions as well as makes for political legitimacy. The recent deliberations on exiting the reliance on coal demonstrate how this governance arrangement can reconcile competing policy ambitions. The 'Commission on Growth, Structural Change and Employment' brought together diverse stakeholders from across the spectrum of energy sector, industry, consumer and environmental/climate activist movement to chart a programme for energy transitioning. Despite the quite different and conflicting positions of participants with regard to the future of coal and the pace of a managed exit from coal, the Commission was able to secure agreement on mapping an exit timetable (Wehrmann 2018; Egenter and Wehrmann 2019).

The Municipality as Agent of Change

There is a still further element in this story of the making of the *Energiewende* pathway engendered by re-municipalisation as an energy democratisation project. In driving the municipal assumption of a role in the energy system, this not only wrought a reconfiguration of the state and its *raison d'être*, municipalities also became a force in pushing ahead with the energy transition.[11] Mirroring developments in Denmark and the United States, the local state became an agent of change cultivating the support of climate citizens and also bringing further pressure to bear on the federal state as well as the *länder* (Baach 2016). This has contributed to consolidating the *Energiewende*'s institutional foundations within the state system as well as its social foundations (Ohlhorst 2016; Elektrizitätswerke Schönau 2017).

Climate citizen politics has generally been a more dynamic and pervasive force in Germany compared with the America, and this appears reflected in the more vibrant role of the institution of the German state across the respective tiers compared to the structure of the state system in America. This provides some salutary lessons for appreciating why the municipality has been a less consequential institutional conduit for progressing the claim for action on climate change in the United States, and we can draw on this in interrogating the role of municipal government in Australia.

[11] State governments, the *länder*, wanted to push back the date for exiting coal and lignite (Wettengel 2020).

In the first instance, it is important to recognise that the local political domain, the municipal or local government, can be a quite vibrant political arena. This is partly the effect of the different dynamics at play in the local political order. The links between the citizens and their elected representatives are more direct and personal. Local government is more accessible and has proven a more progressive instrument in reflecting and responding to the groundswell of community lobbying for action on climate change and this gives voice to the power in community. Well-resourced corporations tend to be less likely to invest their resources in trying to influence the machinations of local politics as they do in the domains of the State or federal state, which is not to say that there is not the potential for local government becoming captive to particular interests (Jessop 1982). On the contrary, the third tier of the state system can be particularly vulnerable to capture, but there appears to be greater scope for the popular will to be projected and accommodated as seems to be the case in the extent to which municipal governments have been responding to community pressure and becoming actively engaged in the development of initiatives to transition from emissions-intensive fuel generation to renewable energy. This can make for a vibrant political milieu that can feature in and influence the politics of counties, cities and municipalities, and which is quite possibly more pronounced in provincial America because of the 'patchwork legacy' of energy providers that resulted from the New Deal's rural electrification programme. The political tradition of 'localism', and its roots in agrarianism and republicanism, which have helped to sustain the municipal utilities and community cooperatives, reinforces this potential for a vibrant community-based politics (Hess 2011, 2013, 2018).[12]

Moreover, notwithstanding the United States being a federated system, the American state structure is set on quite different cultural and organisational foundations. These foundations are more diverse than those in Germany, varying in form and with respect to the authority of the different tiers of the state as these are constitutionally determined. This is evident with the third tier, where authority can be vested in counties, towns or cities and municipal government, varying across the nation and thus in the definition of their function and the various imperatives to which they have to respond and how these articulate across the state system. Unlike the

[12] In addition to the north-west states, other regions with a history of vibrant citizen activism include the north-east states and California, highlighting the variability in political culture across the United States.

articulation of the different levels of the state that one can observe through the lens of the *Energiewende*, the respective spheres of the American state system operate more independently or autonomously of each other. Governance at the level of the State or the municipality functions more independently of the federal state, and this is in part a reflection of the greater purchase of civil society in shaping the policy focus and function of these spheres, and especially municipal government, of the state system.

This lack of correspondence in purpose manifests in the local endeavours to initiate transformation in the energy system facing obstacles that frustrate these endeavours carrying through to the federal political sphere. This blockage also reflects the extent to which the federal state is separated from the thrust of social democratic processes. The state is held to be relatively autonomous of the various social forces that make claims on the state. To maintain its political legitimacy, to manage the range of competing imperatives, the state must be seen as being representative of standing above particular interests and representing the spectrum of competing interests. This exercise of relative autonomy, however much the actuality demonstrates otherwise, severely restricts the scope for citizens to influence policy design and practice. Access through municipal governance to the federal arena is comparatively blocked, organisationally closed to the exercise of a grassroots' democratic voice as government is premised on its ostensible independence (Dryzek et al. 2003). The state generally presents as an edifice of impartiality. Yet the norm has been for the hegemony of the fossil fuel industrial complex to define the measure of state support in defending the energy status quo, and this also tends to constrain the effective reach of the climate citizens' appeal through the office of the municipality. Still, the role of local government should not be presumed to be some invariable instrumental of energy capital. As has been argued here, there is some scope for the municipality to become the institutional conduit for advancing energy democracy.

References

American Public Power Association (2020) 'The complexity of climate change' 11 March.

American Public Power Association (2019a) 'American Public Power Association applauds GREEN Act for fairer energy incentives' 19 November.

American Public Power Association (2019b) 'Environment Issue Brief: Federal efforts to address climate change'.

Baach, S. (2016) 'A local energy transition success story' in institutions' in J. Hoff and Q. Gaussett (eds) *Community governance and citizen-driven initiatives in climate change mitigation* London: Routledge, 130–149.

Baker, D.R. (2019) 'San Francisco's version of a new Green Deal is taking over its power company' Bloomberg 12 March.

Bakker, K. (2013) 'Neoliberal versus postneoliberal water: Geographies of privatization and resistance' *Annals of the Association of American Geographers* 103(2), 253–360.

Blanchet, C. (2017) 'Remunicipalisation of energy systems' *Resilience* 10 July.

Blaylock, D.L. (2019) 'The power of local solutions' Community Engagement 15 March, https://www.publicpower.org/periodical/article/power-local-solutions.

Bozuwa, J. (2018) 'Public ownership for energy democracy: Opportunities for publicly-run energy utilities to revolutionize generation and the grid' Democracy Collaborative 3 September.

Bozuwa, J. (2019) 'Pulling the plug on PG&E' *The Nation* 15 October.

Buxton, N. (2016) 'Empowering out future, Boulder, United States' Energy Democracy 12 December.

Ciampoli, P. (2018b) 'Public power doing more on renewables than stats suggest: Zummo' The American Public Power Association 27 November.

Ciampoli, P. (2018a) 'Public utilities make NREL top 10 lists for green pricing lists' *American Public Power Association* 4 June.

Cumbers, A. (2012) *Reclaiming public ownership: Making space for economic democracy* London: Zed Books.

Cumbers, A., and S. Becker (2018) 'Making sense of remunicipalisation: Theoretical reflections on and political possibilities from Germany's *Rekommumalisierung* process' *Cambridge Journal of Regions, Economy and Society* 11(3), 503–517.

Dalzell, T. (2020) 'California wants to take over PG&E? Learn from Long Island's mistakes' *The San Francisco Chronicle* 19 February.

Ditto, J. (2020) 'The complexity of climate change' The American Public Power Association 11 March.

Dryzek, J., D. Downes, C. Hunhold, D. Schlosberg, and H-K. Hernes (2003) *Green States and social movements: Environmentalism in the United States* Oxford: Oxford University Press.

Dubb, S (2018) 'In Europe, non-profit and local public utilities are seen as key to green future' *NonProfit Quarterly* 9 May.

Egenter, S., and B. Wehrmann (2019) 'Germany to quit coal by 2038, under commission proposal' *Climate Change News* 26 January.

Elektrizitätswerke Schönau (2017) 'Elektrizitätswerke Schönau, Germany: How mothers, teachers and doctors took over the power grid and energy supply', 28 November.

Flemsæter, F., H. Bjørkhaug, and J. Brobakk (2018) 'Farmers as climate citizens' *Journal of Environmental Planning and Management* 61(12), 2050–2066.

Gløjmar Berthou, S.K., and B. Vind Ebbsen (2016) 'Local government of climate change in Denmark: Recasting citizens as consumers' *Journal of Environmental Management and Planning* 59(3), 501–517.

Hess, D.J. (2018) 'Energy democracy and social movements: A multi-coalition perspective on the politics of sustainability transitions' *Energy Research and Social Sciences* 40, 177–189.

Hess, D.J. (2013) 'Industrial fields and countervailing power: The transformation of distributed solar energy in the United States' *Global Environment Change* 23(5), 847–855.

Hess, D.J. (2011) 'Electricity transformed: Neoliberalism and local energy in the United States' *Antipode* 43(4), 1056–1077.

Hockenos, P. (2013) 'Local, decentralised, innovative: Why Germany's municipal utilities are right for Energiewende' *Energy Transition* 28 September.

Hoff, J. (2016) "Think globally, act locally': Climate change mitigation and citizen participation' in J. Hoff and Q. Gaussett (eds) *Community governance and citizen-driven initiatives in climate change mitigation* London: Routledge, 28–53.

Hoff, J. and Q. Gaussett (2016) 'Community governance and citizen-driven initiatives in climate change mitigation: An introduction' in J. Hoff and Q. Gaussett (eds) *Community governance and citizen-driven initiatives in climate change mitigation* London: Routledge, 1–6.

Ison, N., M. Lyons and J. Atkinson (2018) *Repower Australia Plan* Australian Conservation Foundation, GetUp!, Solar Citizens, Environment Victoria, Nature Conservation Council of NSW and 350.org, Melbourne, Australia.

Jessop, B. (1982) *The Capitalist State: Modern theories and methods* London: Basil Blackwell.

Jørgensen, S., and S. Rahbek Pederen (2016) 'Climate ambassador programmes in municipalities: Encouraging climate change mitigation in public administrations and institutions' in Jens Hoff and Quentin Gausset (eds) *Community governance and citizen-driven initiatives in climate change mitigation* London: Routledge, 107–129.

Krick, E. (2018) 'Ensuring social acceptance of the energy transition: The German government's 'consensus management' strategy' *Journal of Environmental Policy and Planning* 20(1), 64–80.

Kuckro (2019) 'Public Power CEO: We must "come to grips" with climate change' *E&E News* 30 September.

Kunze, C., and S. Becker (2015) 'Collective ownership in renewable energy and opportunities for sustainable degrowth' *Sustainability Science* 10, 425–437.

Liccardo, S. (2020) 'There's a better way out of the PG&E bankruptcy' *San Francisco Chronicle* 27 February.

Lovins, A. (2013) 'Lessons from the Energiewende for the US' *Ecologic* 27 November.

Lovins, A. (1977) *Soft energy paths: Towards a durable peace* San Francisco: Friends of the Earth International.

McDonald, D., and E. Swyngedouw (2019) 'The new water wars: Struggles for remunicipalisation' *Water Alternatives* 12(2), 322–333.

Møllenbach, E., and K. Hornbæk (2016) 'Klimafesten: A case study of a municipality's intervention to engage citizens in environmentally sustainable actions' in J. Hoff and Q. Gaussett (eds) *Community governance and citizen-driven initiatives in climate change mitigation* London: Routledge, 215–233.

Morris, C., and A. Jungjohann (2016) *Energy democracy: Germany's energiewende to renewables* Springer.

Morris, J. D. (2020) 'PG&E union resists Bernie Sanders' calls for public takeover' *The San Francisco Chronicle* 23 February.

Moss, T., S. Becker and N. Naumann (2015) 'Whose energy transition is it, anyway? Organisation and ownership of the *Energiewende* in villages, cities and regions' *Local Environment: The International Journal of Justice and Sustainability* 20(12), 1547–1563.

Motyka, M., A. Slaughter, and C. Amon (2018) 'Global renewable energy trends' *Deloitte Trends.*

Ohlhorst, D. (2016) 'Germany's energy transition policy between national targets and decentralized responsibilities' *Journal of Integrative Environmental Sciences* 12(4), 303–322.

Page, S. (2016) 'The second-largest city in the U.S. is on the verge of being 100 percent renewable' *Think Progress* 10 June.

Palazzi, T., and D. Seif (2018) 'Why distributed solar is winning in Texas' *Greenbiz* 28 February.

Partain, S. (2020) 'Working together to address climate change' *The American Public Power Association* 6 March.

Richard, M.G. (2016) 'Top ten green utilities in the US' *Treehugger* 8 May.

Roberts, D. (2019) 'The radical reform necessary to prepare California's power system for the 21st century' *Vox* 1 November.

Roberts, J., F. Bodman, and R. Rybski (2014). *Community power: Model legal frameworks for citizen-owned renewable energy* London: ClientEarth.

Rocholl, N., and R. Bolton (2016) ''Berlin's electricity distribution grid: An urban energy transition in a national regulatory context' *Technology Analysis and Strategic Management* 28(10), 1182–1194.

Roselund, C. (2016) 'Amory Lovins awarded the German Order of Merit for his vision of the energy transition' *PV Magazine* 19 March.

Roth, S. (2019) 'Here's how local governments are replacing California's biggest utilities' *The Los Angeles Times* 25 December.

Rubik, F., and M. Kress (2014) 'Climate Citizens—Analysis of roles, experiences, challenges and opportunities using the example of the citizens of Heidelberg/Germany', Institute for Ecological Economy Research (IÖW), Bergstraße 7, D-69120 Heidelberg/Germany.

San Francisco Priority Check (2019) 'Why does a union oppose the takeover of PG&E?'

Schroeter, R., O. Scheel, O. Renn, and P-J. Schweiser (2016) 'Testing the value of public participation in Germany: Operationalization and a case study on the evaluation of participation' *Energy Research and Social Science* 13, 116–125.

Schryver, Y., and B. Moline (2019) 'Municipalization: Setting the record straight' *Community Engagement* 12 November.

Schumacher, E.F. (1973) *Small is Beautiful: A study of economics as if people really mattered.*

Shemkus, S. (2019) 'Mass. Climate group says municipal utilities moving too slow on clean power' *Energy News Network* 27 February.

Slocum, R. (2004) 'Consumer citizens and the cities for climate protection campaign' *Environment and Planning A: Economy and Space* 36, 763–782.

Smeloff, E., and P. Asmus (1997) *Reinventing electric utilities competition citizen action and clean energy* Washington, DC: Island Press.

Smyth, J. (2019) 'Rural America could power a renewable economy—But first we need to solve coal debt' *Clean Cooperative* 2 July.

Søgaard Jørgenson, M., and S. Rahbeck Pedersen (2016) 'Climate ambassador programmes in municipalities: Encouraging climate change mitigation in public administrations and institutions' in J. Hoff and Q. Gaussett (eds) *Community governance and citizen-driven initiatives in climate change mitigation* Routledge: London, 107–129.

Solar Citizens (2018) 'Renters and Low-income Households Suffer Worse Through Weekend Heatwave', 20 March.

Solar Citizens (2020) 'Supercharge ARENA'.

Trabish, H.K. (2017) 'Join or die: How utilities are coping with 100% renewable energy goals' *Utility Dive* 13 December.

U.S. Energy Information Agency (2019) 'U.S. electric utilities by ownership type (2017)'.

Wagner, O., and K. Berlo (2015) 'The wave of remunicipalisation of energy and supply networks in Germany: The establishment of 72 new municipal power utilities' First fuel now: ECEEE 2015 Summer Study; 1–6 June 2015, Toulon/Hyères, France; proceedings.

Wehrmann, B. (2018) 'Germany's coal exit commission' *Clean Energy Wire* 13 December.

Wettengel, J. (2020) 'Government rejects state demands for changes to coal exit law' *Clean Energy Wire* 8 April.

Witkowski, W. (2020) 'PG&E cuts deal with Cal. Governor Newsom to exit bankruptcy' *Market Watch* 20 March.

World Future Council (2016) 'Energy remunicipalisation: How Hamburg is buying back energy grids' 19 October.

World Future Council (2017) 'Our Hamburg—Our Grid' 2 February.

Wuestenhagen, R. (2012) 'A power struggle in Germany: David, Goliath, and the future of energy systems' Network for Business Sustainability 26 October.

Zummo, P. (2020) 'The path to a more decarbonized future' The American Public Power Association 27 February.

CHAPTER 9

Local Governments as Conduits for Progressing Climate Change Action in Australia

A quite different constitutional organisation of energy governance in Australia has constrained the role of local municipal governments in climate change action, but this has not blocked some quite unique and extraordinary initiatives to mitigate emissions and support the transition to a renewable energy-based economy, and to do so in ways that draw inspiration but from the experiences in Germany and the US. Historically, municipalities had assumed a lead role in the electrification of communities, but with the federation of what had been independent colonies in 1901, constitutional responsibility for energy provision was placed with State governments, and this has been an institutional fetter obstructing municipalities' continuing involvement with energy generation and distribution. Municipalities were, alongside local power companies, pivotal in the electrification of cities. However, this role was circumscribed and became more so as different tiers of the nation state system asserted control of energy generation and transmission, firstly by State governments asserting control over energy generation and distribution, then with the privatisation of State-owned utilities, and more so when the federal government assumed some control over governance with the establishment of the National Electricity Market. Notwithstanding the limiting effects of this architecture, the establishment of the NEM created opportunities for local governments to explore measures that would result in quite robust efforts to hasten the abandonment of coal as the primary energy source.

S. Rosewarne, *Contested Energy Futures*,
https://doi.org/10.1007/978-981-19-0224-6_9

Climate Citizens, Energy Democracy and the Greening of Municipal Councils

Originally, municipalities had a somewhat circumscribed role in electricity generation and supply, working alongside the early initiatives of some fledging private companies. This was the case of the city of Melbourne; for instance, Melbourne, the capital city of Victoria, was among the first of the ventures to introduce electricity to the city, setting up to provide street lighting, replacing gaslighting, and to supply power to a number of establishments. In a move to standardise the nascent power industry, Melbourne City Council took control of some of these private ventures in the early 1890s and commissioned a private company to take responsibility for managing generating and distribution infrastructure. The spatial reach of the Council's supply network was limited with the carving up of the city's electricity market following the privately owned company, the Melbourne Electric Supply Company, successfully negotiating long-term franchises to supply power to a number of other municipalities in the city around the same time as the Council's utility was taking control of the power supply in the designated boundaries of the City.

This institutional mix of public and private power utilities serving distinct markets in the Victorian economy was soon overshadowed by the ascendancy of the State-owned and controlled State Electricity Commission of Victoria (SECV) which was formed in 1918. Victorian power plants had been reliant on black coal transported from New South Wales, but this reliance and the organisation of the power sector was radically transformed following an investigation into Germany's use of its lignite (brown coal) reserves, and the State government concluded that there were parallel advantages to be had from exploiting the brown coal deposits in the La Trobe Valley, in eastern Victoria and which was envisioned as a potential counterpart to the Ruhr. Melbourne City's pre-eminence as a power utility was gradually overshadowed with the founding of the SECV. A similar fate later befell the City Council's principal private counterpart, the Melbourne Electric Supply Company. In the final years of World War One, the State government adopted the plan to extract brown coal and develop what would become a highly intensive and concentrated industrial-scale power generating zone under the management of the State-controlled SECV (CIGRE and AHEF 1996). While it sought to retain a role in supplying power to Melbourne's suburbs, increasing the number of turbines it operated, the City Council could not compete with the cheaper coal-generated electricity being transmitted from the La Trobe Valley. The Council struck a deal with the State Electricity Commission for the grid to draw on the Council's generating capacity to help meet the growing peak load demand. Drawn out as it was, the future of the Council utility was set as the SECV established a monopoly over Victoria's electricity market.

Similar developments unfolded in other States, with municipal councils establishing power utilities and for several of these to be gradually amalgamated under the management of a single entity, such as the Sydney County Council. Following the Victorian model, these were subsequently placed under the management of a State-controlled utility, such as the Electricity Commission of New South Wales in 1950. The Commission later merged with a number of other suburban municipal electricity utilities and renamed EnergyAustralia in 1996.

While none of these organisational developments was subject to popular plebiscite, a distinguishing feature of the subsumption of municipal utilities under state government control was that electricity-consuming communities were not necessarily averse to the concentration of electricity generation and supply in state hands. Apart from complaints about the dust from transported coal and the level of particulate matter emitted into the atmosphere by generating plants, there was very little questioning of state hegemony in the power sector; electricity remained a cheap source of household power by world standards. There was one exception, however. This was the Tasmanian government Hydro Electric Commission which had established itself as a political powerhouse in that State. Its plans to expand hydro generating capacity in the early 1980s, by constructing more dams in the south-west wilderness area of Tasmania, set in train one of the most intense environmental campaigns—the 'No Dams' campaign—in Australian history. The battle was resolved with the outcome of the 1983 federal election when the Labor Party was voted into office, breaking a conservative coalition party hold on federal government, and the Labor government blocked any further dam construction.

Abundant supplies of coal and lignite secured these resources as the primary energy source, and this effectively foreclosed much interest in the exploration of renewable energy technologies. So, despite some expressions of interest in debate on the link between greenhouse gas emissions and the potential for global warming, most notably with the Hawke Labor government's sustainable development agenda of 1990–1991, there was little interest in the subject outside scientific circles and even within the environmental movement (Rosewarne 1992). By the time that interest in the question of climate change was being aroused, emissions-intensive fossil fuel-based industries and the resources sector had been able to define the tenor of public discourse and set the agenda of Australia's Kyoto Protocol's negotiations. The scope for dissenting views on the energy future was severely constrained, and this was reinforced by the projects to privatise state utilities, to liberalise the energy market and to establish a national energy market.

At the same time that the Australian government was deliberating over the Kyoto Protocol, State utilities were being subjected to a process of corporatisation in the 1990s. Anti-privatisation campaigns were launched by unions and consumer groups in Australia in the late 1990s and into the 2000s, which forced the New South Wales government to postpone its privatisation plans and the Queensland government not to proceed with the privatisation of generating and transmission infrastructure. Environmental and climate concerns did feature in the anti-privatisation campaigns, but they were not at the forefront of the struggles to maintain the state's role in electricity generation and distribution. The effect of the liberalisation and privatisation programmes was to foreclose any serious consideration at the local government level of alternatives to a coal-based energy system. The guarantees of returns to those international investors that had successfully bid for the privatised generating plants and the transmission infrastructure limited locked in their futures as pivots in the energy system. The effective marketing positioning of three energy corporations further limited opportunities for climate change activists to bring pressure to bear on municipal councils to become proactive on emissions mitigation and renewable energy in ways that were comparable to their Danish, German and American counterparts.

This void in local engagement soon changed. Despite the constitutional and practical organisational constraints and the federal government's ideological conviction in the coal-industrial complex, climate citizens began to look upon the local political arena and sites where movements could gain a foothold explore novel measures to reduce emissions and promote greater use of renewable sources of energy. In fact, it is highly likely that the Howard Coalition governments' intransigence on climate change policy forced some rethinking among activists to devote energy to working with other levels of the state system—State and local municipal government— as climate citizens where a local political culture provided opportunities for working with councillors who could look to policy and practices to better represent their constituencies and to begin pressuring for policy and practical changes (Bulkeley and Kern 2006). Local councils have proved successful arenas in which councillors card-carrying members the Greens Party and Socialists parties have pressed for progressive initiatives around decarbonising and support for low-income households to ensure inequities are addressed. In the process, the focus of municipal governance has shifted, and municipalities became active agents progressing action to address climate change.

State government constitutional responsibility for energy meant that they would assume a leading role in the climate change policy debate, and Greens representatives elected to State parliaments have played an important role in this. State governments have been responding to the popular support for transitioning to renewable energy. The pace of household rooftop solar panel installation has accelerated because of state support. This support appears to have been moved by two factors. One has been the realisation of the strength of popular opinion regarding the need for action to arrest climate change, clearly evident in the household investments in solar panels and in the very large turnouts for climate rallies. A second was the realisation that a growing number of the coal-fired power plants were on the cusp of reaching the end of their economic life, with some shutting down giving next to no notice and others announcing plans to close in the not-too-distant future while bringing forward planned closure dates. This engendered a sense of urgency to act quickly to ensure that generating capacity would be maintained and made State governments more receptive to entertaining supporting alternative sources of energy. The economics of the renewable energy options helped to extinguish any notion of coal being the principal energy source into the future, and the imperative to ensure energy supply reliability forced the hand of State governments.

Alongside these developments, climate citizens also began to see opportunities where municipalities could contribute to developing decarbonisation ventures. This was to prove a breakthrough moment with the projection of climate politics into the municipal arena beginning to have some positive outcomes. Community activists, and especially environmentalists, had won some council backing to pursue a range of community-fostered initiatives, and many of these took up the cause of climate change to win municipal backing for projects that have resulted in local governments becoming indirectly involved in influencing energy generation patterns. One of the earliest developments, the Moreland Energy Foundation, emerged out of the efforts of a group of environmental activists, who had established permaculture gardens, recycling schemes and other ventures, extending their vision of an ecologically sustainable future to include energy. The Foundation was established in 2000 under the umbrella of the Moreland City Council, an inner-city council in Melbourne. With Council backing, the Foundation has become a vehicle supporting households and businesses to reduce their carbon footprint. It coordinates the purchase of solar panels and hot water systems, cutting their cost to make them a more attractive financial proposition.[1] It has organised fundraising to support the installation of community-owned solar panels on businesses and complexes that enable them to draw on the green power generated on their roofs to help reduce power costs and the community's carbon footprint. And it has served as a model for other community-based action.

Municipalities as Agents of Change: From the Local to Renewable Energy Partnerships

The Climate Council of Australia has acknowledged the leading role local councils, municipalities and shires have demonstrated in responding to the mounting challenge of climate change (2017b). One in five respondents to a 2017 Council survey reported their objective of achieving 'zero emissions' or '100 per cent renewables'. Municipal and shire councils have responded to community pressure to formulate climate change policies and to develop measures that drive emissions mitigation and back greater recourse to renewable energy. Climate citizens have considered the municipality across Australia as an important organisational conduit for engaging at the level of local government measures to reduce their carbon footprint. This has been represented as evidence of the important role that local governments can play in response to community preference for action on climate change.

While acting on behalf of constituents, municipal councils are themselves active agents of change. Some municipal councils have responded to community pressure to follow the lead of households and invest in the councils' own capacity to generate their energy needs by installing rooftop solar panels and heat pumps. Having local councils undertake energy audits pinpoints where energy savings can be made to reduce emissions and support a transition to greater reliance on renewable sources of energy. This has resulted in a range of concrete initiatives, in the installation of solar panels on council

[1] The Foundation was rebranded the Australian Energy Foundation in 2019 to better reflect the broadening of the range of support it provides to strengthen household energy autonomy and underwrite the transition from coal-fired power.

buildings, improvements in insulation of buildings, installation of LED and tendering for the supply of electricity from renewable sources.

Promoting equitable access to renewable energy for low-income households has been one campaign initiative pursued by climate change activists. The principle of climate justice has emerged as a critical feature of the municipal politicking and has resulted in local governments dedicating resources to support the replacement of incandescent light globes with LED lights or the installation of solar panels for low-income households. Council and shire councils have led the way ahead of other polluters yet have been reticence to approach other activists to develop innovative measures to reduce cz. The Melbourne inner-city council, Darebin Council, for instance, was among the first of the councils to introduce assistance programmes for low-income households to enable them to reduce reliance on energy delivered by for-profit energy companies by providing subsidised installation of rooftop solar panels and advising on more energy-efficient practices within the home, including the replacement of candescent light globes with LEDs. Some similar programmes are being organised in conjunction with the Victorian Labor government.

Individual councils have endorsed climate change activist events, such as the recent Climate Emergency protests. More councils sign onto the climate emergency campaign calling for more proactive support of State and federal governments (Kilvert 2019; SBS 2019).

Discussions as to how local governments should respond to the challenges being unleashed by climate change also spawned more profound reflections on the standing of municipal governance *vis-à-vis* other tiers of government. The Australian Local Government Association (ALGA) represented the local calls for action on climate change as evidence of the closeness of community to local government and the local government's awareness of the needs of the community as indicative of the force of democracy at the local level. But, according to the ALGA, there was an urgent need to sanction the legitimacy of this exercise in democracy because the municipal government lacked constitutional recognition, and this could restrict the exercise of local governance and the claims on resources to take action on climate change. This called for sanctioning the political status of the municipality. In fact, ALGA was so confident of its characterisation of local government as an enabling institutional vehicle for the exercise of democratic voices that the Association began lobbying for constitutional reform. It submitted an appeal to a Parliamentary Inquiry into political reform arguing for transformation of the three tiers of the state. The ALGA advocated for a prospective referendum on Australia's structure of governance to include the possibility of citizens being given the opportunity to support constitutional reform that would elevate the standing of local government in the Constitution (ALGA 2012a, b). Lobbying by climate citizens was not the only force that prompted the ALGA request for constitutional reform, but it proved to be an important voice, especially given the factors driving climate change were global in nature and the cost of responding to the deleterious consequences of climate change was likely to be beyond the resources on the local government. Constitutional reform that recognised the local government in the state system would provide the formal right to make a claim on national resources.

The recommendation was not taken up, and the small window of opportunity to write local government into the Constitution passed. This did not, however, bring an end to the municipality as an agent of change. Endeavours to progress the greening of municipal councils have been an important outcome that has been engendered by what elected councils refer to as the democratisation of energy, giving voice to pressure to mitigate emissions and increase reliance on renewable energy.

Partnering as a Climate Change Strategy

The commitment of municipal councils to pursuing action to ameliorate climate change has been reinforced by partnering among councils as well as other levels of government, other public institutions, non-government community associations and private enterprises. The Australian Local Government Association and its State affiliates have proven to be an important conduit for engaging councils in emissions mitigation and progressing renewable energy. The ALGA emphasised the benefits of councils working with local communities that were taking action to mitigate climate change, and the ALGA has been encouraging its members to work alongside and promote grassroots action (ALGA 2018, 2019). The ALGA has been a consistently strong force in endorsing the need for action on climate change acting as a clearing house for ideas and policies to mitigate emissions (ALGA 2018).

Councils have responded to proposals advanced by climate citizens to develop novel ways of consolidating their commitment to a clean energy future. Reflecting the catchcry 'think globally, act locally', climate citizens have built connections that have proved springboards uniting several councils in common cause. Partnerships have brought councils together in alliances to project collective endeavours to embrace renewable energy. The earliest council alliance was instigated by the Moreland Energy Foundation with the formation of the Northern Alliance for Greenhouse Action. This brought together nine councils in 2002 to work as a network for sharing information, coordinating emissions reduction and adaptation activities, and cooperating on the research and development of innovative projects. The Northern Alliance's goals are to 'substantially contribute to the transition to a low-carbon future by delivering effective programs and leveraging local government, community and business action' (Moreland Energy Foundation, n.d.; ARENA 2020). With climate citizens backing in the ensuing few years, networks of regional municipal greenhouse alliances have proliferated across the State of Victoria; Indigo Power (n.d.).[2] The formation of the South Australian Regional Organisation of Councils provides a similar springboard for inter-council dialogue on climate change (South Australia Local Government Association 2021).

Such alliances have translated into a national vision that is being prosecuted by the Australian Local Government Association. For two decades, the Association has been focused on reducing emissions from council operations and supporting the implementation of emissions abatement programmes in communities (ALGA 2017, 2019; Mey et al. 2016).

Partnering with State, and to a lesser extent federal government, has helped consolidate action on climate change (Department of Environment and Heritage, NSW 2016; Department of Environment, Water and Planning. Victoria 2017; (Mondo 2018) Department of Environment, Water and Resources, South Australia, and the Local Government Association of South Australia 2013, 2021; Mazengarb 2020; Department of Land, Water and Planning, Queensland 2017; O'Keeffe 2018a, b).[3]

Such alliances have also begun to incorporate into their clean energy platforms programmes that aim to enhance equitable access to energy (Sansom 2014a; Solar Savers 2017). A partnership between the Central Victorian Greenhouse Alliance (CVGA), Eastern Alliance for Greenhouse Action (EAGA), Northern Alliance for Greenhouse Action (NAGA) and the South East Councils Climate Change Alliance (SECCCA) aims to remove the barriers faced by low-income pensioner (aged and disability) households wanting to install solar. Similar regional groupings have been formed in New South Wales and the State government established an office with the express purpose of encouraging more collaboration and partnerships among local governments to progress action on climate change (Burton 2016; Adapt NSW n.d.).

[2] These include South East Councils Climate Change Alliance, Central Victoria Greenhouse Alliance, Western Alliance for Greenhouse Action, Eastern Alliance for Greenhouse Action, Gippsland Climate Change Network and Goulburn Broken Greenhouse Alliance.

[3] The President of the ALGA is the local governments represented on the COAG, although it is not evident that this provides much of an opportunity to voice local government ambitions.

The Climate Council's *Cities Power Partnership* has provided another forum for enhancing collaboration among local councils and engaging the councils in climate change initiatives. Launched in July 2017, between 1997 and 2009, 233 councils from across Australia joined the *Partnership*'s Cities for Climate Protection campaign, an initiative of the International Council for Local Environmental Initiatives (Climate Council of Australia 2017). In 2019, the *Partnership* organised the conference. An indication of councils of some of the ways in which the agency of the state is being exercised in the battle over climate change policy is demonstrated by the NSW and Queensland governments joining forces to launch coordinated action to pressure the Morrison Government to pursue more decisive measures to cut the nation's emissions (Hall 2021).

Going Global: Consolidating the transition momentum through orchestrating a Global-Local synergy

The Australian Constitution is silent when it comes to municipal councils. They are the hidden tier of the state system, and too often are overlooked when it comes to Australia's international climate commitments. Yet, as is argued here, municipalities have been responding to community pressure, anticipating new, more ambitious emissions reduction benchmarks as has been the case with the Paris Accord (Kilvert 2019; Macdonald 2019). Through the perseverance of climate citizens municipalities have been working together in various alliances to chart a path that is dramatically reducing local governments' carbon profile. The proactive initiatives of municipalities, while of a different nature to those being pursued in Europe and North America, have not been held back by their subordinate status in the Australian state system (Mey et al. 2016). This is intriguing because the climate citizens' *cris de coeur* at the local level appears to be effecting a reconfiguration of the organisation and focus of the local state and its *raison d'être*. The Victorian Greenhouse Alliance is based on a rhizomatic structure in which each of the regional greenhouse alliances or actions—Central Victorian Greenhouse Alliance, Eastern Victorian for Greenhouse Action, Goulburn Broken Greenhouse Alliance, South East Council Climate Change Alliance and Western Alliance for Greenhouse Action, Northern Alliance for Greenhouse Action—is formed by the several municipal and shire councils uniting under a local region's alliance umbrella, bringing 66 councils in partnerships to pursue action on greenhouse gas emissions. This reflects a significant institutional transformation in the organisation of the state system and reveals an interesting shift in the political momentum of climate movements. The trans-municipal boundary alliances have established common cause to drive the momentum of transitioning from reliance on coal-fired to renewable energy sources and a more ambitious emissions reduction target.

The alliances capture the spirit of the catchcry 'think globally, act locally' while drawing on the local to inject substance in the campaign globally. And nowhere is recognition of the state exercising agency better illustrated than in its literal global manifestation for municipalities have gone global in their greening. They have been projecting their commitment to clean energy internationally, becoming a global force in their own right and taking their climate activism to another level, by channelling climate change politics into international arenas (Bulkeley and Betsill 2013). Australian municipal governments have signed onto Cities for Climate Protection and the Local Governments for Sustainability. Australian governments were well represented in such forums (Bulkeley 2013: Ch. 4). The European Covenant of Mayors for Climate and Energy, founded in 2008 and merging with the Compact of Mayors in 2016, brings together local governments committed to accelerating the transition to clean energy (Covenant of Mayors 2019). It now has 11,000-plus members around the world. In the United States the political force of climate politics has found expression in the lateral exchanges forged through local government associations. Over 1000 mayors in the United States have signed the Climate Protection Agreement, which was launched in 2006 and emerged out of meetings of municipal government leaders held across America in the early 2000s and has resulted in a growing number of municipalities committing to 100 per cent renewables (Trabish 2017).[4] The Agreement has become a springboard for engaging local governments from across the world to commit to acting on climate change. Some 25 Australian municipalities had signed the pledge, and more are debating joining (Beyond Zero Emissions et al. 2018; ALGA 2020; Monash City Council 2018; Council of Capital City Mayors, 2019; ICLEI

[4] Twenty-five city mayors in Germany signed the *Making the Energiewende a success story thanks to strong local authorities* pledge in 2013 (German Council for Sustainable Development 2013). New South Wales mayors signed a similar agreement in 2007 committing to lead the response to climate change at the local level (Local Government Association Australia 2007). The Climate Council has recruited cities to join the Cities Power Partnership, which was launched in conjunction with the Local Government Association in 2017 and represents 50 per cent of Australia's population and actively engages with community energy groups (Cities Power Partnership 2019).

Oceania 2019). Most local governments in signing the pledge commit to working with community climate groups (Bulkeley and Betsill 2013). The establishment of transnational municipal networks is a still more exciting illustration of the positive momentum that local government has brought to the design and implementation of measures to accelerate the transition to clean energy (Kern and Bulkeley 2009).

Re-energising Councils: Investing in Renewable Energy Generation

The passing of the Gillard governments Clean Energy Future Package included a provision requiring State and municipal governments to undertake a stocktake of their carbon footprint. Backed by funding, this impelled those that had not directed much attention to conduct an audit of what they had done to date and to reflect on what they could do to address the climate change challenge (McGuirk et al. 2014). Federal government finance was provided for a range of programmes, including to increase energy efficiency and increase reliance on renewable energy. However, unlike their European and North American counterparts, Australian municipal governments have displayed little interest in directly engaging in generating electricity, although they have been supportive of cooperative energy generation enterprises, as is noted in Chap. 10. There is, however, one notable exception, which somewhat surprisingly has been given little fanfare. This is the decision by the Sunshine Coast Council to invest in the 15 MW Sunshine Coast solar farm. Sunshine Coast Council is the first municipality in Australia to offset 100 per cent of its electricity with renewable energy drawn from the Council-owned and operated solar farm (Sunshine Coast Council 2019). The decision to invest in the establishment of the solar farm was based on a simple cost-benefit assessment of the likelihood of ever-increasing power costs. It has been estimated that the investment will save the Council $22 million in power bills over the next 30 years (O'Keeffe 2018a, b). The farm was developed and delivered solely by our council, without any assistance from either the Federal or State governments, and the Council ensured that the cost of building the solar farm would provide employment opportunities within the local region. It was connected to and started feeding power into the grid in 2017.

Several councils in Central Victoria are considering the merits of direct involvement with the intention of hastening the transition to zero-emissions electricity. They have commissioned a study of the 'benefits for

Victorian councils to invest, build, own and operate large scale renewable energy infrastructure to meet their long-term electricity needs' (Central Victoria Greenhouse Alliance 2019). Cooperation across municipal governments has resulted in some quite novel arrangements that deliver clean energy at competitive prices. The formation of partnerships between councils based on consolidating their power requirements to negotiate as a single entity with an established or prospective solar or wind farm has produced noteworthy savings on power bills.

The more popular approach to municipal councils promoting renewable energy has been through collective 'bulk buy' arrangements based on councils forming partnerships and pooling their power needs and contracting with a solar or wind farm to meet the councils' energy requirements. Municipal council Power Purchasing Agreements (PPAs) have, quite literally, changed the landscape. In 2013, the Melbourne City Council began investigating the idea of establishing a consortium that would consolidate the volume of power that they could entertain negotiating in a contract for supply (ACCC). The scaling up of the volume of power being sought would place the consortium in a stronger position to negotiate renewable energy supply with a larger solar or wind farm. Led by the City of Melbourne, 14 organisations from different sectors—and including the Bank Australia, NAB, Australia Post, Citywide, Moreland City Council, City of Yarra, City of Port Phillip, NextDC, RMIT University, The University of Melbourne, Zoos Victoria, the Melbourne Convention and Exhibition Centre and Federation Square—agreed to form a consortium, the Melbourne Renewable Energy Project, and signed a ten-year renewable energy Power Purchase Agreement (PPA) with Pacific Hydro to supply the consortium with clean, low-cost power.

PPAs have become an important feature of the energy system in Victoria. Other States have also embraced the construct. The Sydney City Council was among the first to launch such a venture. Given it is relatively well resourced and had recruited an energy mapping officer to plan the Council's energy options, this leadership has provided a springboard for exploring partnering with other institutions to consolidate the City's commitment to emissions mitigation (Mazengarb 2019). A similar venture was implemented by Monash University with the backing of the Victorian State. Adelaide *City Council* says it will become South Australia's first carbon-neutral local government by the end of this year following the signing of a 'landmark' *renewable* power purchase agreement (Richards 2020; SSROC Projects 2020; Vorrath 2020).

Information about municipal councils negotiating PPAs and the legal requirement for ACCC endorsement is readily available through the Greenhouse Alliances (Point Alliances 2017; Victorian Greenhouse Alliances, 2017). The Climate Change Council has provided some support for councils by guiding municipalities council in exploring measures to reduce their carbon footprint as a means of reining in administration costs.

Power Purchase Agreements have redefined the organisation of the energy market by establishing the durability of exchange relations in the place of the ephemeral nature of the *laissez faire* market and the futures market. Some of the PPAs bring together a sizeable number of entities in alliances that have radically transformed the energy market. The competitive spirit and ephemeral nature of market exchanges have been replaced by collective partnering and enduring contracts negotiated outside the formal arena of the NEM. PPAs enhance the municipal partners' ability to negotiate competitive prices, securing energy at lower cost and reliability in the supply of renewable energy. In contrast with the free market exchanges, the PPAs are increasingly being organised on the basis of a shared confidence as well as a common purpose that breaks with the anonymity and anarchy of the free market.[5]

The significance of PPAs cannot be overstated. In Victoria where municipal climate change alliances have been most successfully developed, the Local Government Power Purchase Agreement (LGPPA) Project brought 39 councils together to begin negotiating a 100 per cent renewable energy procurement option. The organisational structure of the Greenhouse Alliances has been built around 70 of Victoria's 79 municipalities partnering in regional greenhouse alliances that are acting as a springboard for engaging in emissions mitigation and transitioning to reliance on renewable energy. The strategic appeal in PPAs is obvious, as users

[5] At face value such partnerships could be regarded as being in restraint of trade because they fit the mould of a cartel-like formation. PPAs were thus susceptible to be declared as an anti-competitive practice that could be declared illegal by the Australian Competition and Consumer Commission. The Melbourne Renewable Energy Project was the first to test this and it successfully sought exemption from Australia's Competition and Consumer Act, an exemption that was finally granted in 2016 (ACCC 2017). The ACCC authorised the parties 'to establish a joint renewable energy (electricity) purchasing group, conduct a joint tender for an electricity supply arrangement and jointly negotiate the terms and conditions of their electricity supply agreements'. The application had expressly sought protection from legal action for possible breach of the Act in order to negotiate 'Green Electricity Retail Sale Agreements (GERSAs). The PPA was negotiated on the basis that the supplier (individually, or with a developer) 'will use the GERSAs to develop a new renewable energy generation system that will supply renewable energy into the National Electricity Market'.

avoid the cost of investing in the means to honour their commitment to renewable energy while, in negotiating long-term agreements that provide prospective renewable energy generating enterprises with the certainty of a market and sale of energy at a predictable and set price. PPAs have fostered confidence among investors that the sale of the power a solar or wind farm they are generating is assured for a defined period.

The Increasingly Social Form of the Energy Market

The redefinition of market relations consequent upon the negotiation of Power Purchasing Agreements is also transformative in another respect, namely, effecting a change in the socio-economic organisation of electricity markets. The formation of a buying consortium turns the purchase process into a social arrangement.[6] A second project is building on the success of the Melbourne Renewable Energy Project by linking with universities and businesses to accelerate the shift to renewable energy, contracting for the long-term supply of 113 GWh based on a power purchasing agreement (Vorrath 2019, 2020; ACCC 2019).[7]

In a remarkable development, the Darebin City Council has lodged a similar application on behalf of 41 local government councils in Victoria. The Australian Competition and Consumer Commission has approved the application, permitting the Council and partners to act as one buying entity to issue a collective tender. Interestingly, the ACCC's approval of PPAs is transforming the meaning of competition in the market. In the ACCC's oeuvre, competition is not necessarily defined in line with the *laissez faire* construct. Such collective interventions in the market are becoming reasonably commonplace and ever-more ambitious in their reach.[8] The pooling arrangements challenge the legal foundations of the

[6] A somewhat similar model that has gained some traction in California is the Community Choice project. The project aims to authorise the municipality to purchase power on behalf of residents and businesses and use the existing infrastructure to deliver green power to households (Community Choice Energy 2015). The formation of the Clean Power Alliance is a similar scheme, which involves 29 cities committing to buy wind-generated power displacing Edison as a major supplier while using Edison's poles and wires (Roth 2019).

[7] This second group includes RMIT University, Deakin University, Cbus Property, ISPT Property, infrastructure construction group Fulton Hogan, Citywide Asphalt, and major food manufacturer Mondelez International.

[8] In 2019, drawing on this model of a union of enterprises negotiating PPAs that in effect underwrite the establishment of solar and wind farms, some 41 Large Format Retailers joined forces to apply for permission to act as a single buying entity (Large Format Retailers

laissez faire market insofar as sovereign entities, individual consumers, legally constituted corporate entities or legally constituted councils are being replaced by partnering arrangements. Competition law in Australia would normally deem such contractual arrangements to be in restraint of free trade and therefore in breach of the Competition and Consumer Act 2010. However, the ACCC Commissioner has determined such arrangements do not frustrate competition and are acceptable.

This pattern of energy users establishing more social forms of collective contracting. This transformation of energy users' engagement in the market has had limited appeal among energy generators or suppliers. This could be because of the dominant position of the vertically integrated energy corporations. There is one notable exception, the case of Yates Energy Services, a company based in Renmark that has been servicing the efforts of the recently established solar farms that provide a regular source of income for many fruit growers in the region. Yates is acting as a partner and broker. Since 2014, the company has 'developed and constructed over 80 solar farms in the sub 5 MW market, managing 100% of the renewable energy generation offtake into the National Electricity Market through its YES Energy Retail business' (Yates 2020).

Conclusion

The political organisation of climate citizens has taken efforts to decarbonise the economy to a different plane. The union of citizens in common cause has transformed the character of the social relations of energy and modified the institutional framework in which the energy regime is constituted. Whereas households seek to effect the transition to decarbonise by investing in solar panels and effecting a change in energy market relations, the success of their endeavours to progress the transition from fossil fuel dependency remains contingent on the cumulative effects on the multiple decisions by individuals to invest in rooftop solar panels and battery storage systems.

Association 2019). They had determined that 'pooling their electricity demand will allow them to achieve a sufficient scale to access more competitive offers from electricity suppliers than would be possible from individual negotiations' (Large Format Retail Association 2019). While they were only incidentally interested in switching to the purchase of renewable energy, the PPA that they were interested in negotiating was based on energy supplied from solar and wind.

Re-municipalisation, and the various forms this is taking, breaks from the industrial-scale fossil fuel order. The concentrated and centralised model based on energy drawn from coal-fired generators and transmitted across the country is being supplanted by a decentralised order. Taking control of the mode of generating electricity is more than a technical process. It is the corollary of the projection of a political will that can be successfully marshalled and projected to secure a voice in the institutional political arena. Climate citizens, in concentrating their focus on this tier of the state, have succeeded in engaging municipal governments to help advance efforts to meet the challenge of climate change. This has transformed the character of the political project, inhering the projection of climate politics as a democratising process. Moreover, it is a process that injects more authority into the climate movement because the municipality is possessed of the resources that can be directed to enhancing the momentum of the transition. It is also a process of climate politics being socialised and for this more social configuration is to be reflected in climate governance. This is evident in citizens uniting to focus their energies on lobbying the state, becoming personally involved with and engaging the state in the climate change project, and it is more profoundly evident in the alliances that municipal governments are forming. Such alliances are emboldening the municipal ambitions through the connections that are being established, regionally, within the national political arena as well as internationally, radically shifting the dynamics and purchase of climate politics.

This elevation of energy democracy from the local level to the international plane engenders a transformation in the power relations that frame the fossil fuel industrial complex. As with the physical processes of electricity generation and distribution being decentralised, so energy politics is also being decentred. No longer can it be assured that the state, or at least one sphere of the state, can be counted on to partner the international energy corporations and their resolve to maintain the fossil fuel industrial complex. The power relations have been reconfigured and the institutional order recast, and while there can be no certainty that this will displace the hegemony of the emissions-intensive energy behemoths, as we have observed in the instances of Berlin and San Francisco, fossil capital's ability to assert their authority in setting energy agendas is now more precarious. Energy democracy has been placed on the front foot of an invigorated climate citizenry that has successfully petitioned a role for local government to act in the interests of its constituents and contribute to the decarbonisation project.

References

ACCC [Australian Competition and Consumer Commission] (2019) Determination Application for authorisation AA1000455 lodged by RMIT & Ors in respect of establishing a joint renewable energy purchasing group Authorisation number: AA1000455

ACCC [Australian Competition and Consumer Commission] (2017) Retail Electricity Pricing Inquiry: Preliminary report, September

Adapt NSW (n.d.) 'Local Councils and Climate Change'

ALGA [Australian Local Government Association] (2020) 'Oceania mayors call for more support on climate action' 16 October.

ALGA [Australian Local Government Association] (2019) 'Submission in Response to the Climate Change Authority's Updating the Authority's Previous Advice on Meeting the Paris Agreement Consultation Paper' August.

ALGA [Australian Local Government Association] (2018) 'Local councils and communities are taking action on climate change in Australia according to a nationwide report released recently'.

ALGA [Australian Local Government Association] (2017) *ALGA 2014–17 Strategic Plan*

ALGA [Australian Local Government Association] (2012b) Submission to the Joint Select Committee on the Constitutional Recognition of Local Government',

ALGA [Australian Local Government Association] (2012a) 'The case for change: Why Local Government needs to be in the Australian Constitution', November.

ARENA [Australian Renewable Energy Agency] 'Moreland Micro-grid Investigation (2020).

Beyond Zero Emissions, ICLEI and Ironbark Sustainability (2018) *Australian Local Climate Review 2018*

Bulkeley, H. (2013) *Cities and Climate Change* Routledge.

Bulkeley, H., and M.M. Betsill (2013) 'Revisiting the urban politics of climate change' *Environmental Politics* 22(1), 136–154.

Burton, D. (2016) 'Collaboration and partnerships for adaptation' CoastAdapt, National Climate Change Adaptation Research Facility, Gold Coast.

Central Victoria Greenhouse Alliance (2019) 'Large-scale renewables'

CIGRE [The International Conference on Large High Voltage Electrical Systems] and AHEF [The Association for the History of Electricity in France] (1996). 'Electricity in Australia 1880-Present'

Cities Power Partnership (2019) 'A National Program for 2019, Canberra: Climate Council of Australia

Climate Council of Australia (2017) 'Local leadership: Tracking Local Government progress on climate change'

Community Choice Energy (2015) 'What is Community Choice Energy?'

Council of Capital City Mayors (2019) 'Lord Mayors represented at Mayors' climate conference, Seoul' 24 October.

Covenant of Mayors (2019) Covenant Initiative.

Department of Environment and Heritage, NSW (2016) *NSW Climate Change Policy Framework*

Department of Environment, Water and Planning. Victoria (2017) 'Resources for Local Government—Climate Change'.

Department of Environment, Water and Resources, South Australia, and the Local Government Association of South Australia (2013) Local Government Association of South Australia Sector Agreement.

Department of Land, Water and Planning, Queensland (2017) *Pathways to a Climate Resilient Queensland.*

German Council for Sustainable Development (2013) Making the Energiewende a success story thanks to strong local authorities Berlin.

Hall, J. (2021) 'Councils across NSW and Queensland join forces to demand climate action from Morrison Government' news.com.au 3 March 2.

ICLEI Oceania (2019) 'The National Consultation on the Global Covenant of Mayors framework for local government climate action in Australia'.

Kern, K., and H. Bulkeley (2009) 'Cities, Europeanization and Multi-level governance: Governing Climate Change through Transnational Municipal Networks' *Journal of Common Market Studies* 47(2), 309–322.

Kilvert, N. (2019) 'Australian cities are declaring a 'climate emergency', but does that actually mean anything? *ABC News* 27 June.

Large Format Retailers Association (2019) Application for Authorisation Under s88(1) of the Competition and Consumer Act 2010, 6 August.

Local Government Association Australia (2007) Climate Change Mitigation and Adaptation Programme.

MacDonald, J., on behalf of Central Victorian Greenhouse Alliance, Eastern Victorian for Greenhouse Action, Goulburn Broken Greenhouse Alliance, South East Council Climate Change Alliance and Western Alliance for Greenhouse Action, Northern Alliance for Greenhouse Action to the Climate Change Authority (2019) 'Updating the Authority's advice on meeting the Paris Agreement' 29 August.

Mazengarb, M. (2019) 'City of Sydney strikes landmark deal to go 100% renewal' *New Economy* 22 October.

McGuirk, P., R. Dowling, H. Bulkeley (2014) 'Repositioning urban governments? Energy efficiency and Australia's changing climate and energy governance regimes' *Urban Studies* 51(13), 2717–2734.

Mey, F., M. Diesendorf and I. McGill (2016) 'Can local government play a greater role for community renewable energy? A case study from Australia' *Energy Research and Social Science* 21(1) 33–43.

Monash City Council (2018) 'Cities Power Partnership and Pledge Programs' 28 August.
Moreland Community Solar (n.d.) 'Community Solar Portal'.
O'Keeffe, D. (2018a) Councils feel the pinch on climate change' *Government News* 16 August.
O'Keeffe, D. (2018b) 'Councils are "leading action" on climate change' *Government News* 21 May.
Point Advisory (2017) *Energy models for Councils—A decision makers guide* Prepared for the Central Victorian Greenhouse Alliances.
Richards, S. (2020) 'Renewable energy to fully power council' *Indaily* 5 February.
Rosewarne, S. (1992) "The Green Accord", *Capitalism Nature Socialism* 3 (1), 11–17.
Roth, S. (2019) 'Here's how local governments are replacing California's biggest utilities' *The Los Angeles Times* 25 December
Sansom, M. (2014a) Solar saver program gives pensioners gives power over bills' *Government News*
Sansom, M. (2014b) Solar saver program gives pensioners gives power over bills' *Government News* 18 September.
SBS (2019) 'More Australian councils declare a climate emergency' 17 July.
Solar Savers (2017) 'Removing the barriers to solar'
South Australia Local Government Association (2021) *Climate Commitment Action Plan*
Southern Sydney Regional Organisation of Councils (SSROC) (2020) SSROC Projects
Sunshine Coast Council (2019) Sunshine Solar Farm.
Trabish, H.K. (2017) 'Join or die: How utilities are coping with 100% renewable energy goals' *Utility Dive* 13 December
Victorian Greenhouse Alliances (2017) *Discussion Paper: Electricity Procurement in the Victorian Local Government Sector. Aligning Council Money with Council Values* May.
Vorrath, S. (2020) 'COVID-19 Likely to create desire for energy self-reliance: Australia' *Microgrid Knowledge* 23 March
Vorrath, S (2019) 'City of Melbourne leads second "bulk-buy" renewables auction' *Renew Economy* 22 November

CHAPTER 10

Breaking from the Industrial Energy Generation Model Fossil Fuel Complex: Community Renewable Energy Cooperatives

Household investment in rooftop solar voltaic panels provide scope for consumers to pursue a more active role in the organisation of the energy system and to contribute personally to decarbonising, but the structure of the energy system remains relatively unchanged and subject to the power relations that are embedded in the market and the profit imperative. Unlike the emergence of the prosumer household, climate citizens, through their advocacy of municipal utilities, seek to transform the institutional make-up of the energy system to displace the privileging of the market and contest the dominance of energy corporations. In providing an institutional vehicle for hastening the transition to clean energy, the municipality provides the forum for climate citizens' voices to find the space to exercise the democratic impulse in calling for supplanting the emissions-intensive energy regime. But the exercise of this voice can be constrained by the electoral cycle and by the fact that the constitution of the municipal utility, as an apparatus of the state, is subject to the influence of a range of forces, including the definition and reach of the role of the state. The energy market liberalisation project is one such force that has framed the direct engagement of the state in energy generation and distribution. And some of the benefits of municipalities operating utilities, such as the potential of municipal utilities for fostering development within the local economy, cannot necessarily be assured as noted in the instances of major metropolitan city utilities contracting for the supply of electricity with wind and solar farms beyond the municipal boundaries.

S. Rosewarne, *Contested Energy Futures*,
https://doi.org/10.1007/978-981-19-0224-6_10

A more emphatic and substantive exercise of the will of the climate citizenry that seeks to extinguish the opaqueness of market exchanges to transform energy systems has been organised around citizens uniting to establish community renewable energy cooperatives. Albeit limited in number, though more effective in driving the reduction in carbon emissions, renewable energy cooperatives have become a much-celebrated feature of the energy landscape. The establishment of the cooperatives is reflective of a more vibrant and proactive display of climate citizenship, one that is consciously focused on hastening the transition by reducing reliance on fossil fuel-based energy generation by displacing the energy corporations and challenging the role of the state and its sanctioning of the market system and committing to the fossil fuel industrial complex. The imperative for taking control of energy generation and distribution is frequently declared in the trinity of ambitions of democratisation, exercised through the cooperative, decarbonisation, in the substitution of fossil fuel-based energy by renewable sources, and decentralisation, which is based on local generation and distribution of the benefits within the community.

Interest in the deployment of renewable energy technologies as a foundation of such communities began to attract renewed interest in the shadow of the Kyoto Protocol and into the new millennium, especially in the heartlands of the leading fossil fuel industrial complex economies. The inspiration for establishing energy renewable cooperatives lay in the seeds of the 'soft energy' models that were sown in the 1970s. The resuscitation of the 'soft energy' model brought together some older stalwarts, who had held on to the utopian dream of the ideal alternative lifestyle, and a younger cohort, many of whom approached the energy issue from the vantage point of a radical, often anarchist, political persuasion. A second crucial factor was the rapidly falling cost of renewable energy technology. This has proved a game changer and been crucial to the affordability of the renewable energy technology and in getting the new wave of energy cooperatives up and running (Vantsinjan 2017; European Union 2018; IRENA 2018). The revitalisation of interest by the self-proclaimed climate citizenry has been particularly important in rural and regional communities, overcoming the lack of supply reliability and pricing-cost pressures. And while these communities have been the vanguard of investment in a range of new technologies, the formation of cooperatives to organise the generation and distribution of renewable sources of energy reflects the groundswell of support for hastening the radical transformation of the energy system in its entirety.

In engineering the technological transformation, the formation of cooperatives also entails social transformation. Community renewable energy cooperatives transform the energy landscape because, coupled with their *raison d'être* of democratising, decarbonising and decentralising the energy system, they can present a challenge to the established energy regime. First and foremost, the cooperatives bring an end to the community's reliance on the combustion of carbon, Secondly and crucially, the union of these progressive forces has distinguished the initiatives to establish renewable energy cooperatives as being more than simply about reducing emissions. The formation of energy cooperatives is about communities taking control of their destiny, more than simply 'taking control of the bill'. They are about taking control away and decoupling from the fossil fuel industrial complex and a determination to become energy self-sufficient as 'social enterprises', as the northern New South Wales renewable energy cooperative, Enova, has stated (Enova, n.d.). They displace the hierarchical institutional arrangements that enable a few global corporations and some nation states to define the energy order. They dismantle the structure of relations that endows large corporations with the exclusive rights to capture the benefits of generating power and selling it only to those who can afford to pay for the privilege, and they disrupt the distorted and destructive international political economy upon which the fossil fuel industrial complex is founded and which diminishes the integrity and dynamism of communities. The cooperatives break from determinations set in the market to democratise, to establish community control of the enterprise and to link the environmental objectives with social and political transformations. And this community focus is framed by the development of a circular economy based on the benefits of renewable energy generation being re-invested back into enhancing the wellbeing and resilience of the community. The trinity 'democratisation, decarbonisation and decentralisation' is commonly represented as encapsulating the fundamental ambitions of renewable energy cooperatives, and this is often appended with the call for 'climate justice'.

The Origins of Energy Cooperatives

The initiative to break from the industrial capitalist model through the formation of self-sufficient and independent communities can be traced back to the nineteenth century and there has been sporadic enthusiasm over the course of the last century. The cooperative movement experienced another wave of interest in the 1960s and early 1970s in many advanced industrial economies, contesting the preoccupation with unbridled economic growth, resource depletion and increasing pollution, as well as the loss of any sense of community and of citizens recast as self-centred, alienated economic actors. These sentiments aroused a second-wave environmentalism alongside the organisation of a raft of progressive social movements, including the civil rights movement, the women's movement, minority race and indigenous people's struggles and the anti-war movement, and many of the actors in these movements joined forces with others to set up cooperatives or communes. The anti-nuclear power movements had resonance because they raised the broader concern with the ever-expanding scale of the industrial machine, including questioning the industrial-scale energy generation and the associated transmission and distribution networks. Communal modes of organising were often the off-spring of a counter-culture that embraced the practice of 'living lightly on the land', and this engaged the potential of alternative technologies. These technologies were human scale in their organisation, contrasting with the alienating technologies and systems of mass production associated with capitalist enterprise. This was the context in which Amory Lovins' 'soft energy' was welcomed. 'Soft energy' demonstrated that more benign forms of energy generation were feasible and could be fostered in constructing the tapestry of self-sufficient communities.

The contemporary interest in establishing cooperatives has been more consciously motivated by environmental considerations. Climate change has aroused renewed interest in the possibilities of cooperatives and communes demonstrating that it is possible for communities to generate electricity as an emissions-free exercise. Energy, or rather renewable energy, has now become the catalyst placing energy and emissions front and centre of community concerns. The interest in cooperatives is bound up in the search for modes of socio-economic organisation that are carbon neutral. Reducing the carbon footprint has been a uniting force bringing people together in common cause. Needless to say, the formation of renewable energy cooperatives has forced consideration of the other factors that have to be resolved in the making of any cooperative endeavour, so the energy cooperatives confront many of the challenges that their 1960s–1970s cooperatives did. Still there are new challenges that present, including the greater technological skill profiles and capacities required, the legal hurdles that have to be crossed and the financial demands are generally more substantial than was the case with the earlier movements.

In exploring the energy transformation being engineered by the renewable energy cooperatives, we first consider the origins and nature of energy cooperatives and their metamorphoses into renewable energy cooperatives. To better understand the renewable energy cooperative movement in Australia, the focus will follow the lead of two of Australia's foremost advocates of renewable energy cooperatives, Nicky Ison and Jarra Hicks, who received a grant to examine community renewable energy ventures in Europe, North America and India. The focus here, however, will concentrate on the place of community renewable energy cooperatives in the remaking of the energy system in Denmark, Germany and the United States.

Attention then turns to considering the anomalous position of the state in responding to the challenge posed by energy cooperatives and to the maintenance of the emissions-intensive energy regime, a challenge which the energy cooperatives have had to reconcile. This has meant working both with the state, securing its imprimatur, and against the state, by contesting the state's endorsement of a system of production and exchange that is, ostensibly at least, predicated on privileging anonymous economic transactional exchanges. Key to cooperative movement's *raison d'être* is winning state backing for the trinity goals, to support transforming the dominant modes of generating electricity by developing new institutional arrangements that ratify the importance of social exchanges as the necessary means for accelerating the transition to a zero-emissions system (Bilek 2012; Debor 2014; Hess 2013; Hockenos 2013, 2014; Hoffman and High-Pippert 2015; Janzing 2012; Kalkbrenner and Roosen 2016; Klagge and Meister 2018; Levitin 2015; Motyka 2016; Rommel et al. 2018; Weig 2013; Wettengel, 2018; Yildiz et al. 2015).

Climate Citizens Taking Control: Pathways to Community Renewable Energy Cooperatives

Community energy cooperatives are not a new phenomenon. Their significance as an institutional force has waxed and waned over the course of the history of electricity as a preeminent power source. The origins of energy cooperatives can be traced back to the late nineteenth century in the first efforts to modernise by generating electricity as the new, more convenient source of lighting and power. Community cooperatives were established to do what local governments or private investors were reluctant to do, especially in small communities where the overhead costs of investing in the new technology and in the distribution infrastructure made for a risky and possibly uneconomic venture. In those areas where municipalities did not take the lead and establish some generating capacity, communities responded by pooling their resources to invest in the technology to establish generating capacity and/or to negotiate supply and distribution agreements. This was particularly the case in rural areas. In Germany where substantial numbers of energy cooperatives formed, some 6000 cooperatives were established between 1895 and 1932. During this period, coal and oil were the chief inputs used to drive the thermal turbines for generating power and became more so over the course of the twentieth century, although hydro power and wind turbines were not uncommon. In Denmark, for instance, community-owned wind turbines had a long history as a key source of electricity.

Energy cooperatives were promoted as a means of spreading the electrification of communities and industries in rural areas in the United States following the 1930s Depression. Under New Deal programmes designed to lift rural economies out of the lingering effects of the Depression, the federal government's Rural Electrification Administration provided financial assistance and granted loans to fund the establishment of local electricity generating capacity (Benander et al. 2017; U.S. Energy Information Agency 2019; Smyth 2019). The rural electrification scheme, launched in 1937, resulted in the formation of a large number of cooperatives. The great majority were in effect consumer cooperatives that coordinated electricity distribution; some 65 were established as generation and transmission cooperatives. Coal and oil were the principal fuels used to generate electricity for, or by, the cooperatives. The cooperatives have survived to remain a common feature of the energy landscape, with some 812 energy cooperatives operating in 47 states with a disproportionate number concentrated in the Midwest and Southeast. Their continuing and current significance is evidenced in cooperatives operating 13 per cent of the nation's meters (U.S. Energy Information Agency 2019).

The influence of the energy cooperatives in shaping the make-up of energy systems has varied through time, but the 1970s marks a watershed in their orientation. Two developments combined to prompt the first intentional moves to substitute fossil fuel inputs with renewable sources of energy and to reorient the organisation of energy generation and distribution. OPEC's decision to increase the price of oil in 1973 set in motion an energy crisis that forced consideration of the deployment of alternatives to fossil fuels. The crisis also coincided with developing concerns about mounting environmental problems associated with the expansive resource and energy-intensive industrial complexes and prospective resource scarcity. This coupling of the energy crisis with growing concerns with the resource-intensity of industrial capitalism awakened interest in exploring alternative socio-economic models to the fossil fuel intensive, industrial-scale economy. One focus that was attracting considerable attention and which spoke to both concerns was the canvassing of a more community-scale and less resource and energy-intensive socio-economic system in which energy needs could be met with the adoption of 'soft energy', a system smaller in scale and one generating energy from renewable sources. 'Soft energy', and the premises of self-sufficiency on which this was based, was presented as a panacea that could address these two issues. This sparked broader interest in the promise of renewable energy generation.

This was especially prescient for Denmark, which was hit particularly hard by the 1970s hike in oil prices. Oil provided 80 per cent of the nation's energy needs, and the Danish state, municipalities and communities were forced to consider alternative energy sources. The escalating cost of oil sparked fresh interest in small-scale community-owned energy generation and, with an established record of demonstrated technological prowess in wind turbines, Denmark transitioned quite quickly to drawing power from wind turbines, and community cooperatives, some in partnership with municipalities, others with private investors, spearheaded the transition. Denmark's investment in wind power became an iconic model for climate citizens. The conjunction of different interests established a multi-stakeholder-led process that successfully transitioned from the hitherto substantial reliance on oil to demonstrate that the promise of a 'soft energy' future was not some utopian dream. This was clear evidence that community renewable energy cooperatives were both socially and technically feasible and could provide economically practical alternatives to the emissions-intensive fossil fuel-based industrial complex as well as an alternative to nuclear power which was becoming the much-vaunted alternative source of power.

Denmark's energy transition highlighted that there was no particular natural or invariable dominant energy regime. The prevailing energy system was the product of a succession of individuals innovating turbine technologies that enjoyed the backing of their local communities. These developments were cultivated through state endorsement and recognition of the innovative initiatives of individuals and the preparedness to share the technology. The transition to wind power emerged out of a shared reaction of citizens and the state to develop a system over which community control could be strengthened and alleviate the burden that the oil-price hike was placing on the economy. The fossil fuel hegemony was successfully contested with the democratisation of the energy regime.

A similar social and political transformation also unfolded in Germany, where both the oil crisis and development of nuclear power fermented renewed interest in community energy. Much of the momentum for renewable energy cooperatives could be attributed to the conviction in 'soft energy' philosophy and the disdain this engendered in the opposition to the fossil fuel energy industrial complex. The intensive, industrial-scale complex had been constructed under the government of the Third Reich. From the mid-thirties, around the same time that the New Deal project was fostering the establishment of community energy cooperatives in the US, the Third Reich had introduced a range of measures to increase levels of concentration of industry and to scale up the size of industrial enterprises. This included expanding and consolidating generating capacity with the result that the established community energy cooperatives were displaced and electricity generation became dominated by monopoly-like energy behemoths. Coal mining expanded to meet the energy demands of this industrial expansion notwithstanding the ever-increasing cost of extracting coal, which was being heavily subsidised by the German state and the European Coal and Steel Community, and ultimately by consumers.

One consequence of this development was that community energy cooperatives all but disappeared (REScoop 2020). The post-war redevelopment of Germany entrenched this industrial-scale energy model. This was based on consolidating coal as the primary fuel source for electricity generation and conceding the increasing concentration of energy corporations. Municipal governments that owned and operated power utilities also came under mounting pressure to address fiscal challenges and the difficulties the utilities faced competing against the industrial-scale generators that were being supported by the federal state, and this drove a wave of privatisations that reinforced the level of concentration in the energy sector at the same time as power prices kept increasing. Over the course of the 1960s, nuclear power plants, complementing the industrial-scale energy model, were constructed to consolidate the energy generating capacity demanded by the industrial corporations as well as a potential energy substitute for coal. Paradoxically, the turn to nuclear power engendered considerable discord, and had a catalytic effect in igniting a politically relentless anti-nuclear movement (Renn and Marshall 2016). The 1973–1974 inflation in oil prices merely added to the costly energy dilemma confronting Germany and reinforced the moves by citizens to explore alternative energy systems. The 'soft energy' discourse, with its engagement of broader political economy concerns, became a rhetorical force in defining the focus of the anti-nuclear narrative. Renewable energy cooperatives became a popular concrete form for climate citizens to demonstrate their dissent with and challenge the fossil fuel industrial complex.

This shift in climate citizens' oeuvre became more pronounced in the 1990s with the liberalisation of energy markets, a pressure for change that had become the object of energy policy throughout much of the world. Liberalisation had served to remove some of the regulatory shackles that were supposed to be hindering competition among energy providers. The same conventional economic discourse was adopted to prioritise so-called efficient and cost-effective measures for mitigating emissions. Yet, contradicting the promise of the economic theory, market liberalisation opened the way for energy corporations to capture an ever-increasing share of the market, and this made more transparent the inequities engendered by this dominance. Electricity prices escalated, and the energy behemoths were quite resolute in their resistance to investing in more benign systems of generating electricity at a time of increasing popular concern that more positive action needed to be taken to arrest the growth in emissions.

The corporations' resistance served to prompt climate citizens to themselves revisit the potential benefits of assuming direct control of energy generation and distribution, and this effected a dramatic shift in the orientation of the political economy of energy in the latter years of the 2000s decade. Community renewable energy cooperatives became a more significant vehicle for delivering power in rural Germany (Holstenkamp 2015). The state proved to be a positive force in this transition as community-based initiatives also brought changes in the political milieu with legislation being passed to accommodate renewable energy ventures. The passage of the Renewable Energy Sources Act in 2000 was a crucial factor in stimulating the surge in the number and membership of energy cooperatives because the Act established a government-guaranteed feed-in tariff and, in requiring grid owners to purchase energy generated by the cooperatives, prioritised cooperatives' access to the grid (Weig 2013; Balch 2015). There was a surge in investment in cooperatives that accelerated over the course of the decade as the organisational challenges were eased and the economics of operating a cooperative became clearer. Some cooperatives also expanded from generating energy by buying into their local grid, which provided additional cost savings by eliminating 'the middle layers of power suppliers', and this consolidated the transformation of the energy system (Tarhan 2015; German Cooperative and Raiffeisen Confederation 2013).

Foregrounded in this development was consideration of the broader economic consequences of the fossil fuel industrial complex. There were the immediate benefits flowing from climate citizens' initiatives. The state's legitimacy had been thrown into question. There was strong popular opposition to the adoption of the market-based measures designed to put a price on carbon to incentivise emissions mitigation. Pricing carbon did not provide comfort for those who would have to pay a higher price for their electricity and provided no necessary assurance that emissions would be reduced. The state's continuing defence of the low-emissions alternative technology offered with nuclear power reinforced this loss of legitimacy (Renn and Marshall 2016). In fact, given the structure of the energy sector and the time frames required to construct new sources of electricity, it was likely that energy uses would be paying a high premium for 'dirty energy' into the immediate future. By generating clean energy, the renewable energy cooperatives would bring forward emissions mitigation and reduce the deleterious effects that coal-fired power would have for future generations, and thereby addressing the problem of intergenerational equity. Furthermore, because the cooperatives were the product of grassroots' community endeavours, they were more conscious of equity considerations and ensuring that all members of the community should have access to clean energy and not be relegated to the fringes of the energy market to live in energy poverty. The feed-in tariff scheme was extremely popular, and together with state endorsement of energy cooperatives, the greening of state policies contributed to restoring the state's legitimacy stocks.

Building an economic system that internalised and circulated the benefits of local energy generation within the community was integral to the cooperative's broader economic vision. Community energy cooperatives became a means of stemming the loss of economic vigour and the substantial financial drain on the wellbeing of local communities consequent upon energy payments being made to energy corporations headquartered in some far-off location. Community renewable energy cooperatives in Germany complemented the other climate movement forums—the street protests, the carbon-intensive site occupations and the re-municipalisation projects—that were responding to the energy-climate existential crisis. Confronting the role of the state in the constitution of the emissions-intensive fossil fuel complex was a natural corollary of these developments. It manifested in questioning the magnitude of state subsidies that underwrote the construction of industrial-scale generating capacity, be this for constructing fossil fuel-based generators or for nuclear power and which exaggerated the irrationality of the fossil fuel industrial complex.

Community-owned renewable energy cooperatives were held up as a providing a concrete means of producing clean energy, reducing the community's carbon footprint, delivering safer and more reliable energy supplies. A common argument in the advocacy of energy cooperatives has been that cooperatives feature as an essential objective in their *raison d'être* the sequestration of the economic benefits that flow from the cooperative energy venture. The cooperative provides a foundation for strengthening the circular economy, ensuring that the local sourcing of renewable energy will counter the loss of the dynamism of communities consequent upon the wealth that is drained out of the communities as they pay the price of energy supplied by the corporations registered in other localities. Local renewable energy generation and distribution will provide employment opportunities and the reinvestment back into the community that would otherwise be lost, and this helps in providing new momentum in those communities (NRECA 2018b).

In Germany, the number of cooperatives being established each year climbed from 8 in 2006 to 267 in 2011, and there are some 850 functioning today, and much of the growth predated the *Energiewende* legislative initiative of 2010 (Weig 2013; Janzing 2012; German Cooperative and Raiffeisen Confederation 2013). This growth in renewable energy cooperatives highlights the extent to which the momentum of Germany's energy transformation has been 'from below', citizen-driven, political and social life at the level of communities. The grassroots' ventures found expression in the different possibilities for decarbonising and decentralising the energy system, and for democratising decision-making over energy.

Energy market liberalisation has proved to be a somewhat paradoxical backdrop to the formation of community renewable energy cooperatives. On the one hand, liberalisation of markets unlocked the opportunity for the introduction of alternative institutional initiatives into the electricity sector, and states have been supporting such endeavours. In Europe in 2016, following concerns expressed by renewable energy cooperative proponents, the European Commission announced a number of measures to remove legal obstacles to the formation of energy cooperatives (Vantsinjan 2017). The European Union has embraced the important role that community energy can play in increasing renewable energy generating capacity (European Union European Regional Development Fund 2018; European Climate Initiative 2019; Friends of the Earth Europe et al. 2018).

On the other hand, in Europe the consolidation of energy market liberalisation as a developing policy objective has had a detrimental effect on the viability of community energy cooperatives. In Denmark competitive market pressures have resulted in the merging of some cooperatives as they have struggled to meet the competition from private energy providers, and this has resulted in some aspects of their social enterprise objectives being sacrificed (Wierling et al. 2018). In Germany, some of the momentum being generated by climate citizens to deepen and hasten the transition has dissipated following the wind back of feed-in tariffs and the introduction of electricity supply-based bidding systems. Competing for market share at auction has been to the advantage of the energy generation corporations (Werhmann 2017; Wettengel, 2018; Buchsbaum 2019). This change in pricing has prompted some spirited responses from the German Renewable Energies Federation BEE, the German Community Power Association BBE, the German Cooperative and Raiffeisen Confederation and the World Wind Energy Association, who have called upon the German government to provide some relief from the pricing-by-auction system (WindFair 2020). Energy market liberalisation has had a comparable negative impact on cooperatives in the US. The easing of some of the restrictions that limited wholesale market trading to within States to permit cross-border investment by private operators, has created new competitive pressures for energy cooperatives.

The notion that community renewable energy cooperatives should not be excused from being subject to the principle that liberalisation of energy markets is designed to remove the institutionalised market advantages and create a level playing field overlooks the differences in the capacities to shape market processes. Arguments informed within liberal political discourse that renewable energy cooperatives should not be insulated from competition overlook the structure of the network, the transmission infrastructure and the substantially greater magnitude of state financial support provided to the energy corporations and which give the corporations a decided competitive advantage. Moreover, such arguments ignore the fact that the cooperatives are concentrated in the smaller markets in rural and remote areas, the US being the most obvious case, where overhead costs are generally greater. The energy corporations have demonstrated an historical reluctance to service these communities, or until the market expands to a large enough size that justifies private investor's interest. The instances where cooperatives are able to be established beg an interesting question about the constitution of markets and energy supply. Measures engineered by the state to progress the energy market liberalisation that have gone hand in hand with facilitating the formation of cooperatives in rural and regional areas may well have helped to strengthen supply arrangements, but such measures may also be being motivated by a model that sanctions some self-reliant energy enclaves to free large energy corporations from having to service all electricity users and not have to guarantee minimum standard service provisions across the whole network. This seems to be happening in Australia where energy market liberalisation has weakened the service provision requirements that have been the corollary of holding a licence. Energy companies are able to withdraw from servicing areas that are more costly to distribute power, although they have tended to allay concerns about the provision of services by assisting energy cooperatives to become established. A similar organisational change is being debated in California in the aftermath of the wildfires as PG&E navigates reframing of its operation.

The Challenges in Forming Community Renewable Energy Cooperatives: Engaging/Disengaging the State

The formation of a community renewable energy cooperative is no easy feat as a substantial research literature has documented.[1] These investigations highlight how the establishment of cooperatives relies on the endeavours of a highly motivated and determined group of community members

[1] There has been a surprisingly lot of reflection on the challenges that have confronted the German cooperatives, much building on the research led by Walker and colleagues on British renewable energy cooperatives, as well as some research published on Australian renewable energy enterprises. A selection of such research includes Balch (2015), Bauwens et al. (2016), Bilek (2012), Buchsbaum (2019), Centgraf (2018), Hain et al. (2005), Herbe et al. (2017), Holstenkamp (2015), Holstenkamp and Kahla (2016), Ison and Langham (2015), Kalkbrenner and Roosen (2016), Kunze and Becker (2015), Li et al. (2013), McKinnon (2015), Mey et al. (2016), van Summeren et al. (2020), Walker (2008), Walker and Devine-Wright (2008) and Walker et al. (2007, 2010).

who possess the technical, engineering, legal, financial management and organisational skills capable of mapping a proposal and carrying this through to fruition. Those individuals who propose the cooperative must also have the personal attributes that are capable of presenting the idea of an energy cooperative as an appealing proposition, one that will be economically attractive, and they must have the determination to stand up to the likelihood that a commercial energy provider will oppose state approval of the venture. They must also work their way through the labyrinth of development applications and rules and regulatory requirements when applying for authorisation because the energy sector is highly regulated for safety reasons and because of the public good nature of electricity. Unlike approval for the installation of solar panels, which is generally handled by the installation companies or the energy retailers, and the lobbying endeavours to establish municipal utilities where the focus of climate activists is concentrated on petitioning councillors and rallying the support of the community and when the task of managing the energy utility is taken on by the municipality, energy cooperatives rely almost entirely on the volunteer work of its advocates.

Generally, there is limited institutional backing on hand to dedicate the resources required for setting up a cooperative. The costs incurred have to be raised by those advocating the project or, if there is some development support on offer, such as from the state, advocates have to dedicate considerable time and energy in applying for funds. The project is usually dependent on the efforts of that small group of volunteers who have entertained the idea, and who have the time to dedicate themselves voluntarily to marshal interest in the project, work their way through the host of challenges that present from the moment of conception, and raise the funds with which to finance the project. The body of research investigating the establishment of cooperatives has provided a helpful introduction to the challenge, offering many suggestions, and warnings, as to how to address the challenges, although it is generally acknowledged that ventures that are so heavily reliant on volunteers at various, and often all, stages of a venture, that advocates will invariably face some intractable problems.

One challenge that must be confronted is how best to approach and constructively and strategically deal with the state. Community renewable energy cooperatives require the imprimatur of the state at the same time as the state is the institution which is the source of much of what the climate activists advocating cooperatives are campaigning against. State support is necessary because the energy cooperative's formation is conditional, if not dependent, on the state's preparedness to accommodate, even support, their presence (Oteman et al. 2014: 10). In many respects, the renewable energy cooperatives are being established as an organisational form in response to the failure of the state to pursue meaningful action to abolish reliance on fossil fuels. Cooperatives are being formed in opposition to the state as the long history of anarchist movements challenging the legitimacy of the state has demonstrated.

On the other hand, the state is, of course, not a permanently set construct, as can be clearly observed by reflecting on the German state's adoption of *Energiewende*. Many of the climate citizens who were instrumental in the endeavours to establish energy cooperatives based their political philosophy on an anarchist tradition. This tendency also figured prominently in the vibrant and political force of the anti-uranium movement and in the climate movement that built on the opposition to nuclear power. This political tendency challenged the legitimacy of the state regarding it as being complicit in the making of the fossil fuel-based industrial complex. This challenge to the legitimacy of the state was also played out in the concerted efforts to shut down nuclear and coal-fired power plants. The authority of the state was placed into question by the campaigns of civil disobedience and occupations, of transport networks, coal mines. The political momentum of these movements contributed to the dramatic shift in the state's energy policy agenda and its translation into the *Energiewende* (Klagge and Meister 2018; Yildiz et al. 2015; Bilek 2012; Rommel et al. 2018; REScoop 2020). The decision to bring forward the *Energiewende* was shaped by a new determination within the state to legitimise its energy policy, especially in the context of the realisation that the Fukushima meltdown instilled the anti-nuclear movement with a more authoritative voice. Arguments in support of renewable energy gained currency as did new institutional forms of generating energy, and the *Energiewende* commitment to delivering 60 per cent of energy from renewable sources by 2050 and cutting emissions by 80–95 per cent from 1990 levels was reflective of the shift in sentiments of a large proportion of the population. Yet, while some radical political tendencies were more resolute in their determination to cultivate enhanced energy self-sufficiency, through re-municipalisation and renewable energy cooperatives, the German state and industrial corporations retained confidence in the importance of industrial-scale energy generation, and they looked to an alternative energy narrative, one that fostered the manufacture of the new the renewable energy technologies that could provide first leader advantage in transitioning and which would reignite the pace of industrial development (Hillebrand 2013; Buchan 2012).

The political dynamic that framed the energy transition in Germany exposes one of the tensions in the role of the state for community renewable energy cooperatives. As was clearly demonstrated, confronting the challenge of climate change highlighted the contradictory role of the state and how the cooperative movement manages this. The motif of the cooperative movement, democratisation, decarbonisation and decentralisation, is founded on giving voice to breaking free from the established politico-energy system. The ambition is to establish an enclave that built on the foundations of a circular economy and freed of the chains of the state and energy corporations and an operational logic that is not defined in terms of the accumulation of capital and the profit imperative. The cooperative model is constituted as the antitheses of the liberalisation project. Democratisation is a counter to energy generation and distribution as if it is only a transactional process; decarbonisation abandons the notion that emissions should be regarded on the same value chain plane as the energy that is generated, and; decentralisation is about establishing a socio-technical spatial fix that is not bounded by the territorial reach of the state and an accumulation dynamic that the state seeks to underwrite.

Yet, the establishment of energy cooperatives is contingent on their incorporation to comply with regulatory framework that permits their operation (Friends of the Earth Europe et al. 2018). To become established and be permitted to generate and distribute energy, community energy cooperatives have to be legally incorporated. State sanction of the cooperative is a necessary requisite for the cooperative to engage in electricity distribution. In the Australian context the enterprise is required to register as a for-profit business, on terms applicable to established and registered energy providers to operate as a going concern, that is, not running at a loss. It is a common practice for investments in the enterprise to be on the basis of a share issue, but unlike the private corporation the democratisation goal is embodied in the constitution and practices of the enterprise. The norm is to limit shares issued to an individual to prevent member participation in the cooperative being compromised by a controlling shareholder. Members will be rewarded with a return on their investment, usually set at a predetermined level that does not forego the generation of a surplus that can be used for social and/or environmental enhancement. To avoid having this surplus (or profit) generating requirement becoming the overriding goal and overshadowing the cooperative's substantive mission, the enterprises are generally managed through a two-tier organisational structure. The entity registered as a for-profit business, operates as if it is a private investment enterprise, preparing and issuing annual reports like its counterparts, issuing a dividend to its community investors, its shareholder members, but operationally it is responsible to the cooperative entity. Financial surpluses are allocated to the managing parent, the actual incorporated energy cooperative, which invests these funds back into community ventures in line with its stated mission (Enova, n.d.).

Community renewable energy cooperatives must therefore operate in a manner that is consistent with a nation's legal codes. In practice, state imprimatur of cooperatives has been critically important to their development. Indeed, the establishment of community renewable energy cooperatives could not, and mostly has not, proceeded without some measure of direct state support. The formation of cooperatives is contingent on the states having established the legal structures that regulate their formation and their management and, crucially, the governance arrangements that permit an energy cooperative to engage in the generation and distribution of electricity. Moreover, cooperatives generally have to rely on the cooperation of the owners and operators of transmission infrastructure because, as much as it is their ambition to break free of this infrastructure, in most cases the distribution of renewable energy is dependent on access to the grid network, and this access can only work if it is governed by state-determined regulations.

State support has also extended to the financial underwriting of energy cooperative ventures. In Denmark, the state has been crucial in supporting the development of community renewable energy cooperatives. Many of the wind turbine projects were developed with the backing of the national state or municipal governments (Friends of the Earth Europe et al. 2018; Oteman et al. 2014; Cumbers 2012; Martin 2017). In the United States, it was mostly the federal government that advanced funds to back the formation of cooperatives with the aim of redressing the low level of electrification of rural communities. In Australia, state support has been of immeasurable assistance in helping energy cooperatives get off the ground. With the first community renewable energy cooperative, the Hepburn Wind farm in north-west Victoria, government seed funding proved a significant factor in turning what was an idea into a concrete enterprise. Enova, the community renewable energy cooperative operating in the

Northern Rivers area of New South Wales, has been awarded several funding grants (Energy Source & Distribution 2021), as has Totally Renewable Wangaratta in north eastern Victoria. Several other cooperatives have received seed funding from federal state as well as from local municipal governments. Interestingly, ARENA has supported the organisation of annual conventions of renewable energy cooperative advocates, and the Victorian government financed the preparation of a 'how-to' manual (Lane and Hicks 2017).[2] The federal government has a competitive grant scheme which awards funds to successful applicants to support for energy efficient equipment, energy generation and storage systems, and energy audits (Department of Business 2015). Such support clearly serves the purpose of the state being seen to be representing a broad array of competing energy and climate positions by the nation's constituents, while other motivations are considered in Chap. 10.

Community Renewable Energy Cooperatives as Political Project

An important aspect of the re-emergence of community renewable energy cooperatives has been the formation of associations representing the collective interests of the cooperatives. The peak associations in the US and Europe have emerged as critically significant lobbyists giving voice to the cooperative goals, providing some direction for the development of established and newly forming cooperatives, and securing financial and legal support. Such associations are not new. The National Rural Electric Cooperative Association in the US was formed at the start of American involvement in World War Two to provide a unified voice for the rural cooperatives and to lobby the federal government (NRECA, n.d.). While member cooperatives have been locked into the fossil fuel intensive energy regime, they have been increasingly receptive to transitioning to renewable energy, and the National Rural Electrical Cooperative Association has encouraged this in signalling that the future of cooperatives lay with

[2] For instance, grants have been awarded for specific community renewable energy projects—Frontier Impact Group partnering with Trading as Community Power Agency and Embark Australia; the Moreland Energy Foundation—and organising and strategic forums—the National Community Energy Strategy; UTS Institute for Sustainable Energy in partnership with Community Power Agency, Embark; the National Community Energy Strategy Regional Australia's Community and Regional Renewable Energy Program.

renewable energy. In 2004, it announced that it would support the uptake of solar generating capacity by cooperatives. In fact, the actual uptake by cooperatives has far exceeded expectations, and the momentum of members' interest in transitioning has clearly shifted in favour of replacing coal with renewables, especially as the cost of solar has declined (Rocha 2019). Responding to pressure from members and growing consumer expectations, an increasing number of community cooperatives are now committed to reducing emissions and transitioning to renewable energy technologies.[3] One report observed that this interest is going 'viral', and that the number of co-op members has been increasing as households embrace clean energy and community cooperatives' share of the renewable energy market has been growing (NRECA 2018a).

A quiet energy revolution is underway in America, manifest in the more-than fourfold increase in cooperatives' solar power generating capacity over 2015–2018. A change in the Association's leadership, a leadership that has confidence in the future being renewable energy, will likely consolidate this shift (NRECA 2018b).[4] The majority of power generated by cooperatives now comes from low-emissions and zero-emission sources, and the cooperatives-coal partnership is being further contested in locations where the coal remains king.

But what appears to be the revitalisation of the peak association is not just about renewable energy. It is as much about reasserting the role of cooperatives in enhancing their place in the life of communities. This is reflected in the management of cooperatives being subject to democratic processes and reminding the world that they are more than just energy providers (Hoffman and High-Pippert 2015; Johnson and Lewis 2017; NRECA 2017; Peters et al. 2018). Their purview extends beyond energy to demonstrate commitments to strengthening the resilience of community through supporting health and education facilities, promoting equity with respect to access to energy and equality of employment opportunity, contributing to the development and dynamics of the local economy and enhancing the environment. While not all are as committed to the goal of

[3] Coal is the source of 40 per cent of energy generated and distributed by the electricity coops; natural gas 25 per cent and nuclear 15 per cent, with hydro delivering 10 per cent and other renewables 8 per cent. Investment in solar power has resulted in quite a spectacular increase in the amount of power generated, from 35 MW in 2013 to 96 MW in 2018.

[4] In line with this transitioning, NRECA's partner organisation, the Tri-State Generation and Transmission Association has set a goal for half of its generating capacity to be based on renewable sources by 2024 (Holly 2020).

decarbonisation, a growing number of NRECA's members do now regard transitioning to renewable energy as key to the goal of social progress (Hatlestad et al. 2019; Rocha 2019; NRECA 2018b).

Peak bodies representing the energy cooperative cause and environmental non-government organisations have played an even more active role in Europe. Lobbying pressure of European state institutions by Friends of the Earth Europe, REScoop and the European Renewable Energy Coalition have been particularly effective in garnering the support of the European Parliament for community renewable energy (European Climate Initiative 2019; Friends of the Earth Europe et al. 2018. In 2016, the European Commission, having come to the conclusion that cooperatives could play a crucial role in Europe's transition to renewable energy, announced a number of measures to remove legal obstacles to the formation of energy cooperatives (Vantsinjan 2017). The European Union has embraced the important role that community energy can play in increasing renewable energy generating capacity (European Union European Regional Development Fund 2018). The lobbying and publicising roles of the sector's peak associations and the concrete transformations being wrought in Germany and the Danish model have emboldened similar movements across Europe, including Britain (Rossetto 2020). This has recently involved out-reach programmes to foster energy cooperation in accession countries. Some of these activities are being supported by the German Federal Ministry for the Environment, Nature Conservation and Nuclear Safety, which is indicative of strategic links being formed with municipalities. These have been particularly constructive in strengthening the viability of cooperatives where this gives access to municipal-governed grids. The pursuit of these objectives provides crucial lessons for the relative latecomers to renewable energy cooperatives, such as Australia, and they are being heeded.

There have been several community renewable energy cooperative initiatives launched in Great Britain, some independent, some with state support, and many emerging out of municipality-focused movements. The example of Nottingham becoming the first British municipality in 75 years to establish a power utility, Robin Hood Energy, has set a precedent that climate change activists are drawing upon to pressure other councils to follow this lead (Dubb 2018). The energy cooperative movement is receiving a boost from the formation of renewable energy cooperatives in the United Kingdom (Mayo 2012). For example, Community Energy Scotland was registered as a charity in 2007 and has been promoting the benefits of non-profit community energy schemes and assisting with the development of projects. It grew out of the successful renewable energy cooperative, the Highlands and Islands Community Energy Company, which was formed in 2004 (Community Energy Scotland 2020). Community Power Cornwell, formed in 2008, presents an interesting case study in the successful expansion of community energy cooperatives because in addition to a number of up-and-running projects and more in the pipeline, it purchased the private company West Country Renewables Limited to consolidate the position of community renewable energy in the region (Community Power 2020) The Institute of Welsh Affairs is pressing the merits in involving community as an essential step in promoting clean energy (Jones 2018; Buckland-Jones 2019).

Emerging from this focus has been the normalisation of community renewable energy cooperatives as a feature of the energy system. These developments have contributed to shifts in state policies as is demonstrated in Britain's Department of Business, Energy and Industrial Strategy supporting the trial of a community energy project in 2013, and the subsequent release of its Community Energy Strategy (Department of Business, Energy and Industrial Strategy 2015). In Britain, governments are offering financial support for community energy projects in Cornwall, Scotland and Wales, and this support can be critical in providing the initial start-up capital to get the venture underway. Springing up in other locations, Ireland's renewable energy cooperative retailer is now operating. It emerged out of Templederry Wind Farm and has partnered with several other Irish renewable energy cooperatives (Community Power 2020).

These developments are also translating into the development of international collaboration. Transborder connections are most developed in Europe (Interreg Europe 2018). The NRECA has participated in initiatives to forge global links. The formation of the International Renewable Energy Agency in 2011 following two years of planning and the doubling in the members of states that are members of the IRENA have helped normalise the standing of community renewable energy cooperatives in a global transitioning programme (IRENA 2018).[5]

[5] There are 161 members of IRENA, including Australia, Denmark, Germany and the United States.

Conclusion

The enduring force of climate change politics has been translating into concrete community-based action as climate citizens establish a common cause in partnerships to invest in renewable energy technologies, take control of and transform the energy system. Community energy cooperatives are not a new phenomenon. They can be as old as the beginnings of the electrification of cities and industry. What is new is that over the course of the last decade and more, there has been a transformation in the *raison d'être* of cooperatives as they lead the transition to relying on renewable sources of energy. This transformation has not been simply a technological process. Established cooperatives, such as those in America, the reinvigorated cooperative movements such as has occurred in Germany and Europe, and the recent interest in renewable energy cooperatives in Australia, are forcing changes to the social relations and political constitution of the energy system. Decarbonisation lies at the heart of this project, and this is predicated on giving voice to climate citizens through the democratisation of energy such that communities can determine the make-up of the energy system. Fostering the decentralisation of the spatial context of the system has become an important objective in underwriting decarbonisation and democratisation. Together, these goals are guiding principles establishing the means for securing energy self-reliance and self-sufficiency. They speak to the challenge of decarbonising, of reducing the community's carbon footprint and the broader dynamic of the community's ecological footprint. And these principles inhere in the social fabric that constitutes the cooperative and the community whose interests it represents. Energy democracy becomes the metaphor for climate justice.

The momentum of this movement has gathered in pace and intensity as the climate challenges become increasingly transparent and in the face of the resistance of global energy corporations and states to taking steps to mitigate emissions. Record temperatures and heatwaves, highlighted by the ravages of climate change, especially as these have manifest in the wildfires in California in 2018 and 2019, Greece in 2018, Portugal in 2017 and Spain in 2019, and an extraordinary 2019–2020 season of bushfires in Australia, have propelled communities into appreciating the benefits of local control of electricity generation and distribution and the urgency to decarbonise. The threat and increasing frequency of extreme weather events that are increasingly associated with climate change are also challenging the engineering integrity and reliability of the industrial-scale energy systems and reinforcing the social and political merit in the energy cooperative movements endeavour to recast energy systems as well as the global goal of net-zero emissions.

References

Balch, O. (2015) 'Energy co-ops: why the UK has nothing on Germany or Denmark' *The Guardian* 2 October.

Bauwens, T., B. Gotchev and L. Holstenkamp (2016) 'What Drives the Development of Community Energy in Europe? The Case of Wind Power Cooperatives' *Energy Research & Social Science* 13: 136–147.

Benander, L., D. Anagrita Horowitz and I. Baker (2017) 'New Economy Energy Cooperatives bring power to the people' in D. Fairchild and A. Weinrub (eds) *Energy Democracy: Advancing Equity in Clean Energy Solution'* Washington D.C.,: Island Press, 195–201

Bilek, A (2012) 'Revitalizing Rural Communities through the Renewable Energy Cooperative' Washingt0n D.C.: Heinrich Böll Stiftung'.

Buchan, D. (2012) The Energiewende – Germany's gamble, Oxford: Oxford University Press.

Buchsbaum, L.M. (2019) 'Germany renewable energy cooperatives struggle as markets collapse' *Energy Transition* 19 July.

Buckland-Jones, S. (2019) Community and local ownership of renewable energy in Wales' Institute of Welsh Affairs 6 March.

Centgraf, S. (2018) 'Supporting civic engagement in German energy cooperatives—Transdisciplinary research based on the reflection of individual needs' *Energy Research and Social Science* 44, 112–121.

Community Energy Scotland (2020) 'Community Energy Futures'.

Community Power (2020) 'Ireland's first community owned Renewable Electricity Supplier'

Cumbers, A. (2012) *Reclaiming public ownership: Making space for economic democracy* London: Zed books.

Debor, S. (2014) 'The socio-economic power of renewable energy cooperatives in Germany: Results of an empirical assessment' Wuppertal Paper no. 187 (April) Wuppertal Institute for Climate, Environment and Energy.

Department of Business, Energy and Industrial Strategy, United Kingdom (2015) 'Community Energy'.

Dubb, S. (2018) 'In Europe, non-profit and local public utilities are seen as key to green future' *NonProfit Quarterly* 9 May.

Energy Source and Distribution (2021) 'Network-connected battery storage trial to begin in Queensland' 21 March.

Enova (n.d.) 'Enova Community is the not-for-profit arm of Enova Community Energy'.

European Climate Initiative (2019). 'Clean energy for all Europeans package'.

European Union European Regional Development Fund (2018) *Renewable Energy Communities: A policy brief from the policy learning platform on low-carbon economy* August.

Friends of the Earth Europe, energy cities, Greenpeace and ReScoop EU (2018) *Unleashing the Power of community renewable energy*. Brussels.

German Cooperative and Raiffeisen Confederation (Deutscher Genossenschafts- und Raiffeisenverband e. V.) (2013) https://www.cleanenergywire.org/experts/german-cooperative-and-raiffeisen-confederation.

Hain, J.J., G.W. Ault, S.J. Galloway. A. Cruden and J.R. McDonald (2005) 'Additional renewable energy growth through small-scale community oriented energy policies' *Energy Policy* 33 (9), 1199–1212.

Hatlestad, E., K. Rock and L. Veazey (2019) *Rural Electrification 2.0: The Transition to a Clean Energy Economy* Minnesota: CURE.

Herbe, C., V. Brummer, J. Rognli, S. Blazejewski and N, Gorecki (2017) 'Responding to policy change: New business models for renewable energy cooperatives – Barriers perceived by cooperative' *Energy Policy* 109: 82–95.

Hess, D.J. (2013) 'Industrial fields and countervailing power: the transformation of distributed solar energy in the United States' *Global Environment Change* 23(5) 847–855.

Hillebrand, R. (2013) 'Climate protection, energy security, and Germany's policy of ecological modernisation' *Environmental Politics* 22(4), 664–682.

Hockenos, P. (2014) 'Germany's revolution in small batch, artisanal energy' *Foreign Policy* 31 October, https://foreignpolicy.com/2014/10/31/germanys-revolution-in-small-batch-artisanal-energy/.

Hockenos, P. (2013) 'Local, decentralised, innovative: Why Germany's municipal utilities are right for Energiewende' *Energy Transition* 28 September.

Hoffman, S.M., and A. High-Pippert (2015) 'Community solar programs and the democratisation of the energy system' Proceedings of the European Consortium for..., 2015—ecpr.eu.

Holly, D. (2020) 'Tri-State G&T announces plans to phase out coal generation' NRECA 20 January.

Holstenkamp, L. (2015) 'The Rise and Fall of Electricity Distribution Cooperatives in Germany' (September 1, 2015). Arbeitspapierreihe Wirtschaft & Recht Working Paper Series in Business and Law 22/No. 22, Leuphana Universitåt Lümburg.

Holstenkamp, L., and Kahla (2016) 'What are community energy companies trying to accomplish? An empirical investigation of investment motives in the German case' *Energy Policy* 97: 112–122.

Interreg Europe. (2018) Renewable Energy Communities: A policy brief from the Policy Learning Platform on low-carbon economy August.

IRENA [International Renewable Energy Agency] (2018) *Community Energy: Broadening the ownership of renewables* Abu Dhabi.

Ison, N., and E. Langham (2015) 'National Community Energy Strategy' UTS.

Janzing, B. (2012) 'Energy cooperatives are booming in Germany' *DW* 6 July.

Johnson, D., and A. Lewis (2017) 'Organizing for Energy Democracy in rural electric cooperatives' in D. Fairchild and A. Weinrub (eds) *Energy Democracy: Advancing Equity in Clean Energy Solution'* Washington D.C.,: Island Press, 93–112.

Jones, Calvin (2018) *The Economic Costs and Benefits of Renewable Energy Transition in Wales*, Institute of Welsh Affairs.

Kalkbrenner, Bernhard J. and Jutta Roosen (2016) 'Citizens' willingness to participate in local renewable energy projects: The role of community and trust in Germany' *Energy Research and Social Science* 13(1), 60–70.

Klagge, B. and T. Meister (2018) 'Energy cooperatives in Germany—an example of successful alternative economies?' *Local Environment: The International Journal of Justice and Sustainability* 23 (7), 697–716.

Kunze, C., and S. Becker (2015) 'Collective ownership in renewable energy and opportunities for sustainable degrowth' *Sustainability Science* 10, 425–437.

Lane, T., and J. Hicks (2017) *Community engagement and benefit sharing in renewable energy development: A guide for renewable energy developers* Environment, Land, Water and Planning, State Government Victoria.

Levitin, M. (2015) 'Community energy model is speeding US moves to renewables' *The Guardian* 8 May.

Li, Wen Li, Janine Bermel, Harald Schach and Werner Konold (2013) 'Transitioning to community-owned renewable energy: Lessons from Germany' *Procedia Environmental Sciences* 17, 719–728.

Martin, B. (2017) 'Communal ownership drives Denmark's wind revolution' *Green Economy* 20 September.

Mayo, E. (2012) 'The communities taking renewable energy into their own hands' *The Guardian* 7 January.

McKinnon, John (2015) 'Australia's National Community Energy Strategy' Australian Environmental Makers Network 11 June.

Mey, F., M. Diesendorf and I. McGill (2016) 'Can local government play a greater role for community renewable energy? A case study from Australia' *Energy Research and Social Science* 21(1) 33–43.

Motyka, M. (2016) 'Unlocking the value of community solar: Utilities find opportunity in the inevitable growth of distributed energy', Deloitte Trends.

NRECA [National Rural Electric Cooperative Association] (2018a) [National Rural Electric Cooperative Association] (2018a] *A Solar Revolution in Rural America* July.

NRECA [National Rural Electric Cooperative Association] (2018b) 'Waging a comeback for small town America' 11 December.

NRECA [National Rural Electric Cooperative Association] (2017) 'America's Electric Cooperatives Fact sheet'.

Oteman, M., M. Wiering and J-K. Helderman (2014) 'The institutional space of community initiatives for renewable energy: a comparative case study of the Netherlands, Germany and Denmark' *Energy, Sustainability and Society* 4 (11), 1–17.

Peters, M., S. Fudge, A. High-Pippert, V. Carragher and S.M. Hoffman (2018) 'Community solar initiatives in the United States of America: Comparisons with—and lessons for—the UK and other European countries' *Energy Policy* 121, 355–364.

Renn, O. and J. Marshall (2016) 'Coal, nuclear and renewable energy policies in Germany: From the 1950s to the "Energiewende"' *Energy Policy* 99, 224–232.

REScoop (2020) 'The energy transition to energy democracy'.

Rocha, V. (2019) 'Rural revitalization: Applying co-op know-how to rebuilding communities' NRECA 1 January.

Rommel, J., J. Radtke, G. Jorck, F. Mey and Ö. Zildiz (2018) 'Community renewable energy at a crossroads: A think piece on degrowth, technology, and the democratization of the German energy system' *Journal of Cleaner Production* 197(2), 1746–1753.

Rossetto, N. (2020) 'Community led renewable energy initiatives: A multi-faceted and evolving phenomenon' Florence School of Regulation, 30 March, https://fsr.eui.eu/community-led-renewable-energy-initiatives/.

Smyth, J. (2019) 'Rural America could power a renewable economy—but first we need to solve coal debt' *Clean Cooperative* 2 July.

Tarhan, M. D. (2015) 'Renewable Energy Co-operatives: A review of demonstrated impacts and limitations' *Journal of Entrepreneurial and Organizational Diversity* 4(1), 104–120.

U.S. Energy Information Agency (2019) 'U.S. electric utilities by ownership type (2017)'.

van Summeren, L.F.M., A.J. Wieczorek, G.J.T. Bombaerts and G.P.J. Verbong (2020) 'Community energy meets smart grids: Reviewing goals, structure, and roles in Virtual Power Plants in Ireland, Belgium and the Netherlands' *Energy Research and Social Sciences* 83, May.

Vantsinjan, J. (2017) 'Europe, renewables and the future of community energy' *Coop News*.

Walker, G. (2008) 'What are the barriers and incentives for community-owned means of energy production and use?'

Walker, G., and P. Devine-Wright (2008) 'Community renewable energy. What should it mean?' *Energy Policy* 36(2) 497–500.

Walker, G., P. Devine-Wright, S. Hunter, H. High and B. Evans (2010) 'Trust and Community: Exploring the meanings, contexts and dynamics of community renewable energy' *Energy Policy* 38(6), 2655–2663.

Walker, G., S. Hunter, P. Devine-Wright, B. Evans and H. Fay (2007) 'Harnessing community energies: explaining and evaluating community-based localism in renewable energy policy in the UK' *Environmental Politics* 7(2), 54–82.

Werhmann, B. (2017) 'The Energiewende's booming flagship braces for stormy times' *Clean Energy Wire* 14 June.

Wettengel, J. (2018a) 'Citizens' participation in the *Energiewende*' *Clean Energy Wire* 25 October.

Wierling, A., V.J. Schwanitz, J.P. Zeiß, C. Bout, C. Candelise, W. Gilcrease, J.S. Gregg (2018) 'Statistical evidence on the role of energy cooperatives for the energy transition in Europe' *Sustainability* 10 (9), 1–25.

WindFair (2020) 'Open letter to the German energy minister calls for more community power', 12 February.

Weig, A (2013) *Please, in My Backyard: How renewable energy cooperatives advanced citizen involvement in the German energy transition*, Washington D.C., Heinrich Böll Foundation and The Climate Network.

Wettengel, J. (2018b) 'Citizens' participation in the *Energiewende*' *Clean Energy Wire* 25 October.

Yildiz, Ö., J. Rommel, S. Debor, L. Holstenkamp, F. Mey, J.R. Müller and J. Radtke (2015) 'Renewable energy cooperatives as gatekeepers or facilitators? Recent developments in Germany and a multidisciplinary research agenda' *Energy Research and Social Science* 6(1), 59–73.

CHAPTER 11

Renewable Energy Cooperatives Challenging the Liberalised Energy Economy in Australia

There has been a sudden burst of enthusiasm in the idea of renewable energy cooperatives. The unprecedented record hot temperatures that prevailed through 2019 and culminated in the devastating bush fires that obliterated many communities over the summer of 2019–2020 has moved many to consider the most sensible strategy for rebuilding. The regularity of the sustained heatwaves demonstrated that climate change was already having an impact, and this was resulting in changes in building regulations. Royal Commissions conducted to review previous conflagrations exposed one hazard was connected to the power networks that delivered electricity to homes and businesses. There was the potential for power lines to ark and set ablaze the surrounding flora. The premier source of greenhouse gas emissions, the industrial-scale coal-fired power system, was now recognised as a potential source of ignition. The need to rebuild is now accepted as more than an exercise in the rebuilding of dwellings. How power is to be delivered safely to the communities has become an integral feature of the reconstruction of the devastated communities, and one feature of this has been a mushrooming of interest among residents in areas that felt the full force of the wildfires. Along the south-east coast of Australia and forested retreats, the idea that electricity self-reliance would make for a safer environment has attracted considerable appeal. Of course, this is a subject that is attracting attention around the world.

This interest in renewable energy and especially in renewable energy cooperatives is not unique to Australia, but it did come at a time when

S. Rosewarne, *Contested Energy Futures*,
https://doi.org/10.1007/978-981-19-0224-6_11

renewable energy cooperatives were being established in a growing number of countries in the first decade of the new millennium and there were increasing experiments in replicating model cooperatives being explored in Australia. One of the spin-offs of this interest in cooperatives was the potential for cooperatives set on becoming self-sufficient to take the lead in energy self-sufficiency through the adoption of renewable energy technologies. Two climate activists, Nicky Ison and Jarra Hicks, were successful in applying for a grant to investigate energy cooperatives in Europe, North America and India. This reflected a burgeoning movement advocating the benefits of such cooperatives as one potential positive means of confronting the challenge of climate challenge. As noted in the previous chapter, the idea of cooperatives is not new. Community cooperatives, and more specifically communes, have been an experiment in social and political organisation that dates back over a century, the most celebrated being the group of socialists, and followers of William Lane, who migrated to Paraguay to form the 'New Australia'. The more recent cooperative movement of the late 1960s and early 1970s, with the establishment of communes, built on the confidence aroused at the Nimbin Aquarius Festival counter-culture festival held in 1972 (Hastwell 2016; Faulkner and Stickells 2019). They drew on the experiences being celebrated in the United States and Europe, as well as India, and attracted some luminaries including Jim Cairns, the former socialist and Treasurer in the Whitlam Labor government (VC Cairns 1993, 1999). Many of these communes and cooperatives did not stand the test of time, although a number still exist and reside in the sub-tropical areas of north-east New South Wales and south-east Queensland. While these were not specifically formed as a part of a move to reduce their carbon footprint, they take inspiration from Lovins' 'Soft Energy' philosophy.

Origins of Renewable Energy Cooperatives

There is not a great deal of evidence of organised advocacy of community energy cooperatives. The first initiative coincided with commencement of work on the Hepburn Wind Farm with the establishment of Embark as a fund-raiser for community-based renewable energy projects. Following their successful research trip, climate activists, Nicky Ison and Jarra Hicks set up the Community Power Agency in 2011, and it was incorporated as a not-for-profit workers' cooperative in 2015. Their continuing commitment to the cause of renewable energy has been a springboard for successful grant applications that have led to the launch of various ventures to promote the uptake of renewable energy.

Another project specifically designed to raise funds to support proposed renewable energy projects was the Community Own Renewable Energy Network Australia, formed in 2012 by a group of South Australians determined to divest from fossil fuel-based sources of energy. CORENA mobilised interest by inviting voluntary donations to enable communities to 'collectively help pay for practicable renewable energy projects' (CORENA 2021). CORENA provides interest-free loans to support the installation of solar panels, mostly in non-profit community centres. There have been over 30 projects funded to date, mostly less than 7 KWh solar panels, in child-care centres, community medical clinics, community schools and more ambitious projects such as Totally Renewable Yackandandah, although larger systems are now being funded. The centres pay back the loans drawing on the savings made from the energy generated by the solar panels, and the funds are recycled into investing in new projects. Funds are advanced in line with the agreed project and control remains in the hands of the borrowing centre.

In the mid-2010s, one lobby group, Embark, reported that there were some 100 or so cooperatives established or in the process of being established. This represented a bit of a catch up with overseas trends because at much the same time that the number of renewable energy cooperatives were being established in Europe and American, there was a diminution in community interest in establishing cooperatives generally. The enthusiasm for cooperatives that was a feature of the 1970s and 1980s had dissipated, and in the first decade of the new century the number of cooperatives declined by almost a third. One explanation for this was that compliance costs were increasing, although the more likely explanation was the dominance of neoliberal ideology which influenced attitudes towards cooperatives. This was the case with mutual societies many of which were reformed as private companies, resulting in members' interests in the enterprises being converted into shares at highly inflated values (Balnave and Patmore 2012). The scope for establishing energy cooperatives was severely constrained. Apart from a lack of expertise, the dominance of State energy utilities and regulations restricting the setting up of enterprises to generate and distribute energy meant that there were all but no examples of successful community energy cooperatives to draw upon. The corporatisation and privatisation agendas of the 1990s that brought an end to States' direct involvement in energy generation and distribution were other discouraging factors, although these would come to influence the decisions to explore setting up community renewable energy cooperatives. It was the burgeoning of interest in energy cooperatives in Europe that inspired the first moves to establish energy cooperatives in Australia.

The first substantial initiative in Australia was the Hepburn Wind cooperative venture. Hepburn Wind (n.d.) was the first truly community renewable energy cooperative formed as an association in 2007 following discussions over a couple of years and key proponents having investigated cooperative-operated wind farms in Europe. A small group of concerned climate citizens had set the cooperative movement in motion. It was also the first to establish a network of members drawing on the energy generated by the wind turbines and distributed through a micro-grid. A small group of concerned climate citizens had set the cooperative in motion, devoting considerable effort to discussing and engaging the local community in the development of the project. Highlighting the benefits to the local community and beyond became a critical feature of this process. Members of Hepburn Wind's Board of Management volunteer their time and energy. All have a professional occupational background, and some have been actively involved in advocacy work around renewable energy and climate change.

Located in the rural community around Daylesford, 100 kilometres north-west of Melbourne, the cooperative commenced construction of the wind farm in 2010 after a lengthy process of formalising the legal entity to operate the farm and raising the capital through a well-received share issue, as well as securing State government assistance and some funding from the local shire (see: https://www.hepburnwind.com.au/). The cooperative negotiated a 25-year lease with a local landholder to host the turbines and, after securing necessary development approvals and meeting all the regulatory requirements, began construction of the towers and installation of the wind turbines in 2010. The Farm was connected to the grid in 2011. The Wind farm now has 2000 members who have invested in the project and who are mostly drawn from the local community. Supporters were able to purchase green energy from 2015 so it effectively operates as a micro-grid.

Hepburn Wind operates according to 'values of self-help, self-responsibility, democracy, equality, equity and solidarity' and in line with the principles that inform the global community of renewable energy cooperatives: of 'voluntary and open membership, democratic member control, member economic participation, autonomy and independence, education, training & information, co-operation among co-operatives and concern for the community'. The project has served as an important 'how-to' model for other community ventures. It has an active management that is focused on generating other environmental and social projects in the Daylesford-Hepburn local area, and, after meeting costs and paying shareholder dividends, this is backed by dedicating a proportion of its surplus revenue to funding other environmental and community projects.

Hepburn is proceeding with the addition of a 5–6 MW solar farm, after having been awarded a $500,000 grant from the State government's Community Renewables Program. The farm will generate sufficient electricity to power up to 2000 homes.

Several communities in the Northern Rivers area of New South Wales were developing keen interest in the possibilities of becoming zero-carbon and energy self-sufficient (Blakester 2014; North Coast Energy Forum 2016). Among one of the most ambitious projects has been Enova which was established to function as a clean energy retailer aggregating energy generating capacity from local renewable sources. The Enova community energy project in the Northern Rivers could be considered the most vibrant and largest, enjoying 'overwhelming consumer support' (Enova Energy). The enterprise was the fruition of discussions among a small group of professionals, some of whom had retired to the area, having departed from a major metropolitan city to enjoy the sub-tropical climate. They had joined forces in common cause in participating in the intense struggles against fracking in northern New South Wales, campaigns that exposed some of the challenges linked to capital-city based energy companies making decisions to the detriment of the local region.[1] This led to an ambition for the community to exercise more control over how power was generated and distributed, and more generally enable locals to take some control over the region's electricity market. The concept of a local renewable energy retailer that was community focused, that could demonstrate to the local community that generating renewable energy was viable and that Enova could harness this energy and, operating as a non-profit enterprise, could oversee the distribution of the energy to supply local needs. More public and open discussions involving local communities were formalised through the North Coast Energy Forum, which was established in 2010 to canvass how to reduce reliance on energy imported from outside the region (Blakester 2014) and progressed further during the Community Energy Conference in Canberra held in July 2014.

[1] The area has a long history of environmental activism and hosts a number of alternative life-style communities that have their origins in the late 1960s and 1970s, many inspired by debates around 'soft energy'.

Seed funding, awarded by the NSW State government Office of Environment and Heritage, helped to finance a feasibility study, which showed a strong business case to justify proceeding with the concept (Latimer 2018). In 2015, Enova was granted an Energy Retail Licence, 'conditional on us raising $3 million capital to cover prudential guarantees, fixed operating costs and cash flow requirements' (Enova Energy n.d.-a).[2] This was crucial because the existing market was becoming dominated by corporate energy providers that provided the 95 per cent of the areas electricity supply, and whose headquarters were based outside the region and, it was calculated, were draining $300 million from the local economy every year. Building on the organisational and electrical engineering, legal management and other skills of the community members volunteering to spearhead the project, the group determined that they could make a difference to energy and community wellbeing by drawing on existing local renewable energy generating capacity and the surplus being generated within the area, expanding this capacity by encouraging others to invest in renewable energy technologies such as solar panels, and developing a grid network that linked this capacity (an arrangement that was negotiated with the State-owned regional utility Essential Energy).

The shared commitment to the 'soft energy' pathway provided the glue that saw the incorporation of Enova Energy, the trading division of Enova Community Pty Ltd., in 2016. This was based on setting up 'a community-owned company ... [that] aspire[s] to [be] a commercially and socially vibrant company'. It was determined to promote community-owned and operated renewable energy systems based on building more self-sufficient, ecologically and socially sustainable communities and establishing economic sovereignty. The development envisioned a future in which greater energy self-reliance and self-sufficiency would engender a more resilient and buoyant economy and provide more employment for locals.

Enova set about promoting the uptake of green energy, providing incentives in the form of higher-than-market feed-in tariffs for renewable energy generators to focus on local market opportunities, providing advice on solar power and storage options. A critical objective of the project has been to plug the leakage of locally generated wealth to the head offices of the corporate energy retailers located in distant cities by exercising control

[2] There was considerable interest in investing in Enova, within the local area and from beyond the region, and Enova set a limit of the purchase of shares from investors outside the region, ensuring that 70 per cent of shares would be held by locals.

over the generation of (renewable) energy and distribution. As one of the key figures involved in setting up Enova, Tony Pfieffer observed: 'We're taking back control. This is creating a circular economy, keeping money in the region' (Latimer 2018).[3] Enova has enjoyed 'overwhelming community support', built up through a team almost entirely of volunteers, and successfully raising close to $4 million capital from shareholders between August and December 2015, with 75 per cent of these resident in the Northern Rivers area. The project has obviously garnered strong support from outside the region given the shareholder profile.

The Enova project is motivated by more than simply establishing an economic enclave by retaining energy money within the region. Strengthening the circular economy is envisaged as a means to an end, to enhancing the equitable distribution of energy and other resources within the region, ensuring energy access to all, empowering the local communities, reducing the region's carbon footprint and developing a regional economy dynamic through the local generation and deployment of energy to strengthen the local community. As is the case with Hepburn Wind, Enova is concerned to give back to the community, and its charter commits its retail arm, Enova Energy Pty Ltd., to channel 50 per cent of its profits into its non-profit arm Enova Community. Enova Community plans to take solar power to households throughout the community. It sets tariffs and payment options that ensured equitable access to electricity throughout the region and has supported low-income households to install solar panels. Enova promoted the project of local power generation and retail distribution as being relatively autonomous from the grid by building the network as a micro-grid.[4]

Another quite successful and iconic venture that is attracting much interest, though with somewhat different in origin, is the Totally Renewable Yackandandah (TRY). TRY was established in the small rural town of Yackandandah, in north-eastern Victoria. It developed organically, out of the recognition of a shared outlook on life across the community. This was demonstrated in a range of ventures, including the installation of rooftop solar panels, recycling programmes, various energy efficiency measures through to permaculture gardening projects that were framed by the determination by families to reduce their ecological footprint and become

[3] See: https://enovaenergy.com.au/about-us/.

[4] Electricity is currently transmitted through the grid network that is managed by the State-owned regional energy infrastructure and retailer Essential Energy.

more self-sufficient and sustainable. One turning point was the agreement to form a collective to take on the running of a local store that was flagged for closure in order to ensure that essential services that it had been providing continue to be maintained. Another followed several households installing rooftop solar panels and 14 of these deciding to join forces set up a micro-grid. The TRY Perpetual Energy Fund was established in 2014 as a repository for donations and the Fund would make loans to households and local enterprises to support efforts to increase energy efficiency, such as energy-efficient lighting, and to support investment in solar panels. This soon led to the installation of panels on public buildings as a cost-saving measure. A large solar and battery system was installed at Yackandandah's fire station, which can operate in isolation in the event of a power outage. In 2015, the local health service and aged care facility installed 130 kW of solar panels on its roof. In the process, TRY was incorporated with the eventual aim of establishing a micro-grid that would serve the local community, achieve energy sovereignty, and set a goal of 100 per cent renewable energy by 2022 (TRY 2021; Mondo 2018). Approximately 60% of the houses in the town have solar panels, and almost every community-owned building does as well, and by 2035, it will have saved several hundred thousand dollars in power bills.

TRY' strategic vision mirrors that of Enova. It is committed to the democratisation, decarbonisation and decentralisation trinity and to embedding this in the goal of climate justice.

The State Called into Play

In contrast to American and European counterparts, the history of peak bodies providing technical, legal and organisational assistance to support the establishment of community energy projects in Australia, let alone lobbying on behalf of community renewable energy cooperatives, has been quite recent. It has also been much more limited in scope, relying on the enthusiasm of a few individuals. This has had a quite interesting implication for the political momentum of endeavours to promote community renewable energy ventures as one component in the decarbonisation of the economy. Individual community renewable energy projects appear to have been much more forceful promoters of cooperative ventures.

Reference has been made to ARENA providing financial support that enabled the community energy forum, the National Community Energy Strategy, held in Canberra in 2014, and additional funds to back a couple of follow-up events. The Energy Strategy provided a springboard for the establishment of the peak body, the Coalition for Community Energy. It organises annual gathering to bring interested parties and climate citizens together. C4CE now has over 100 members who have established renewable energy enterprises or are in the process of or contemplating establishing renewable energy cooperatives.

Recent efforts to provide assistance to interested communities exploring the possibility of setting up a community energy cooperative have been an important development. Federal financial support has been a start, with ARENA assisting the momentum of the community energy movement by funding the drafting of the *Finance and Funding Toolkit for Community Energy and Small-Scale Solar* in 2017. The Victorian government's launch of the *Community engagement and benefit sharing in renewable energy development: A guide for renewable energy developers* has also been an important development. But, such guides, which were drafted by the sector's leading advocates, also highlight the comparatively limited development of community renewable energy cooperatives in Australia. Most of these initiatives, including the drafting of guides, reflect the energetic and determined endeavours of a small cohort of climate citizens proselytising the benefits of community renewable energy cooperatives. To their credit, they have been exploring the many opportunities to recruit backing for establishing renewable energy projects, although this has meant promoting ventures, such as solar gardens, that are not consistent with all elements of the trinity democratisation, decarbonisation and decentralisation. In some respects, the successful energy cooperative projects have proved more effective champions in engaging communities in considering establishing cooperatives, and this highlights a unique aspect of the politics of renewable energy in Australia.

In Australia, ARENA has provided funds for advocates of community energy projects to conduct annual forums, the first, the National Community Energy Strategy, being held in 2014. ARENA provided financial support to meet the cost of the *Finance and Funding Toolkit for Community Energy and Small-Scale Solar* in 2017. The Victorian government commissioned the *Community engagement and benefit sharing in renewable energy development: A guide for renewable energy developers* (Lane and Hicks 2017). State governments have made grants to several of the iconic renewable energy cooperatives, including the Hepburn Wind farm, Enova and Totally Renewable Yackandandah.

Yet, care must be taken in not overstating the measure of state support for cooperatives. The clear lesson to be learnt from the German experience is that the state places much greater emphasis on the industrial-scale energy corporations, and this is evident in Australia in the organisation of the coal-industrial complex and its metamorphous into the fossil fuel industrial complex and the industrial-scale transmission grid and distribution networks that cannot readily accommodate cooperatives. By the same token, there is cause for exercising some suspicion of those instances where energy companies support an energy cooperative to take over responsibility for distributing power as is the case of Enova.

Climate Political Strategising: Building a Rhizomatic Structure and Mosaic of Community Energy Enterprises

Notwithstanding their conviction and their commitment, the relatively low-key character of community energy cooperative on the Australian political plane appears to have translated into a quite distinctive and intriguing development. As success stories, individual energy cooperatives' proselytising seems to have injected more of the momentum into impelling communities to explore or set up renewable energy cooperatives. Drawing on the experience at the grassroots level, the actual cooperative venture is proving a stronger force in engendering interest than are the poorly resourced renewable sector's nascent advocates.

Enova's community outreach endeavours illustrate this story. It is quite active in pursuing this endeavour in what can be best described as a rhizomatic projection. Enova was established as an energy collector, sourcing most of its energy from solar photovoltaic solar panels installed on the rooftop of households, farm buildings, other businesses and community groups in the area. By recruiting more members and supporting the installation of solar panels, Enova has been able to increase its supply capacity. Its energy enterprise relies on access to the grid, which is based on an agreement for distribution rights it has negotiated with the regional State-owned energy utility, Essential Energy. The development of the enterprise has concentrated on consolidating a relatively self-contained energy platform in which supply and distribution flows through a mini-grid which is connected to the national grid network. But this mini-grid functions relatively autonomously from the larger grid. It is a rhizome-like structure, the node being constructed on a centrally organised platform with its root and shoot system conduits for supply and distribution. The mini-grid structure is sufficiently flexible to enable it to become a means of spreading its roots and shoots, tentacles reaching into the hinterland, sweeping up more and more supply, and distribution, capacity, and displacing the for-profit flows of electricity.

Working with local communities both within the Northern Rivers and the immediate vicinity, Enova has recruited ventures to partner in developing similar renewable energy networks. According to Enova's CEO Felicity Stening, the enterprise is gradually spreading its presence, rhizome-like, in building grid-scale energy storage capacity and peer-to-peer trading to enhance energy self-sufficiency (Energy Source and Distribution 2020).

Enova's strategic goal, by localising energy flows or decentralising the energy system disrupts, is challenging the hegemony of the fossil fuel, emissions-intensive, industrial-scale energy regime. It is 'purposefully and actively disrupting the energy model that is reliant on polluting energy travelling along long-distance networks of poles and wires', argues Stening (Pordage 2019). It is both economic and efficient: the twentieth-century model is 'outdated and inefficient, as power is lost along the way' in the industrial-scale transmission network. This is the 'soft energy' model in operation. The model is based on not only progressing the object of decarbonising communities, it's also about injecting a dynamism and resilience in local communities and to displacing 'a model that sucks money out of local communities and directs it into often foreign-owned private enterprises'.

This rhizome-like strategy is manifest in Enova's endeavours to expand its own footprint within the Northern Rivers area and beyond. Enova is enlisting other local communities to adopt the model, recently partnering with Repower Byron, Zero Emissions Byron and COREM to change established energy production and use patterns and to move towards 100 per cent renewables. It is consolidating the model.[5] But the rhizomatic model is anything but parochial in purpose. Enova is seeking to engage others to follow in its footsteps. It is broadening the reach of energy generating capacity and the micro-grid as an integral feature of its vision of renewable energy self-reliance and equitable energy access. It originally defined its place in the energy system as an exemplar of what is possible, 'a community-owned company ... [and it is now a model that] aspire[s] to [be] a commercially and socially vibrant company that can be copied by communities across Australia' (Pordage 2019). It is pursuing this in dialogue with other prospective renewable energy enterprises and advancing the broader

[5] The Community-Owned Renewable Energy Mullumbimby is a not-for-profit, located inland from Byron Bay.

challenge to the corporate-dominated, coal-fired, base-load generation and national grid electricity distribution model (Enova Annual Report 2019). Enova has proceeded to facilitate the adoption of the model 'by communities across the country and in the longer term, produce enough renewable energy locally to meet all its customers' energy needs' (Pordage 2019). It is recruiting partners in Sydney, Central Coast, Newcastle and Wollongong, with plans to expand into south-east Queensland.

Totally Renewable Yackandandah is mirroring Enova's rhizomatic strategy to progress the take-up of renewable technologies. TRY plans to become 100 per cent renewable by 2022. Its progress has prompted the Yackandandah community to work with neighbouring communities to develop similar projects. It has adopted a similar model to Enova except with one big difference. Its proselytising approach has been gaining momentum, but it has taken a somewhat different trajectory. TRY has partnered with another community renewable energy cooperative, the retailer and certified social enterprise, Indigo Power, to develop what it describes as a 'mosaic of community energy groups' (Quaglio 2019).

Indigo Power was established specifically to develop a model system, an energy hub that would enable the pooling of members' generating and sharing or distributing capacity through linking members by means of a community-centred mini-grid with a back-up battery storage system. The proven working hub model could act as a template and be duplicated in other communities (Indigo Power n.d.). The energy hub with its mini-grid-like tentacles operates in a similar manner to Enova's rhizomatic model. The Indigo model is more set on establishing community hubs that are energy self-sufficient, as much as is possible independent, to avoid having to partner with a for-profit grid company, which would have to be the case in Victoria given the State's grid has been completely privatised. (Enova has an arrangement with the State-owned Essential Energy to connect to the grid.)

In both instances, the local energy hubs remain connected to the Australian energy network, and in the event of there being insufficient power generated within the hub to meet community demand, ENOVA and TRY, or any other community hub that is established, could draw power from the national grid. However, the ambition is to avoid such recourse because the mini-grid system provides the infrastructure for managing energy flows, with reasonable incentives provided in the form of the feed-in tariffs to incentivise cooperative members in Enova and Indigo-TRY to feed excess solar power into the hub's pool and to permit their member and other neighbours to draw from that pool if need be (Grogan 2020; Indigo Power n.d.). Indigo has committed to investing 50 per cent of its profits back into the community.

The partnership between TRY and Indigo has been supported by substantial Victorian government grants, philanthropic funding and other donations, as well as members' direct contributions in meeting the costs of solar panel installations. The current focus of planning is fixed on increasing battery storage capacity.[6]

In adopting the goal of expanding community investment in renewable energy, TRY has set the goal of 'working hard across Australia to reform the national energy system' (Quaglio 2019). The partnership between TRY and Indigo has injected increasing momentum into the recruitment of communities to commit to advancing the goals of democratising, decarbonising and decentralising by forming community renewable energy cooperatives and organising energy hubs. A crucial feature of this strategy is that the hubs become the means for defecting, or disconnecting, from the national electricity grid, albeit a qualified defecting because they remain connected to the national grid in order to be able to draw on power when there is not sufficient power being generated or stored within the hub. In the north-east Victorian region, one of the five distribution networks owned by private corporations, the establishment of TRY represents a challenge to the monopoly control by Chinese-Singapore-owned AusNet Services over the flows of energy.

The installation of solar panels on the roof of the Old Beechworth Gaol marks the journey of Beechworth down the totally renewable energy path, and of the mosaic coming into fruition, but it is not alone. Indigo Power and TRY have formed the North East Community Energy Network as a springboard to engineer an energy transition across north-eastern Victoria. Indigo is partnering with 15 other communities in the development of community energy hubs in north-east Victoria with plans to expand into southern New South Wales (Beck 2019b). In providing organisational and technical assistance to help other communities in the region establish community renewable energy and establish their energy hubs, TRY and Indigo Power are, in effect, ramping up the challenge of this monopoly by creating an archipelago of energy islands in a sea of energy flows (Indigo

[6] TRY has received approximately $700,000 in Victorian government grants, philanthropic funding and donations, which has been matched by household infrastructure investment of over $1.1 million.

Power n.d.). Enova is doing much the same in north-eastern New South Wales.

The rhizomatic character of this expansion is also being matched by seemingly more spontaneous stirrings of interest in energy cooperatives, an emerging new energy mosaic. Anecdotal evidence points to a mushrooming of proposals to set up renewable energy cooperatives in the wake of the bushfires that ravaged communities over the summer of 2019–2020. The destruction of infrastructure and the loss of reliable supply, including telecommunication links, appears to be having a catalytic effect in prompting communities on the South Coast of New South Wales to explore the advantages of decentralising and democratising the energy system as the means for rebuilding the infrastructure that helps bind communities and for revitalising the local economies, echoing the reactions to the wildfire events in California.

One estimate maintains that there are over 100 cooperatives operating in Australia. Some 80 per cent are organised in rural communities and are being touted as a means of rejuvenating the local economies (Daley 2020).

Charting New Pathways for Engaging Renewable Energy Cooperatives

In a nod to the Yackandandah project, the independent member for the federal electorate of Indi, Helen Haines, is proposing to introduce a bill to set aside funds to establish a new Australian Local Power Agency that would sit alongside ARENA and the CEFC specifically to support the setting up of community energy cooperatives (Haines 2020; Mazengarb 2021). The ALPA would provide 'financial support for the construction of medium-scale projects with community ownership models, and a requirement for large-scale projects to offer local communities the opportunities to take an ownership stake'. Such funding could transform the region of North-East Victoria, which takes in Yackandandah other several rural communities that are exploring following the WAY model to establish community energy projects, into a community-owned and controlled dynamic renewable energy power hub.

One other important development that has emerged out of the rhizome-like or mosaic clean energy model that has recently attracted interest and brought climate citizens out of the shadows of the energy system, and which should be mentioned, has been the establishment of solar gardens which provide opportunities for those whose housing arrangements or residential location preclude them from installing solar panels or joining a renewable energy cooperative. Modelled on some ventures in the United States, solar gardens allow individuals or households to invest in a solar panel array that has been set up with the express purpose of attracting and expanding investment in renewable energy. One scheme that has proven particularly attractive, and the first to be established in Australia, is being managed by Enova (Enova n.d.-b). An Australian-first innovation, the farm is being undertaken in collaboration with North Coast Community Housing (NCCH), COREM and Splendour in the Grass. The 100 solar panels, which will generate up to 35 Kw, are situated on the Community Housing roof. The array will substantially reduce the Housing's energy costs and the savings invested back into community projects, and this is further affirmation of Enova's commitment to advancing community welfare (Wallen 2019; Enova 2019).

Solar gardens complement the rhizomatic model, although those investing in the ventures take on the garb of second-class climate citizens, silent partners, because they do not have a say in the organisation of the ventures. The solar gardens open up the opportunity to contribute to the decarbonisation goal. Investors are paid a standard dividend, and in some other cases solar garden investors can have the power generated treated much as feed-in tariff and have this managed through their utility account. Hepburn Wind has a similar scheme, and in both instances the cooperatives have set caps on the magnitude of investment that a non-member of the cooperative can make in order to preserve the principal focus of the cooperative. ARENA has funded a project undertaken by UTS's Institute of Sustainable Futures to investigate the viability of solar gardens (ARENA 2018).[7]

So attractive have solar farms been to investors, share issues are being snapped up in no time, and other community energy projects have followed the development by also establishing up solar gardens (Bainbridge 2017). Some are located in rural communities, such as in the case of Swan Hill (https://www.swanhill.vic.gov.au/.../sustainable-living/mallee-sun-community-solar-garden). Supporters of community renewable energy

[7] 'ARENA CEO Ivor Frischknecht said initiatives such as this are an important step in giving the consumer more options when it comes to their energy bills.' As AGL has demonstrated, the solar garden model provides an opportunity for energy retailers to expand their access to renewable energy (AGL n.d.).

projects have been enlisting industrial enterprises and retail centres that possess expansive, unutilised and exposed rooftops to dedicate the space to installing solar panels. These have generally occurred in contexts where a local climate action group sees an opportunity and in instances where the rooftop owner does not have the necessary capital or organisational capacity to launch into installing a solar panel array. Moreland Community Solar, for example, having received a positive reception to its proposal for installing panels then proceeded to invite people in the local community to join others and become involved in a solar venture. It negotiated the installation of solar panels, the rate of the tariff to be paid, which will generate savings for the host, and the rate of return for investors while meeting the objective of emissions mitigation (Elmerhebe 2017; Moreland Community Solar n.d.; ARENA 2020).[8]

Conclusion

Renewable energy cooperatives are driving radical transformations in the political economy of the communities in which they are located. The generation, garnering and distribution of electricity are changing the energy dynamics as well as socio-economic relations. Coop members acquire a voice that empowers them to be able to make more transparent and determine the place of energy in the community and how the value generating capacities of the renewable energy technology can be deployed to advance the wellbeing of the whole community.

At the same time, the establishment of cooperatives disperses the generation of electricity, and this is occurring not only because the provision of power is being established within these communities but also because the cooperatives are developing opportunities for those who do not reside in the area to invest in surrogate cooperative electricity ventures. The establishment of solar gardens, for instance, is proving to be a very appealing and popular investment opportunity. The gardens are engaging a broader range of people who, for example, do not have the option of investing in rooftop solar panels or who have exhausted the extent to which they can invest in panel installation on their roofs. The opportunity to invest in solar gardens allows an expanded commitment to emissions reduction which further displaces fossil fuel generated electricity, reducing

[8] They are increasingly having to compete with commercial ventures leasing roofs for the installation of solar panels (Fehring 2017; Schlesinger 2019).

both the national and the personal carbon footprints of investors who are thus reducing their carbon footprint, as well as extending the reach of energy democracy given that solar gardens generally provide the space for investors to express their views on the operation and distribution of the benefits of the solar garden.

Still, it has to be accepted that renewable energy cooperatives are making but a small dent in the armoury of the industrial-scale coal-fired generation and transmission system although the development has substantial symbolic significance in eroding the legitimacy of the coal-based electricity system.

References

AGL (n.d.) 'How you could save with solar—even without solar panels'.

Australian Bureau of Statistics (2012) 'Regulation of Cooperatives in Australia', 1301.0—Year Book Australia.

ARENA [Australian Renewable Energy Agency] (2020) 'Moreland Micro-grid Investigation'.

ARENA [Australian Renewable Energy Agency] (2018) 'Solar gardens to bring rooftop solar to all Australians' 17 May.

Bainbridge, Amy (2017) 'Investors snapping up community energy projects, with some selling out in minutes' *ABC News* 30 April.

Balnave, N., and G. Patmore (2012) 'Rochdale consumer cooperatives in Australia: Design and survival' *Business History* 12(2) 986–1003.

Beck, E (2019b) 'On Country renewable energy partnership gets green light' *ABC Goulburn Murray* 9 July.

Blakester, A. (2014) North Coast Energy Forum Sustainable Living Armidale, May.

Cairns, J. (1999) *On the horizon: a cultural transformation to a new consciousness* Narre Warren East, Vic.

Cairns, J. (1993) *Towards a new society: A new day has begun*, Melbourne.

CORENA (2021) Community-funded projects tackling the climate emergency.

Daley, M. (2020) 'The amazing world of our thriving co-operatives, especially in the regions' *The Fifth Estate* 28 September.

Elmerhebe, T. (2017) 'Baker Maison goes green as solar panels come on' *The Telegraph* 8 May.

Energy Source & Distribution (2020) 'New community battery for regional NSW' 11 March.

Enova (n.d.-a) 'Our story' https://enovaenergy.com.au/about-us/our-story/.

Enova (n.d.-b) 'The Community-owned model'.

Enova (2019) 'Community projects'.

Faulkner, H., and L. Stickells (2019) 'Bodhi Farm and other counterculture communities live like the 1970s never ended' *ABC News* 15 December.
Fehring, R. (2017) 'Developers turn to energy supply' *The Australian Financial Review* 28 September 2017, 43.
Grogan, M. (2020) 'Blog: The power to rebuild' *Totally Renewable Yackandandah* 16 January.
Haines, H. (2020) Unlocking community energy in Australia, 15 May.
Hastwell, A. (2016) 'Life on a hippie commune 40 years on' *ABC News* 6 May.
Hepburn Wind (n.d.) Hepburn Wind Community Energy.
Indigo Power (n.d.) 'Everything you need to know about becoming Indigo Powered'.
Lane, T., and J. Hicks (2014) *Best Practice Community Engagement in Wind Development*. ACT Government Environment and Planning Directorate, Canberra.
Lane, T., and J. Hicks (2017) *Community engagement and benefit sharing in renewable energy development: A guide for renewable energy developers* environment, Land, Water and Planning, State Government Victoria.
Latimer, Cole (2018) 'Microgrid rise sees the light of day' *The Sydney Morning Herald* 20 January.
Mazengarb, M. (2021) 'Haines proposes $483m Local Power Agency to help communities buy in to wind and solar' *Renew Economy* 22.
McKinnon, John (2015) 'Australia's National Community Energy Strategy' *Australian Environmental Makers Network* 11 June.
Mey, F., M. Diesendorf and I. McGill (2016) 'Can local government play a greater role for community renewable energy? A case study from Australia' *Energy Research and Social Science* 21(1) 33–43.
Mondo (2018) 'Mondo and the Victorian Government's Microgrid Demonstration Initiative' at Moreland Community Solar (n.d.) 'Community Solar Portal'.
North Coast Energy Forum (2016) 'Transitioning the North Coast.'
Pordage, C. (2019) 'The first community-owned social enterprise energy retailer' *Utility* 28 October.
Quaglio, A. (2019) 'Community Energy Hub launch' *Totally Renewable Yackandandah* 29 November.
Schlesinger, L. (2019) 'RAM generates solar income stream' The Australian Financial Review 26 March, 34.
TRY (2021) Totally Renewable Yackandandah.
Wallen, S. (2019) 'Solar garden unlocks the benefit of solar' *The Echo* 13 December.

CHAPTER 12

Holding onto Fossil Fuels

In the decades following the negotiation of the Kyoto Protocol, successive Coalition governments have hidden behind the idea that the liberalisation of the energy market and the unfettered *laissez faire* market economy more generally would inevitably see Australia embrace a low-emissions, or even zero-emissions, future. Yet, a distinctive feature of the definition of the National Electricity Market was that it was formed according to a set of rules that were based on an ideal construct, on a conventional economic representation of a perfectly competitive, *laissez faire*, market that would function as a discrete entity. This was conceived as a closed system in which coal is extracted to fire the turbines that generate electricity, powering the nation and becoming embedded in products and wellbeing, while the emissions being generated are explicitly excluded from this flow. This is an extraordinary qualification of the definition of the energy market given that energy generation produces such a significant proportion of greenhouse gases. It is an omission that beggars belief given that the deliberations over the definition of the market occurred simultaneously to the global community coming together in an endeavour to arrest the increasing concentrations of emissions. And yet, despite the mounting concern that the global community is being confronted by the possibility of runaway climate change and the international pressure for action is becoming more desperate, successive Coalition governments have hardly wavered. Indeed, its commitment to a continuing future with coal is being compounded by its determined efforts to add another fossil fuel, gas, to its

S. Rosewarne, *Contested Energy Futures*,
https://doi.org/10.1007/978-981-19-0224-6_12

developmental arsenal. Notwithstanding the Rudd Labor government's attempts to introduce a price on emissions and the Gillard government's short-lived *Clean Energy Policy* package, the Labor Party appears to be on the same page as the Coalition.

This presumption that the liberalised energy economy would operate according to the laws of competition, efficiently and sustainably, optimising the benefits for all participants, has been found to be wanting. The calls and protests demanding action continue, and, as has been argued here, various enterprises have been pursuing, or have pursued, a range of endeavours to take control by generating their power needs and committing to zero-emissions power generation, and this has prompted some defection from the liberalised energy market. Moving beyond the market and constructing new modes of organising the generation and distribution of electricity, often resulting in new forms of social organisation, is proving a more radical transformation. The momentum of this transformation is unlikely to abate given the Coalition's Prime Minister Morrison aligning with China, India and Russia in refusing to strengthen the 2030 emissions reduction target at the COP26 deliberations or to set methane reduction goals.

The debate on how to respond to the challenge of climate change has been a brutal one. This is most obviously evident in the bitter personal battles that have resulted in the frequent toppling of party leaders. Five Prime Ministers have had their political careers cut short primarily because they have not been able to secure the backing of their party or the electorate for their approach to climate change governance. Two Coalition Prime Ministers have been voted out of public office because most of the constituents in their electorate could no longer accept the Coalition's resistance to formulating a climate change policy. The consequence has been a policy vacuum which has created much investment uncertainty, and this is despite a general consensus among key government departments as well as international agencies that there is a new urgency for action to be taken. Surprisingly, the surge of investment in renewable energy technologies has continued unabated, although in the absence of meaningful policy this presents its own problems. The policy vacuum persists, and in tracking the developments that have occurred over the two-plus decades of the NEM, it is appropriate to reflect on what is holding back a policy package that would support the passage to a clean energy future and see Australia committing to the net-zero Paris emissions target that so many OECD member states have signed on to. We first consider the policy positions that are on the table at the moment and this could well be because there is a void that many want to be resolved. Taking stock of what has underpinned this policy vacuum and where the mapping of climate change policy has got to is a salutary exercise because it is important to consider just how much progress there has actually been made towards meeting the Paris commitments. We introduce the analysis by considering the state of policy formulation and its weak foundations, focusing initially on the different official models that chart future energy system scenarios before reflecting on the current manifestation of the dirty politics of fossil fuels, and the extent to which this informs the role of the state and the conviction in the merits of the *laissez faire*, free market energy economy.

Resetting Policy: The *Integrated System Plan* and the *Post-2025 Energy Market* Versus a *Technology Investment Roadmap*

In addressing a Minerals Council Parliamentary Dinner prior to flying out to join the G7 meeting held in Cornwell in June 2021, the Prime Minister sought to placate those criticising the government's resistance to legislating an emissions reduction policy, reminding the assembled with the oft-repeated observation that Australia will meet its commitment 'in a canter'. Conscious of the criticism for refusing to commit to a net-zero-emissions target, Morrison parodied Frank Sinatra, announcing that Australia would not be pressured into setting an emissions reduction target in line with the Paris Agreement of net-zero emissions by 2050. It would be *Doing It My Way*: 'I call this the Frank Sinatra approach. We're going to do it our way in Australia, the Australian way' (Morrison 2021). This, of course, begs the question as to what is the Australian way.

There are, in fact, two emissions reduction policy mappings being canvassed, and as has been outlined here a range of other strategies adopted to reduce energy generated emissions as well as consolidate the place of coal in the political economy and expand gas extraction. The Coalition has released a provisional policy of sorts, outlined in a Ministerial Statement: the *Technology Investment Roadmap Discussion Paper* released in May 2020 (Department of Industry, Science and Energy Resources 2021: 28). Its counterpart is reflected in two policy positions commissioned by the COAG Energy Council, the Australian Energy Market Operator's *Integrated System Plan* released in July 2020 and the companion piece, the

Energy Security Board's *Post-2025 Market Directions Design* released at the beginning of 2021.

Both the *Roadmap* and *ISP* and *Post-2025 Market Design* reiterate the now longstanding NEM principle that the free market should provide the institutional umbrella under which policy should come into play. Both pretend to resuscitate this privileging of the *laissez faire* market while setting some parameters on the restructuring that would have to occur to meet emissions and transitioning targets. They each are not so much advocating an emissions reduction path as responding to the surge in investment in renewable energy technologies that are transforming the energy system and the need to respond to the transition from reliance on coal. The two proposals posit mapping a path for an orderly transition, and both are being implemented by default. They emphasise that the state should assume a minimal role, setting some parameters, and that transitioning should be enabled by unleashing the full force of the free market. In important respects this conviction in the free market is subterfuge.

The *Roadmap*'s embrace of free market rhetoric is more duplicitous because it argues that, unlike the Coalition's political opponents, its proposal is not about the state intervening to pick winners, which is exactly what it does seek to do. While it does draw on some of the scenario mapping undertaken by AEMO and the ESB, the *Roadmap* shifts the emphasis back to some long-dismissed possibilities, such as gas, carbon capture and storage, and soil sequestration[1] although it does introduce some novel technologies that are being explored. Its future scenario, for instance, includes hydrogen, but its overwhelming preoccupation is with defending the place of fossil fuels in the energy mix.[2] Recourse to increasing reliance on gas is being represented as critical to achieving supply reliability and affordability and, with this, the *Roadmap* argued for a substantial reorientation of incentive systems.

This complex of state and fossil capital is evident in the *Technology Investment Road Map* for the most part presumes that private enterprise

[1] The idea that simply burying carbon emissions could suffice as an emissions reduction strategy had proven to be an expensive and dismal failure, and here was the government trying it on again. The history of Coalition emissions reduction policy suggests this is a desperate effort to avoid doing anything that entails actually cutting emissions.

[2] Intriguingly, the preoccupation with technological solutions and the commitment to 'technology neutrality' mirrors the focus of the Chief Scientist that was articulated in the *Finkel Review*. See also Finkel's 2021 *Quarterly Essay* 'Getting to Zero: Australia's Energy Transition'.

will lead investment in existing technologies. The *Road Map* holds up the private sector as the principal driver of technology deployment and, should the *Roadblock* ever see the light of day as policy, the state's role is largely going to be focused on removing roadblocks, enabling consumer choice and supporting the emergence of the best enabling environment. The *Road Map* is relying on the state inviting capital 'to bring forward solutions to solve our technology challenges ... [to] never crowd out private investment, and deployment of commercially mature technologies should be left to the private sector' (Department of Industry, Energy and Resources 2021: 28). This is reflected in reconfiguring the Ministerial Reference Panel into a permanent fixture—the Technology Investment Advisory Council—to advise the Minister on Australia's Low Emissions Technology Statements. The Council is chaired by Finkel and includes several prominent businesspeople.[3]

This is the context in which, over the course of 2020, the Coalition expressed its concern with renewable energy sources displacing coal- and gas-fired power plants. It saw that the rapid substitution of dispatchable power generated by the combustion of coal by renewable energy sources would result in a less reliable and secure supply of energy and threatened to have the state invest in building a gas plant should the private sector not invest in new capacity to make good the loss of dispatchable capacity (Clarke 2021b).[4] Although there would be instances where the state will make strategic investments, particularly where there is clear market failure, the *Roadmap* envisages a more systematic engagement of the state in advancing the interests of fossil capital.

To be frank, the *Roadmap* is a bit of a concoction, lacking in conceptual coherence, made worse by it being but one arm of an energy policy that is taking shape alongside the adoption of a range of other funding-support measures, mostly for fossil fuel-related enterprises. The launch of the Underwriting New Generation Investments (UNGI) policy is one such example, promoted as being designed to deliver affordable and reliable power, and which is an obvious case of the government stepping in to pick winners. The Grattan Institute has found the funding support for UNGI

[3] Among those appointed are Macquarie's chief executive officer Shemara Wikramanayake, Grant King, the Business Council of Australia's Alison, Environmental investor and new appointment to ARENA Justin Punch, Deutsche Bank's Steven Skala and Clean Energy Regulator Chair David Parker.

[4] Snowy Hydro was granted $6 million and awarded the contract to build a gas-fired plant in the Hunter Valley, in the New South Wales coal belt.

projects is preventing more viable projects from being built and which could generate power sooner (Morgan 2019).

The second policy emphasis is based on modelling the likely future energy mix by AEMO and ESB. They are unequivocal in regarding the transition to renewable energy as pivotal to future energy scenarios, and that coal and gas are no longer considered credible and sustainable sources of energy. Their joint concern is to chart the orderly transition to renewable sources of energy and to signal how the construction of solar and wind farms will be managed and how the new generating capacity will be connected to the grid in a guided manner. Building on the existing grid and distribution network infrastructure and the likely timing of dispatched generating capacity being retired, the *Integrated System Plan* (*ISP*) outlined a staged mapping of the physical infrastructure that would have to be augmented to accommodate the increased reliance on and the different dynamics of distributed energy resources and variable energy resources. Drafted by the Australian Energy Market Operator, the *ISP* developed a model of the system that incorporated the whole energy system and the full range of the different generating technologies and transmission and distribution infrastructure. This framing was designed to minimise as much as was feasible the need for new investments while ensuring that all demand would be met. At its core was a strengthening of the grid to accommodate distributed energy resources. This conceded the possibility that it would be more economical for the energy requirements of more remote communities to be better developed as stand-alone generating systems. It also envisaged consolidating the National Electricity Market with the construction of more interconnectors to enhance power flows across State borders when supply shortages in particular locations became apparent and thereby also reducing the need for investment in expanding generating capacity to meet the occasional peak-load demand pressures.

The *Post-2025 Market Design Directions Paper* builds on the *ISP* by consulting with industry participants on possible measures to deal with the multiple ways in which the electricity market is being disrupted by the narrow focus of government incentives to encourage investment in renewable technologies and the effects of the greater reliance on distributed energy resources and the displacement of dispatchable power. Confronted by an increasingly complex energy system, the ESB mapped responses to smooth the transition across short to medium and longer time frames. Its immediate objectives were to put in place essential services measures that provide the means for countering the problems of frequency irregularity and system strength variability arising with intermittent power generation from households and solar and wind farms. The variability in power flows highlighted the importance of inertia in the effective operation of power stations, and the ESB contended that there should be some reckoning of this. The development of service markets for market-based measures was considered to provide effective, transparent and structurally consistent responses accessible to all participants in the energy system. The establishment of such market-based measures would help to ensure the smooth and efficient functioning of what was becoming a more complex and diverse energy system. Some of these service markets could be formed organically with entrepreneurial spirit anticipating the demand for new energy services or such service markets could be set up by the state realising the need for a specific market mechanism, the logic of which is generally informed by reference to conventional economic theory.

The ESB identified some major changes to the national energy market that would make the market a more pervasive feature of the energy system. For instance, it envisaged much greater reliance on demand side management as a vehicle for optimising energy use, rewarding energy users for forgoing energy use during periods of high peak use, and thereby avoiding investment in additional generating power to meet a limited number of occasions when demand was testing the limits of supply. Engineering stability in a system undergoing change has led to the design of market-based measures to forestall the problems of congestion caused by a surfeit of power being fed into the grid at times of peak generation, recommending retailers be incentivised to act as the gatekeepers of reliability. A proposed Retailer Reliability Obligation that mandates them ensuring 'the right mix of [generation] capabilities are invested in' is proposed to introduce greater stability into the NEM.

As complementary plans for action, the *ISP* and the *Post-2025 Market Paper* stress the importance of promoting unfettered market processes. This does not preclude the role of the state. The intention, indeed, the practice established to date, involves the state, or more precisely the Energy Market Operator and the Australian Energy Market Commission, assuming an active and interventionist role in energy system governance, coordinating and pacing the accommodation of new generators into the NEM.

The Dirty Politics of Fossil Fuels

The policy vacuum has been even more exposed by the increasing investment in renewable sources of energy. Dr Kerry Schott, who was commissioned to head the ESB's brief to explore strategies for dealing with the consequences of this, has been quite forthright in declaring the policy impasse has seriously compromised the reliability and security of the system, reducing the National Energy Market to a state of anarchy (Latimer and Hannam 2018; Potter and Macdonald-Smith 2018; Iggulden 2018; Clarke 2021a). Her observations have been endorsed by Energy Networks Australia which argued that the pace of the increase in renewable energy generation had exceeded the capacity of the distribution network's capacity to accommodate this additional flow and this was causing congestion and forcing the system into lock-down (Edis 2020; Macdonald-Smith 2019).

If there was any indication that the policy vacuum and the uncertainty this was causing was going to be resolved, it was clear that the federal Minister for Energy and Emissions Reduction, Angus Taylor, was not going to be the person who could garner sufficient popular support for progressing his policy agenda. His petulant demeanour and disregard for constructive dialogue with State and Territory Energy Ministers was on full display when he called an abrupt end to a meeting of the Council of Australian Governments (COAG) Energy Council held in December 2018. He closed the meeting to block an attempt by 'the New South Wales Energy and Climate Minister, Don Harwin, to restore an obligation to reduce emissions in national energy policy' (Carabott 2019; Ludlow 2019; Mazengarb and Parkinson 2019). Harwin's request exposed Taylor's manoeuvres in formulating a policy that would include a future for coal and gas in his *Roadmap* and the characterisation of the *Roadmap* as being 'technology neutral'. It would be almost another calendar year before the Minister was prepared to convene a meeting of the Energy Council, and that was only after Victoria's Energy, Environment and Climate Change Minister, Lily D'Ambrosio, requested a meeting because the Council was long overdue in responding to reports it had commissioned. The reports included the AEMO's first iteration of the *Integrated System Plan*, the Australian Energy Market Commission's proposed energy supply reliability obligation reforms, the ESB's consultation outlining reform options for the *Post-2025 Energy Market Design* and the emerging plans to fund research and development of renewable hydrogen. In the interim there had been a federal election which returned the Coalition to office and the government had released its draft energy roadmap while Angus Taylor was concentrating on trying to negotiate bilateral agreements with the Coalition governments of New South Wales, Tasmania and South Australia (Elsworthy 2020; ABC News 2021; Australian Associated Press 2021).

The COAG Energy Council met in November 2019, but the possibility that this would mark a new era for progressing action on energy and climate change was soon overshadowed by the COVID-19 pandemic and by a significant change in the reporting format of the Energy Council. The Prime Minister dissolved the COAG and set up the 'National Federation Reform Council' (or the National Cabinet) in its place, introducing a 'Cabinet-in-confidence' gag on the Ministers attending. With this confidentiality requirement and the federal Minister's decision to suspend Energy Council meetings for almost all of 2019 and reconstitute the Council in the National Cabinet as the Energy National Cabinet Reform Committee (or the Cabinet Energy Committee), the possibility of progressing agreement on energy and climate policy through the Cabinet Energy Committee was effectively quashed. State and Territory energy ministers' participation in the Cabinet Energy Committee was by invitation and a ring of secrecy brought an end to transparency in its dealings (Mazengarb 2020g, 2021b; Macdonald-Smith and Ludlow 2020a). The onset of the COVID-19 pandemic dramatically underscored the improbability of constructive dialogue around emissions reduction policy.

The Coalition sought to take control of the energy and climate change debate by using the pandemic as a springboard for emphasising the opportunities that could be grasped in mapping an economic recovery plan. The Coalition established the National COVID-19 Coordination Commission chaired by industrialist Nev Power to focus on managing responses to the pandemic. The Commission was renamed the National COVID-19 Commission Advisory Board in July 2020. It operates within the Prime Minister and Cabinet framework and reports to the Prime Minister. A second committee, also led by industrialists, Grant King and Andrew Liveris, was set up to develop a post-pandemic economic recovery plan (Morgan and Long 2019). Andrew Liveris, former chief executive officer of Dupont, was appointed to chair the COVID Commission manufacturing industry task force and he wasted no time promoting gas as a springboard for economic recovery (Brissenden 2021). The conduct of both committees has largely been kept under wraps, and the thrust of their contributions to debate on the future shape of the economy has been to prioritise the importance of expanding exploration and extraction of gas, to reforming the gas market to elevate gas as the pivot on which the recovery would be pinned (Brissenden 2021).

This dovetailed neatly with the *Technology Investment Roadmap Discussion Paper*, which foresaw gas playing 'an important role in balancing renewable energy, ramping up and down to match supply and demand'. The *Roadmap* envisaged setting a short-term objective of integrating investments in renewable energy to support affordability, reliability and security, and the medium-term goal was to match renewable capacity with dispatchable generation delivered by 'clean coal' and gas (Department of Industry, Science, Energy and Resources 2020: 28). The Minister sought to garner legitimacy for the *Roadmap* by setting up a body, the Technology Investment Advisory Council, that brings together Finkel with several prominent business people to advise the Minister on Australia's Low Emissions Technology Statements.[5]

It is not clear how the COVID-19 management and economic recovery committees engaged with the development of the *ISP* process or the *Post-2025 Market Directions Paper* or for that matter the Coalition's proposed *Roadmap*, although media releases in which the Prime Minister and the Energy Minister stressed the importance of gas as a transition fuel indicating that the committees were fully on board with the Coalition's fossil fuel agenda (Brissenden 2021; Wilkinson 2021). This common sense of purpose was reflected in the Coalition negotiating bilateral agreements with State governments involving special billion-dollar funding deals on the condition that the States develop more gas capacity. The Minister made it clear to the Victorian government that there would be no special deal with it while ever its moratorium on terrestrial exploration was in place,[6] and this uncompromising attitude left Queensland complaining that it had not received comparable financial rewards for developing its gas supply capability (Karp 2020; ABC News 2021; Australian Associated Press 2021; Towell et al. 2020; Macdonald-Smith and Ludlow 2020).[7] The Coalition government has still not formulated a developed energy

[5] Among those appointed are Macquarie's chief executive officer Shemara Wikramanayake, Grant King, the Business Council of Australia's Alison, Environmental investor and new appointment to ARENA Justin Punch, Deutsche Bank's Steven Skala and Clean Energy Regulator Chair David Parker

[6] The Victorian Labor government relinquished, announcing in March 2020 the intention to end the moratorium on onshore exploration (Rollason 2020).

[7] Victoria had lifted restrictions on offshore exploration in 2018 (Tchetvertakov 2018).

policy that incorporates emissions mitigation measures or goals,[8] though it has maintained its rhetoric on the importance of dispatchable power to achieve reliability, security and affordability and the need for energy corporations to invest in either coal- or gas-fired plants. The Coalition continues to be backed by the fossil fuel proponents who have become more zealous in projecting the rights to explore and extract coal and gas.

The jockeying over the design of the future increasingly centred on the status of gas in the energy mix and this had morphed into a decidedly unpleasant process. Taylor was reported to have personally pressured AEMO's chief to amend the *ISP* to list gas as being in the future energy mix while his Department Secretary contacted ESB chair to encourage her to resign (Brissenden et al. 2021). Both Zibelman and Schott stood their ground and made clear that Australia was being presented with two distinct options: one informed by economic theory that was rooted in the market economy, the other that was politically driven coloured by political loyalties and which would revive the role of an intervening state. As if repenting his sins and endeavouring to rebuild his political stocks, the Minister for Energy and Emissions Reduction reconfirmed his conviction in the free market, declaring that ad hoc decisions were a thing of the past (Ludlow and Macdonald-Smith 2019; Williams 2021a).[9]

Bringing the State Back in? An Administrative, Coordinating or Formative Role in Energy Governance?

The tension that engulfed debate on the shape of the extent to which the state should intervene in the liberalised energy market could be partly attributed to the fact that there were all but no precedents for informing how to integrate several discrete State markets, constructed around

[8] The Coalition has announced plans to promote research on 'clean' hydrogen development in several regional zones and more funding for carbon capture and storage as have several States.

[9] This recanting did not extend to the Coalition government because the Minister for Resources and Northern Australia vetoed a decision by the Northern Australia Infrastructure Facility to provide financial support to a proposed wind farm. The Queensland government has since announced it will provide funds for the wind farm (AAP 2021), and the Federal Environment Minister has blocked approval of the massive wind and solar project, the Asian Renewable Energy Project, in the Pilbara area on the grounds that it would negatively impact on a Ramsar declared migratory bird habitat (Morton 2021a; Garvey 2021).

state-owned monopoly utilities, into one entity and establish from scratch a market to operate on the basis of a quite different *raison d'être* to those that had previously prevailed.[10] Some foundations had been laid, and were initiated with the passing of legislation that established a set of rules and formalised the principles that would govern the operation of the market. The policy impasse engendered some tension between the Commonwealth and the State governments with respect to their respective roles in shaping the development of the NEM. These were not easily resolved given that the Constitution deemed energy a State responsibility. For governance arrangements to be consistent across the NEM, the first step was to get agreement on common legislative agenda that would be pursued in each member state. The *National Electricity (South Australia) Act 1996* became the model template, duplicated in other States and the governing architecture began to be out in place. The Act incorporated the National Electricity Code, later the National Electricity Rules, which mandated the principle of the competition-driven energy market. The National Electricity Market Management Company Limited (NEMMCO) had been formed to oversee the electricity market participants. This included drafting the rules promoting and maximising competition and commercial freedom, combatting monopoly rent, and enabling market participants to exercise the greatest amount of commercial freedom, maximising efficiency and avoiding special treatment (AEMC 2005).

Reacting to the policy impasse in the federal domain, State governments reasserted their role in designing measures to support the transition of State economies to place greater reliance on renewable energy technologies. This drew a spirited backlash from the Coalition government, the most striking example of this being Prime Minister Turnbull's and Treasurer Josh Frydenberg's criticism of South Australia's Premier, Jay Weatherill, for overseeing a dramatic uptake in wind and solar power and which the Prime Minister and Treasurer allege contributed to a State-wide blackout (Norman 2017). But it was the moves within the Coalition government spearheaded by the coal stalwarts to pressure energy corporations to resist winding back reliance on coal that the issue of the state intervening in the liberalised NEM that became an enduring feature of a bitter debate. The pressuring of energy corporations questioned the integrity of the Coalition and its commitment to the unfettered free market in the government's evolving energy policy. Energy Minister Taylor announced

[10] The privatisation of water management systems was an exception.

that the government was prepared to assume a direct role in generating dispatchable electricity if the private sector did not, a proposal that was clearly at odds with the quarter-century project to liberalise the energy market.[11] The Minister threatened to intervene in the energy market, to punish energy generators that did not comply with government demands to make good the loss of dispatchable energy when closing power plants. The proposed intervention was clearly not consistent with the National Electricity Law that mandated competition as a priority (AEMC 2015). Zibelman[12] expressed her concern that unilateral intervention in the market would compromise the efficiency of the sector:

> What we can't afford to do anymore, is not make a decision and continue to whinge to each other about not making a decision. … We need to either reform the market and regulations to allow for efficient market responses or decide that we can't and that we need to go towards a more regulated and local market. (Mazengarb 2020d; Murphy 2020b)

Zibelman's clear preference was for resurrecting the dynamism and integrity of the NEM and for government and industry combining their resources:

> 'to have the National Electricity Market work[ing] efficiently to deliver more lower-emissions power supply' rather than 'return to more heavy-handed state-based regimes'. (Ludlow and Macdonald-Smith 2020)

Zibelman proceeded to discuss the need to acknowledge that the energy system had undergone a radical transformation and that renewables and batteries, alongside gas and subject to its price, would define the immediate future and that incentives—'the right kind of market signal'—should be concentrated on developing dispatchable or firming capacity.

ESB chief Schott had no qualms in expressing her concerns when the government signalled an intention to block plans to decommission the Yallourn power plant. Schott warned the Minister for Energy and

[11] The Coalition government had bought the Victorian government's interest in state-owned Snowy Hydro to establish a controlling interest, while the New South Government retained its interest. Snowy Hydro would subsequently be awarded $6 million to construct a gas-fired generation plant in New South Wales.

[12] Zibelman announced her intention to resign from her role at AEMO in December 2020.

Emissions Reduction that such interventions are 'not helpful' (Mazengarb 2021a). Schott was more sanguine about the future, arguing that:

> 'comprehensive change to the market design' could remove the need for governments to intervene in the market to fill gaps when coal or gas plants close. ... If we can get the market measures in place outlined in the report, governments may not need to intervene at all. But comprehensive change to the market design is needed. (Toscano et al. 2020)

Taylor responded to such observations by suggesting that 'policy makers want to force out coal generators prematurely' and called for them to say so 'upfront'. This was a spurious contention and all the more so given his position to break with 20-plus years in the making of the NEM. From an administrative role—getting the right kind of market measures in place to ignite the self-correcting 'invisible hand'—Taylor's threat to have the state invest in generating capacity amounted to a radical refashioning of the role of the state. Such intervention exposed just how politicised policy had become in the government's determination to secure the future of fossil fuels. According to Schott, it made 'very little commercial sense', and nor did the initiatives to include gas as an integral feature of the future energy mix, and she questioned the integrity of a party that was ostensibly the 'champion of free enterprise' (Murphy 2021; Murphy and Karp 2021; Parkinson and Mazengarb 2020; Baxter 2020).[13] It was no wonder that peak business councils and prominent corporations were expressing disquiet with the direction of Coalition policy.

The basis of the differences between the Coalition government and the AEMO and ESB lay in the principles that were set when the NEM was being designed. The energy market governance architecture was an unusual creation. The two key agencies that were established to take over from the founding agencies in 2005—the Australian Energy Market Commission and the Australian Energy Regulator—and, in 2009, the Australian Energy Market Operator established to succeed the NEMMCO were better resourced, with the industry paying a levy to cover the costs of governance. There was no ambiguity about the expected focus of the principles that should inform their vision of the energy market. It was *laissez faire* fundamentalism, clearly specified in the Australian Electricity Rules,

[13] The Labor Party's energy and emissions reduction policy is all-but the mirror image of the Coalition government's policy (Parkinson 2020)

and staff were imbued with the ideal of the rule of the competitive market. The three agencies were responsible to the Council of Australian Governments and required to report to the Energy Ministers. AEMO was designed as an independent agency to oversee the operation of the energy market and to represent the interests of participants (AEMO 2020). It is wholly funded by industry participants and does not receive any financial backing from government, and this makes for a unique set-up. Although this financial arrangement could be regarded as pressuring AEMO to act in the interests of the member participants, its constituency is quite diverse with interests across the spectrum of the energy value chain, including energy generation, transmission, distribution and retailing, and includes established operators and newcomers that have just recently invested in the sector. It is mandated to provide 'full representation of relevant parties … [and support]. Extensive involvement by market participants through advisory committees and consultation processes.' Crucially, and in contrast to the clientelist relations that had framed Coalition governments, AEMO is an independent agency while it was equally beholding to the rules that made absolute clear that enhancing competition was the unequivocal priority of governance (AEMO 2020).

Complementing this architecture was the Energy Security Board, which was established in 2017 on a three-year term specifically to coordinate the implementation of the Finkel Report. When the Report was abandoned, ESB was tasked with developing the National Energy Guarantee, and following the dumping of the NEG, the ESB was commissioned to take the lead in developing the *Post-2025 Market Directions Design*. It—and Schott as the ESB's Executive Officer—was under no illusions that it was governed by the same objectives as the AEMC and the AEMO, to promote competition.

As noted, free market stalwarts Zibelman and Schott had no qualms in standing up to Minister Taylor to defend the cause of competition, but it would be overly simplistic to accept their characterisation of the options as a choice between the 'heavy-handed state regime' and a perfectly competitive, self-managed, market 'if we can get the market measures in place … [which would] remove the need for governments to intervene' (Toscano et al. 2020).[14] This contention clearly downplayed the hands-on role of

[14] There were more colourful caricatures of state intervention by free market stalwarts declaring intervention as a 'socialist approach' (Ferguson and Brown 2020). By way of contrast, the Parliamentary Report *Powering Our Future* had recommended strengthening the

the AEMO in directing the connection of solar and wind farms to the grid and coordinating investments across the NEM. AEMO was also working with State governments in planning the development of renewable energy zones. It had aroused the wrath of some investors who had proceeded with developing solar and wind farms on the presumption that they would be automatically connected to the grid but were blocked by AEMO from doing so to avoid grid congestion. There are reports of AEMO intervening in the market to determine which generators are given favoured treatment. Solar farm operators have questioned the impartiality of AEMO in blocking solar farms feeding energy into the grid while permitting wind farms to deliver energy in order to optimise the utilisation of the wind technology. The AER undertakes a zealous policing of the rules, so whichever way the subject was approached, the functioning of the liberalised energy market was contingent on a unique architecture of governance.

Notwithstanding the protestations of Zibelman and Schott that the government should be clear about the preferred governance form for the NEM it was either market-based or state-determined, but it could not be both. Their modelling of the future shape of the energy market presumed the need for some market oversight and channelling in order to deliver energy reliability and security. This did not necessarily contradict the rhetoric about unfettered competition being the paramount and defining feature of the liberalised energy market. The state was a necessary institutional force to enable the competitive impulse to flourish.

However much AEMO advocated the perfectly competitive market in preference to state governance, the actual scope of AEMO interventions placed the Operator in the statist camp. In issuing directives that determined the location of new investments in solar and wind farms, coordinating development and exercising the authority to approve new generating enterprises being connected, competition was subordinated to the administrative and coordinating decisions being made by AEMO. While there were few barriers blocking investments to enter the energy system by establishing new wind and solar farms, the multiple steps that had to be taken made for a drawn-out and often anarchic process. Investors would firstly have to identify a suitable site and then contract with landholders to lease land on which the farm would set up operations, and this could be a haphazard venture, dependent upon the vagaries of the state of pastoral or

governance architecture by establishing a single authority, an overall planning body (Australia 2017).

agricultural enterprise and the landholder's need or interest in supplementing income. Agreement on the terms of the land lease had to be finalised before a development application to state planning authorities could be considered before finance was organised. Once the development application was approved, investors could proceed to contract with developers, and this has proved a complicated process, compounded by delays with a large number of developments and their approval, and resulted in labour shortages and processing logjams, and it is only when the construction work was completed, and the energy farm operational, would it make sense to apply to AEMO for the right to access and feed power into the grid. AEMO then has to determine if there is sufficient network capacity to accommodate the additional power that is on offer (de Atholia et al. 2020). At each of the respective steps, investors face a sense of urgency in that once one stage of the process has been resolved, there is pressure to gain another relevant authorities' approval before moving onto the next stage of the development because the ultimate goal is generally to produce electricity and feed this into the grid to realise the value of the investment.

The competitive pressure can be manifested across the different stages of development because not all investments are launched with the objective of developing a project that the investor plans to hold onto in perpetuity. In fact, a common feature of the development of the renewable energy sector is that the different stages of a project will be undertaken by a succession of enterprises, and when this is linked to development approval processes, the anarchic character of the energy industry can become quite pronounced. Given the relatively disorganised nature of the project development, one that fits the liberalised model, the final approval may not necessarily be forthcoming because, over the course of the project being developed, there have been other solar or wind farms constructed, supplementing and testing generation capacity which, in those instances where grid connection is imminent, AEMO determines what is in the interests of order with the consequence that some approvals are withheld.

A similar filtering or rationing process restricting access to the grid has framed AEMO's management of the longer-term consequences of excess generation capacity and oversupply of distributed resource electricity. AEMO's active role contradicts its declaration of being committed to a minimalist role for the state. Just as Taylor's *Roadmap* claims to not being a case of choosing winners when this is what government policy is all about. Zibelman's and Schott's criticism of the Energy Minister's intervention was somewhat contrived and an oblique way of questioning the clientelist tendencies of the energy state. Their oeuvre could accommodate a conception of the state as being an essential feature of the liberalised energy market, but they lacked the analytical tools to develop a critique that exposed the functioning of the clientelist state.

AEMO's interventions in public discourse emphasise the relative merits of the market *vis-à-vis* the state in the context of States assuming a dominant and defining role in the structure and organisation of the market. AEMO became more interventionist than its pronouncements on the market versus the state indicate. With the competition regulator, the ACCC, hovering in the background monitoring the market, and especially prices, the market is not the unencumbered institution that is supposedly the model for the NEM. Reflection on this is coloured by one of its chief advocates, and the free market operator, AEMO hides behind the veneer of the *laissez faire* model.

Beyond the State, Markets and Competition

As is the case with the inadequacies in the problematising of the state, there has been a tendency to regard the market and competition, the force that is represented as giving life to the market, as straightforward, unproblematic constructs. This was made clear in the Minister for Energy and Emissions Reduction taking issue with the roles of the Australia Renewable Energy Agency and the Clean Energy Finance Corporation as institutions that were compromising the spirit of competition and the energy market liberalisation project more generally. Taylor contended that ARENA and the CEFC, in providing financial start-up support for new ventures that were novel and did not have ready access to finance, gave renewable energy projects an unfair competitive advantage over fossil fuel-orientated projects. Actually, the terms of the ARENA and the CEFC were quite transparent and consistent with the liberalisation project because their support assumed the form of market-like incentives and the decision-making processes that filtered funding applications were transparent (unlike some of the funding support schemes for fossil fuel projects that were being made behind closed doors). The Minister was not overly concerned about ensuring a level playing field, after all the fossil fuel industry was massively subsidised. He was pursuing an entirely different agenda of wanting to provide ventures in fossil fuel-related projects with access to CEFC and ARENA financial backing. The Minister justified his efforts to relax the eligibility rules for applying for financial support in terms of creating a level playing field for the full gamut of energy market participants.[15]

The respective parties overlooked a more fundamental issue in making clear their defence of the free market system and the role of the state. Indeed, their differences on another fundamental issue were not all that at variance, especially after Taylor stated publicly that ad hoc interventions in the market were a thing of the past, as we have noted. Both failed to expose the limited meaning of competition when considering the critical element in climate change politics, namely, emissions and the failure to incorporate a reckoning of the cost of environmental externalities associated with the generation of emissions. Zibelman and Schott could be excused for not taking up this failure because in their case the void can be explained by the design of an energy governance architecture that explicitly excludes an accounting of emissions, but not really excused. To their credit, they were focused on smoothing the transition to a renewable energy future. The Coalition's failure, and Morrison's and Taylor's more particularly, was the result of them being the instruments of the coal-industrial complex. There is, however, no excuse for the ACCC's failure to address the subject of greenhouse gas emissions in its investigations into the failure of the liberalised energy market to deliver a more economical supply of electricity and gas.

The ACCC was established as an independent agency whose function is to oversee an economy in which anti-competitive practices are dismantled. Any combination of factors that has the potential for diminishing competition is subject to its investigative powers. One demonstration of this arose when the Melbourne City Council approached several cultural and other enterprises in Melbourne's central business district to form a consortium that could negotiate power purchasing agreements with renewable energy suppliers and/or generators. By partnering with other energy users, the Melbourne Renewable Energy Project (MREP) consortium would permit the scaling up of the quantum of power supply that could be negotiated, and this could provide scope for electricity being supplied at a more economical price than might otherwise be possible. The consortium, however, could be regarded as establishing a cartel that could be held to be in restraint of trade and thus subject to a Commission ruling vetoing the

[15] Faced with the prospect of legislation being rejected by the Senate, Taylor indicated his intention to make the change by regulation, and he was advised by the chair of the Senate committee that this would be illegal Morton (2021b).

arrangement. Fearing a possible breach of the Competition Act, before proceeding to negotiate 'Green Electricity Retail Sale Agreements', the City Council sought ACCC advice on the legality of the consortium.

In an intriguing decision, the ACCC authorised the formation of the consortium and its right to proceed with negotiating the supply contract. The Commission approved the application because of a composite of factors, including the 'environmental benefits'. The PPA was approved on the understanding that the supplier (individually or with a developer) 'will use the GERSAs [Green Electricity Retail Sale Agreements] to develop a new renewable energy generation system that will supply renewable energy into the National Electricity Market' (ACCC 2018). The Commission also imposed some conditions on the operation of the energy generator that would otherwise be contractually extrinsic. An essential feature of the approved PPA was that it locked in the terms of the consortium's contract as to how the supplier should go about its business, the required form of generation. In imposing conditions on the supplier, the competition construct became infused with a more substantive meaning. Competition was recast to include an environmental obligation that reached beyond the norm of commercial and pricing practices.

If this was not a radical departure from the norm, the Commission's inquiry into retail energy pricing reinstated the quite narrow definition of competition that had been established with the formation of the NEM governance architecture. The inquiry was focused on unravelling the factors that had resulted in sustained price inflation and compromised '*Australia's competitive advantage*'. The Commission took issue with the costs associated with the incentives provided to households to subsidise the installation of photovoltaic rooftop solar panels and the payment of feed-in tariffs. These incentives were designed to increase the proportion of electricity being generated from renewable sources and reducing emissions in the process. The Commission justifiably criticised the inequities in these arrangements because the cost of the measures was incorporated into retail prices identified as 'environmental costs' which were borne by all users (2018). However, there was no counter-argument put forward to applaud these particular 'environmental costs' that lauded the benefits accruing from the reduction in household emissions, let alone any reference to the costs associated with emissions-generated environmental externalities. It was as if in the corridors of the ACCC some doors opened to entertain the Coasean principle while, for others, the field of environmental economics had never been entertained.

Erased from the price inquiry was any interest in the foundations of competitive advantage. This was a *Groundhog Day* moment, a repeat of the story that had been carried to the Kyoto Protocol negotiations. Emissions as an intrinsic factor in coal- and gas-fired power generation received no mention in the ACCC report. The competition that engendered economic advantage did so only because it was sanitised of the effects of emissions. The ACCC economists who prepared the report were as much the prisoners of the coal-industrial complex as were those in AEMO, the AEMC, the AER and the ESB.[16] They are captive of a reified economic orthodoxy, completely shorn of any consideration of the cost of environmental externalities associated with emissions. This is now of some considerable significance because Australian commodity exports that do not incorporate a reckoning of the emissions generated in the course of their production face the prospect of a surcharge being imposed on commodities imported into the economies of the European Union following the European Parliament's decision to endorse a proposal to introduce a 'border carbon adjustment mechanism' as part of the EU Green Deal (Simon 2021).

While the debate on the future trajectory of the NEM railed over the relative significance of renewable sources of energy *vis-à-vis* fossil fuels, neither the AEMO plan and the ESB *Discussion Paper* nor the Coalition's *Technology Investment Roadmap* envisaged any fundamental change in the organisation of the energy industry or in the basic structure of the value supply chain. They both were wedded to a market that was built on industrial-scale generation, and a transmission and distribution network that was centralised and key to the organisation of the resource processing industry. Although there were some differences in the emphasis of the models, most explicitly with respect to the role of the state, when one

[16] One other explanation that might explain the ACCC's disregard for considering the import of emissions when assessing competitiveness and the inflation in electricity prices is that it was not all that concerned with the concentration of ownership in the energy industry. The ACCC chief had approved AGL's 2007 takeover of the Torrens A and Torrens B generators in South Australia, giving AGL an 80 per cent control of demand and 70 per cent control of the retail sector. The ACCC warned that approving the takeover would 'give AGL the ability and incentive to "bid strategically" and raise prices in the wholesale market significantly', and in early 2008 this is exactly what AGL did: increasing wholesale and retail prices (Parkinson 2016; Leitch 2017). More generally, the ACCC chair appears to have been somewhat slapdash in setting competition benchmarks for the energy sector (Parkinson 2016; Leitch 2017).

looked beyond the rhetoric of a supposed shared commitment to a *laissez faire* market, both models overlooked what was driving decentralisation. They presumably could not consider this because they regarded recentralising the energy system as crucial to the stability and security of supply. It is also likely that this commitment to centralised control could be explained by just how much industry policy and the support for energy-intensive industries were intertwined, tied to the centralised model. The resource processing industries of aluminium and steel industries had been forged on the foundations of an industrial scale and centralised energy generation and distribution system that provided energy at substantially discounted prices. Governments were reluctant to allow these industries to disappear, and as much as the *Technology Investment Roadmap* envisaged a future liberated from the heavy reliance on electricity with the development of 'green steel' or an aluminium industry producing 'green aluminium, governments, federal and State, were reconciled to the fact that these industries' futures were contingent upon maintaining the supply of power at substantially discounted prices and at considerable cost to taxpayers (Tomago Aluminium Company Pty Ltd. 2017; Taylor and Tehan 2020; EUtility 2020; Williams 2021b).[17] A centralised energy system was crucial to enabling subsidised electricity.

The AEMO-ESB modelling presumes that every effort is made to consolidate renewable energy generation in reasonably contained regional zones. This will minimise the extent to which the transmission grid network requires augmenting. The modelling also anticipates that the renewable energy network be consolidated by the addition of at last three new high-voltage interconnectors that would enable increased transmission between Tasmania and the mainland, Victoria and South Australia and New South Wales and South Australia and an additional link connecting Snowy Hydro to western Victoria and south-west New South Wales. This would provide the capacity for transmitting power across the NEM to meet the regional variations in peak-load demand and thereby lessen the need for constructing still more generating capacity. This is not going to

[17] As such large users of energy—Tomago is the single largest energy consumer in New South Wales and Alcoa the largest electricity consumer in Victoria, each consuming 10 per cent of the respective State's energy—aluminium producers argued that in signing demand management agreements they were now critical actors to ensuring the stability of the energy system (Potter et al. 2019; Taylor and Tehan 2020). The federal government set aside $76.8 million over four years to meet the cost of the supply and demand management scheme (Department of Industry, Science and Energy Resources 2021).

be inexpensive, and it is a model of energy development that is loudly applauded by the transmission corporations whose status in the electricity value supply chain is preserved. Regulating the flow of electricity in this manner would not obviate the need for greater capacity to accommodate energy generated from renewable technologies. For example, Snowy Hydro, which had been consolidating energy generated by wind and solar farms and repackaging, and 'firming' this, to supply commercial and industrial users maintained that unless more transmission capacity was installed, it would lose this market opportunity (Macdonald-Smith 2021a). Financial support from the CEFC and State governments has been crucial in backing the identification of zones in which renewable energy ventures could be established in NSW, Victoria and Queensland, although the number of zones and their location did not correspond with what the AEMO had negotiated as regional politics came into play. No time was wasted in moving forward on the development of the transmission projects (Toscano 2021).

Initiatives to consolidate household generation have also been a high priority, with their capacity being regularised through the formation of virtual power stations or other aggregating businesses. Instead of this giving households some control over how they meet their energy needs, the AEMC with AEMO support is modifying market rules that allow for varying the value of feed-in tariffs across the day, reducing the tariff paid during the afternoon when solar panels were most productive and introducing mechanisms that centralise control over household generating and storage capacity to determine when the power is fed, or blocked from being fed, into the network. In the *Roadmap*, and more especially so with the *ISP* and the *Post-2025 Market Directions Design*, the task at hand has been one of marshalling and consolidating the export of power into the established infrastructure. The AEMC recommended a more radical arrangement that would entail households being charged for exporting power into the grid. Crucial to the effectiveness of these proposals would be the installation of the smart technology that captured details of energy flows, the emerging pressures and the means for processing this data. These technologies are integral to providing the AEMO with the ability to assert control over the flow and storage of energy, to recentralise the system. The plans are entirely consistent with the *Technology Investment Roadpath* and, in fact, several of the proposals have been incorporated into the drafting of the *Roadpath*.

The installation of industrial-scale batteries is also proving critical to reproducing this industrial-scale and centralised energy governance model, and the distinctive feature of this has been the initiating role of State governments. This is despite the potential that batteries could also become the means for greater independence of communities. The South Australian government set the pattern in 2017 when it negotiated the installation of a Tesla battery with Elon Musk, in partnership with French renewable energy company Neoen. The battery complex at Hornsdale in the mid-north region of the State was then the largest lithium-ion battery in the world at 129 MWh and 100 MW. It proved an overwhelming success in providing grid reliability and its capacity has since been increased to 150 MW. Neoen has approval for a $500 million project at Crystal Brook and a $3 billion Goyder South project, near Burra 194 MW.

Other States soon followed installing utility-size battery storage systems to enhance reliability. The New South Wales Liberal government announced funding support for four new big battery projects in the State as it flicks the switch on the transition from coal to a grid dominated by 'on demand' renewables and storage (Parkinson 2020). With support from ARENA, the NSW government is proceeding with TransGrid which will own and operate the 50 MW/75 MWh grid-connected lithium-ion battery to be installed at its Wallgrove substation in western Sydney (Foley 2020). In each instance, the back-up systems are predicted to attract more investment in renewable generating capacity. In 2020, the Victorian announced its intention to construct one of the world's biggest batteries, a 300 MW system in Geelong (Toscano et al. 2020). Victoria's Powercor is also proposing up to 20 big batteries scattered around its network to support the increase of wind and solar and boost grid security (CNEgypt 2021).

Established energy companies have been investing in battery storage, and a particular focus of their investments has been to locate storage systems within the shadows of generators that are connected to transmission infrastructure, saving considerably on the cost of transitioning. Origin Energy, for example, has announced its intention to build a 700 MW battery at its Eraring coal-fired power plant in what would be one of the world's largest battery installations, four times the size of the Hornsdale battery (Hutchen and Terzon 2021). The battery would enable continued use of the infrastructure after the scheduled decommissioning of the plant in 2032. Neoen has plans to construct a 500 MW battery at the site of the former EnergyAustralia Wallerawang power plant. EnergyAustralia was backed by the Victorian government and ARENA to construct a 50 MhW battery to be located in Ballarat and Gannawarra, the first to work in tandem with a wind farm (Filatoff 2021). AGL and the New South Wales government have lodged a scoping report for a grid-scale battery system to be located alongside the existing plant at the Liddell Power Station, which is part of the 850 MW multi-site integrated battery system AGL is planning on developing by 2024 (Hartmann 2020). A similar rationale has informed AGL's decision to proceed with its plans for installing a battery at Loy Yang in Victoria, and the company has lodged development plans for a battery installation connected to the Torrens Island power station in Adelaide (Macdonald-Smith and Evans 2021; Hartmann 2020). AGL has partnered with the Singaporean energy venture, Vena, to construct what will be the largest battery storage facility in Queensland, the 100 MW/150 MWh near Wandoan in the State's south-west (Maisch 2020). Queensland's PowerLink has been working with ARENA to identify how utility-size batteries can introduce stability in an energy system in which intermittent sources of power can engender disorder (CNEgypt 2021). The Queensland Premier has announced five regional locations across Queensland, in Hervey Bay, Bundaberg, Townsville, Yeppoon and Toowoomba, to host a large-scale, network-connected battery storage trial aimed at supporting the state's continual uptake of renewable energy (Booth 2021). The ACT is planning a network of batteries across the Territory, including 250-megawatt-plus Big Canberra Battery project (Lowrey 2020; Brewer 2021).

Supporting the development of pumped-hydro storage is regarded as another power source that can contribute to an industrial-scale system. Snowy 2.0 has been promoted as fulfilling this role, underscored by grid augmentation to connect with renewable energy zones in New South Wales and Victoria. The federal government's allocation of $94 million to support the construction of the proposed Marinus Link interconnector between Victoria and Tasmania enables Hydro Tasmania to become the 'battery of the nation' (ARENA 2018). ARENA is also supporting the development of pumped-hydro schemes as back-up storage (Energy Source and Distribution 2021b).

Whereas there is considerable effort being invested in developing infrastructural support to capture distributed energy resources as a feature of industrial-scale transmission and distribution system, the accommodation of smaller-scale renewable energy has not attracted anywhere near the same interest. Indeed, the scope for a more comprehensive deployment of batteries to provide 'fast frequency response service and voltage control' for system security is limited by regulation (Bucci 2021). However, the AEMC is considering changes to market regulations to marshal energy flows to and from the grid involving small, distributed energy resources, which would include households and community energy storage facilities. The reach of such changes may first have to win the support of the Australian Energy Council and corporate energy companies which have voiced their concerns that State initiatives supporting local battery storage systems were compromising the security of the national energy market (Toscano et al. 2020).

These belated initiatives finally give some recognition to just how much the energy system has changed, although the Coalition government and the AEMO-ESB have remained transfixed. The energy governance architecture has also been captured by the endeavour to work towards realising a market modelled in the image of the ideal *laissez faire* construct. Yet, an increasing number of participants are organising their participation and investment in the transition to renewable energy by abandoning the individual-orientated commercial transactions through the market by exploring extra-market engagements or defecting from the market. The energy system has been decentralised as engagements in the generation and distribution of electricity are becoming more social in form. The high rate of household investment in rooftop solar voltaic panels and batteries is reflective of the socialisation of energy, with energy corporations endeavouring to recapture their customers by involving them as 'partners' in virtual power plants. The community-based initiatives are being reinforced with the introduction of community-based battery systems. Municipal local governments have become a conduit for citizens to play an active part in fostering the transition while councils partnering with one another have been injected an extraordinary impetus into investments in renewable energy through the negotiation of Purchase Power Agreements based on establishing enduring relationships with new renewable energy ventures that are now able to borrow on the basis of long-term revenue flows. The formation of energy cooperatives reinforces this transformation in the energy system.

Conclusion

The Coalition government's determination to not commit to an emissions reduction target reflects a resolve to not jeopardise the development of the resources and energy-intensive industries. This is embedded in the psyche of those stalwarts in the Coalition who continue to regard the science of climate change with scepticism and outright denial. The extent to which government policy has been captured by the minerals industry is quite extraordinary, but this is not just the product of very effective lobbying on the part of the industry, it is also a consequence of just how unimaginative and lazy the Coalition has become in not seeing beyond the immediate short-term rewards of cheap fossil fuels. These can be economic, but there can be political gains in the pro-fossil fuel stance insofar as it is an appealing message that can attract the votes of constituents who reside in areas where mining could be threatened by the closure of mines and returned pro-coal candidates in the last federal election giving the Coalition a one-seat majority as government. The macroeconomic vision of the fossil fuel-based economy being the impetus underwriting the dynamism of the economy has become embodied in the oft-repeated claim that mining coal fuels the operation of energy-intensive industries as well as a commodity export that regularly excites the observation that Australia is an 'energy superpower'. This has had several successive renditions, firstly as the largest exporter of coal, now the world's largest exporter of gas and the rush to engage hydrogen as a possible fourth. The fifth and outlier is the proposal to build what would be the world's largest solar farm, the $20 million, 12,000 hectare, 30 GW Sun Cable farm in the Northern Territory. In its original form the plan was to construct a transmission cable to export power to Singapore and Indonesia, but the developers are considering that the production of clean hydrogen utilising the green electricity to be the more economic prospect.

The proposed renewable energy project does not appear to have aroused anywhere the enthusiasm that prompted the then-Treasurer Scott Morrison to wield a lump of coal, to launch into a 'show and tell' exhibition in Parliament. Accusing the members of the Opposition Labor Party of having an ideological and pathological fear of coal, Morrison has not resiled from this, and it is this conviction in coal that has defined the Coalition's policy intransigence. And, as we have observed in reflecting on the fossil fuel industrial complex, the commitment to coal has morphed into a complex that is also invested in the promise of gas. The government envisages five new gas fields. The largest, the Beetaloo Basin shale gas reserve in the Northern Territory, could become one of the world's largest gas fields (Vivian 2021). The federal government will have invested over $224 million of public monies in the private venture by the time the project is up and running, and concerns expressed about the scale of the gas-take, with up to 117 million tonnes of greenhouse gases being emitted each year, have been dismissed with the retort that emissions will be effectively managed, although this assurance awaits the detail that would be provided in a yet-to-be prepared environmental impact statement. The potentially deleterious impact that a project of this magnitude has on the viability of other energy ventures has been completely overlooked and this serves to highlight the contradictions in the Coalition's energy development agenda.

The Coalition's longstanding commitment to the development of the resources sector and to coal has been based on a contrived argument that Australia has a natural comparative advantage because of the rich endowment of the fossil fuel. This is an argument that is being taken up with gusto by the gas lobby and the Ministers for Energy and Resources. However, the translation of this comparative advantage into a competitive advantage and the substantial export market relies on ignoring the environmental externalities, the greenhouse gas emissions, that are generated in the process of mining coal and producing electricity. The integrity of the theory that informs the logic of comparative advantage is abandoned in the process of disregarding the environmental externalities. Moreover, the sleight of hand that informs the translation of the theory of comparative advantage into a competitive advantage is built on the principle that it is the unfettered free market that gives free rein to the competitive advantage of coal in the Australian economy. The coal-industrial complex hides beyond the rhetoric of the *laissez faire* model, of making the case for minimising the role of the state, and capital, in the constitution of this free market, but, as has been argued here, this argument is also contrived. The free market's operation is contingent on the active agency of the state in cementing the conditions that engender the materiality of competitive advantage. The state's resistance to taking action to mitigate emissions, in effect, hides behind the rhetoric because this enables the capture of the value that is embodied in the fossil fuel while sidelining the costs that are generated.

References

ABC News (2021) 'Scott Morrison reveals $1b energy deal with South Australian government' 18 April.

Australia, The Standing Committee on the Environment and Energy (2017) Inquiry into modernising Australia's electricity grid, *Powering our Future*, December.

AAP [Australian Associated Press] (2021) 'Scott Morrison signs $1b deal to shore up energy reliability in South Australia' *The Guardian* 18 April.

ACCC [Australian Competition and Consumer Commission] (2018) Restoring electricity affordability and Australia's competitive advantage Retail electricity pricing inquiry, June.

AEMC [Australian Energy Market Commission] (2005) *Australian National Electricity Rule.*

AEMC [Australian Energy Market Commission] (2015) *Australian National Electricity Rule.*

AEMO [Australian Energy Market Operator] (2020) *Integrated System Plan.*

ARENA [Australian Renewable Energy Agency] (2018) 'Battery of the Nation—A Concept Study of Tasmanian Pumped Hydro'.

Baxter, T. (2020) 'Scott Morrison's transition plan is a dangerous road to nowhere' *Climate Council of Australia* 5 February.

Booth, E. (2021) 'Battery storage trial underway in Queensland' *Energy* 29 March.

Brewer, P. (2021) 'Wooing process begin ACT's big batteries supplier' *The Canberra Times* 17 April.

Brissenden, M. (2021) 'Fired Up' *Four Corners ABC* 12 April.

Brissenden, M., A. Donaldson and L. Day (2021) 'Government accused of pressuring experts who questioned its gas-fired recovery plan' *ABC News* 12 April.

Bucci, N. (2021) 'Renewables plus batteries offer Australia the same energy security as coal, research finds' *The Guardian* 29 March.

Carabott, M. (2019) 'State government's angry at Canberra's failure to call COAG energy policy meeting' *Leading Edge Energy* 22 July.

Clarke, M. (2021b) 'Federal Government threatens to build gas plant if electricity sector doesn't replace retiring coal-fired power stations' *ABC News* 14 September.

Clarke, M. (2021a) 'Blistering assessment gives Australia 'just months' to fix nation's energy security' *ABC News* 5 January,

CNEgypt (2021) 'Powerlink looks to battery storage to help solve grid stability problems' 2 June.

de Atholia, G. Flannigan and S. Lai (2020) 'Renewable Investment in Australia. *Bulletin*, Reserve Bank of Australia, March.

Department of Industry, Energy and Resources (2020) Growing Australia's Hydrogen Industry.

Department of Industry, Science and Energy Resources (2021) 'Securing Victoria's energy system' 19 March.

Department of Industry, Science, Energy and Resources (2020) Growing Australia's Hydrogen Industry.

Edis, T. (2020) 'Australia had a real future as an energy superpower' *The Canberra Times* 30 January.

Elsworthy, E. (2020) 'NSW strikes 'landmark' deal with federal government, Greens call it 'climate criminality" *The ABC News* 31 January.

Energy Security Board (2021) Post-2025 Market Directions Design, January.

Energy Source and Distribution (2021b) 'ARENA commits $47m to Genex Power's Kidston Stage 2 PHES project' 24 March.

EUtility (2020) 'Alcoa Portland aluminium smelter invites bid for electricity supply' 30 September.

Ferguson, R., and G. Brown (2020) 'PM's 'socialist approach' on climate' *The Australian* 6 February: 4

Filatoff, N. (2021) 'NSW coal country gets a big-grid balancing battery' *PV Magazine* 11 January.

Finkel, A. (2021) 'Getting to Zero: Australia's Energy Transition' *Quarterly Essay* 81.

Foley. M. (2020) 'NSW to get a big new battery as renewable power surges' *The Sydney Morning Herald* 23 October.

Garvey, P. (2021) 'Backlash over axed $48bn renewables plan' *The Australian* 22 June: 7.

Hartmann, I. (2020) 'AGL's massive big battery investment *Energy Magazine* 17 August.

Hutchen, G., and E. Terzon (2021) 'Largest battery in Australia to be built at ageing NSW coal-fired power plant' *ABC News* 13 January.

Iggulden, T. (2018) '"Anarchy": Whither the government's energy policy' *ABC News* 12 October.

Karp. P. (2020) 'Morrison strikes $2b deal with NSW to boost gas supply' *The Guardian* 31 January.

Latimer, C., and P. Hannam (2018) '"I haven't left anger yet": energy architect slams policy "anarchy"', *The Sydney Morning Herald* 10 October,

Leitch, D. (2017) 'Wivenhoe pumped hydro: the big little plant that didn't' *Renew Economy* 23 February.

Lowrey, T. (2020) 'ACT Labor promise Canberra-wide network of energy batteries if re-elected' *ABC News* 30 September.

Ludlow, M. (2019) 'Taylor finally calls COAG meeting—in two months' *The Australian Financial Review* 24 September.

Ludlow, M., and A. Macdonald-Smith (2020) 'Zibelman's call to arms to reform the grid' *The Australian Financial Review* 23 November.

Ludlow, M. and A. Macdonald-Smith (2019) 'Energy Minister to cut deals with "collaborative" states' *The Australian Financial Review* 18 November.

Macdonald-Smith, A. (2021a) 'Energy users the losers without more transmission: Snowy' *The Australian Financial Review* 25 May:

Macdonald-Smith, A. (2019) 'Plan targets saving as solar strains grid' *The Australian Financial Review* 22 July, 19.

Macdonald-Smith, A. and S. Evans (2021) 'AGL adds to big battery but hurdles loom' *The Australian Financial Review* 24 March, 13.

Macdonald-Smith, A., and M. Ludlow (2020) 'Future tense for Schott's Energy Security Board' *The Australian Financial Review* 29 July, 8.

Macdonald-Smith, A., and M. Ludlow (2020a) 'Vic, NSW under the pump as Qld fumes' *The Australian Financial Review* 29 January.

Maisch, M. (2020) 'New giant battery creates base for more renewables in Queensland' *PV*

Mazengarb, M. (2020g) 'ESB left in limbo after stalemate in 'confidential' energy ministers meeting' *Renew Economy* 20 July.

Mazengarb, M. (2020d) 'Anything but wind or solar: Taylor's technology plan is "road to nowhere"' *Renew Economy* 22 September.

Mazengarb, M. (2021c) 'Morrison government moves to strengthen secrecy around energy ministers meetings' Renew Economy 2 September.

Mazengarb, M. (2021b) 'Transparency lost as Taylor seizes control of now "secret" energy minister forums' *Renew Economy* 18 March.

Mazengarb, M. (2021a) 'Schott says Taylor's threat of intervention on Yallourn closure "not helpful"' *Renew Economy* 11 March.

Mazengarb, M., and G. Parkinson (2019) 'Taylor finally calls COAG meeting, but will he face up to clean energy future' *Renew Economy* 24 September.

Morgan, E. (2019) 'Government should abandon its power play on energy says Grattan Institute' *The ABC Business* 8 October.

Morgan, E., and S. Long (2019) 'Government should abandon its power play on energy says Grattan Institute' *The ABC Business* 8 October.

Morrison, S. (2021) 'Remarks: Minerals Council Parliamentary Dinner' *Transcript* 3 June.

Morton, A. (2021b) 'Liberal-led committee says that Angus Taylor's planned changes to renewable energy could be illegal' *The Guardian* 18 June.

Morton, A. (2021a) 'Coalition vetoes funding for wind and battery farm in northern Queensland' *The Guardian* 6 May.

Murphy, K. (2021) 'EnergyAustralia confirms new gas plant in Illawarra after Morrison government threatened to intervene' *The Guardian* 5 May.

Murphy, K (2020b) 'Australia must stop "whingeing" and make a decision on national energy market, operator says' *The Guardian* 23 November.

Murphy, K., and P. Karp (2021) 'Australian energy board chair says gas-fired power plant in Hunter Valley doesn't stack up' *The Guardian* 30 April.

Norman, J. (2017) 'Watch South Australian Premier Jay Weatherill shirtfront Josh Frydenberg over energy policy' *ABC News* 16 March.

Parkinson, G. (2020) 'The insanity of Angus Taylor's fossil fuel technology road map' *Renewable Economy* 29 February.

Parkinson, G., and M. Mazengarb (2020) 'Morrison's refusal to move on climate, and the "prosperity doctrine"' *Renew Economy* 3 February.

Parkinson, G. (2016) 'ACCC says it's OK for big utilities to exploit the market power' *Renew Economy* 25 August.

Potter, B., and A. Macdonald-Smith (2018) 'Rooftop anarchy' *The Australian Financial Review* 13 October, 17.

Potter, B., A. Macdonald-Smith and S. Evans (2019) 'Alcoa ordered to power down as AEMO takes action amid Victoria, SA heat' *The Australian Financial Review* 25 January.

Rollason, B. (2020) 'Victoria gives green light to conventional onshore gas exploration but bans fracking' *ABC News* 17 March.

Simon, F. (2021) 'European Parliament backs plan to price carbon at EU's border' *EURACTIV* 7 April.

Taylor, A., and D. Tehan (2020) 'Securing Victoria energy system' Joint media release the Minister for Energy and Emissions Reduction with the Minister for Education and Member for Wannon, *Dan Tehan*, 14 December.

Tchetvertakov, G. (2018) 'Victoria approves further offshore gas exploration and sails into early criticism' *Small Caps 18* May.

Tomago Aluminium Company Pty Ltd (2017) Tomago Aluminium Company Pty Ltd—Submission to the ACCC Inquiry into Retail Electricity Supply and Pricing' 12 July.

Toscano, N. (2021) 'TransGrid funds 900km power link to move cheap renewables' *The Sydney Morning Herald* 1 June: 12.

Toscano, N., M. Perkins and A. Prytz (2020) 'Victoria bets on big battery near Geelong to bolster grid reliability' *The Sydney Morning Herald* 5 November.

Towell, N., B. Preiss and N. Toscano (2020) 'State's onshore gas ban a federal deal breaker' *The Age* 31 January.

Vivian, S. (2021) 'Commonwealth tips another $173 million into Beetaloo Basin gas reserve insists emissions targets are on track' *ABC News* 21 January.

Wilkinson, M. (2021) 'How the gas lobby captured Morrison's Glasgow response' *The Saturday Paper* 30 October.

Williams, P. (2021a) 'No more ad hoc power plans: Taylor' *The Australian* 25 May, 20.

Williams, P. (2021b) 'Alcoa's Portland smelter's survival hangs on power deal' *The Australian* 19 March.

CHAPTER 13

The Irrevocable Transition to Renewable Energy: Coal's Death Knell

Australia's transition to a future where energy is increasingly being generated from renewable sources has been one of the world's surprising success stories. South Australia, criticised just four years ago for moving too quickly and creating an unreliable energy system, is now being celebrated for its leadership, having managed to meet all its energy needs from renewable sources in October 2020, albeit for one hour. It is currently meeting 60 per cent of energy from renewable sources and is set on becoming an energy exporter. Other State and Territory governments have set deadlines, some comparable and some less so, but this begs the question as to what the future shape of the structure and organisation of the energy market is going to look like.

The enduring resistance of the Coalition governments to formulating and implementing an energy and emissions policy can be attributed to the hegemony of the coal-industrial complex. The Labor Party, following its flirting with the policies of the Rudd and Gillard governments, has been reluctant to outline a climate change policy frame for fear that this would provide its critics with too large a target, which could frustrate its election prospects. The politics of climate change has been quite attenuated. Debate tends not to canvass much beyond reflections on the Coalition government's *Roadmap* and the various complementary and piecemeal policies, such as carbon sequestration, that have been adopted, or the recommended infrastructure developments advocated in the AEMO

S. Rosewarne, *Contested Energy Futures*,
https://doi.org/10.1007/978-981-19-0224-6_13

Integrated System Plan and the ESB's *Post-2025 Market directions design*. The Coalition's *Roadmap* is essentially wedded to a 'business as usual' approach to the NEM and the defence of coal, although the centrality of coal in this narrative has at times been overshadowed by the promotion of gas as a transition fuel. The *ISP* and the *Post-2025 Market* capture what they envisage is the future shape of the structure of the industry and map pathways that would inscribe some order in this by mapping how the increased supply of distributed energy resources and variable energy resources could be incorporated within the NEM to ensure reliability and integrity of the market. Notwithstanding the ideological conviction in the market as the most efficacious institutional structure for reconciling the supply of and demand for this essential commodity, the AEMO and ESB adopt a realist appreciation of the future shape of the market that is also quite pragmatic.

In constantly coming up against the conservative resistance to formulating and implementing an emissions reduction and renewable energy policy, it is not uncommon to overlook the surge in investment in renewable energy and the transformation in the way we do energy. Likewise, it makes sense to look beyond the arrogance of a Coalition prime minister who had no qualms in breaching parliamentary protocols in his 'show and tell' with a lump of coal and in parodying Frank Sinatra in asserting 'We're going to do it our way in Australia' (Morrison 2021).

A musicologist he is not. Nor a shrewd politician because quite clearly, like the Coalition he pretends to lead, he is out of touch. There has been a remarkable shift in the tenor of climate change politics and the clamour for transitioning that is having some considerable traction. The AEMO has now stated that coal as the primary energy source is destined for the scrap heap with over 60 per cent of plant capacity expected to be retired within two decades (AER 2021). The *ISP* predicts that renewable energy using the existing transmission infrastructure will provide 90 per cent of energy requirements by 2035 and the newly appointed chair of AEMO has indicated that the energy system should be capable of being 100 per cent renewable by 2025 (Westerman 2021), although the system will require some dispatchable power back-up with this being replaced by greater reliance placed on battery and pumped hydro for firming.[1] This is affecting the focus of capital as well as forcing a significant reorientation in

[1] Energy Networks Australia contends that the displacement of coal will occur over a longer time frame (Narayan 2019).

how key departments within the state system are now approaching climate change concerns. There has been a radical shift in how we think about and do energy with the elevation of renewable energy defining the economic trajectory of the nation. In place of the policy thrombosis that has characterised the Coalition government's response to the climate change challenge for the last two decades, a phalanx of climate citizens, state institutions and energy capital has been reconstituting the political economy of energy. This is as much a social transformation as it is a technological transformation involving new ways of doing energy and casting coal to the scrap heap. Paradoxically, the politics of climate change continues to be captured by conservatives while the realpolitik tells a different story that recognises the physical and social transformation that is underway.

The Conservative Obstruction in the Renewable Energy Contest: A Last Gasp?

The conservative opposition to transitioning from a fossil fuel driven economy to one in which renewable energy prevails has proved a bitter and divisive war of attrition. Within the federal political sphere, conservative factions within the Coalition have been able to dictate the terms of the defence of the fossil fuel industry,[2] resulting in a policy vacuum that, industry analysts observe, has created a level of uncertainty that has discouraged capital from investing in upgrading the fleet of coal-fired power stations for the last decade. There has been no new investment in coal-fired generation since 2012 (AER 2021). The coal lobby has been extremely well organised in mustering local communities to show their support for coal as the mainstay of regional communities. In the lead up to the 2019 federal election, a march through regional Queensland organised by climate activists to draw public attention to the Adani coal mining project was successfully rebuffed by coal defenders. So persuasive has this conservative narrative been that the Coalition's chief political opponent, the Australian Labor Party, looks equally impoverished when it comes to the climate change political stakes. Fearful of a negative reaction in key electorates to any policy initiatives that would question the future of coal as a key input in the Australian economy and reduce the number of

[2] A successful leadership challenge within the National Party, the junior partner in the conservative Coalition, has hardened the National's demand that the government invest in more coal-fired power stations (Foley and Toscano 2021).

constituents prepared to signal their support for Labor, the party has become silent on the subject, a far cry from the halcyon days of the Rudd and Gillard Prime Ministerial leadership.

The reality is that renewable energy is displacing coal as the energetic force and doing so quite rapidly, and this presents a policy conundrum for the Coalition government's defence of the coal-industrial complex. The *Roadmap* provides some sense of direction but, as noted, it is a messy concoction of focuses, some elements of which have been informed by the AEMO's *Integrated System Plan* and the ESB's *Post-2025 Market design*, others that hark back to the days when governments were in search of the easiest means for promoting whatever would permit continued combustion of coal. Nominated solutions include carbon capture and storage or encouraging and providing financial incentives to reward land-based carbon sequestration. The integrity of these measures has been seriously questioned, and some, such as the carbon credits awarded for tree-planting missions to produce carbon offsets, are more than overshadowed by the immensity of the scale of deforestation that has followed the Queensland and New South Wales governments removing restrictions on land clearing. The one measure designed to penalise enterprises that do not meet the emissions reductions benchmarks under the terms of the Emissions Reduction Fund Safeguard Mechanism has never been implemented, and the government has indicated that it is considering an incentive system that would be based on carrots replacing the Safeguard Mechanism stick to align with the carbon sequestration reward system (Baxter 2019; Jotzo 2020). Santos has been the first emitter to be rewarded, with a federal grant provided as it proceeds with the trial of sequestering CO_2 in the in an exhausted oil void in central Australia (Thornhill 2021).

The Coalition has resisted any measures that would impact on the international competitive advantage of the resources sector. Not only has it not penalised any enterprise for not meeting its baseline level of emissions, but it is also rewarding those that do, and this adds to government largesse in providing financial backing to a host of fossil fuel ventures in a last gasp at defending the hegemony of fossil fuels. The rate of support exceeds all G-20 economies with not a due diligence assessment in sight (Bloomberg NEF 2021). Among the most egregious allocations has been the government's support for increasing the pumped hydro capacity of the Snowy Hydro or Snowy 2.0[3] and the $600 million awarded to Snowy

[3] The government-owned Snowy 2.0 venture is proving an extraordinarily expensive proposition. At an estimated cost of $10 billion, the economics of expanding Snowy Hydro

Hydro to construct a gas power plant at Kurri Kurri.[4] The grant of $21 million to Empire Oil and Gas, in spearheading exploration of gas and trial fracking in the Beetaloo Basin, could turn a step in the government's 'gas-led recovery into an emissions upsurge' (Knaus and Cox 2021). Most of the funds dedicated to these projects are expensive exercises in ideologically based hubris, often aimed at pork-barrelling the voting constituents. And when opposition to some of these schemes is aired by climate governance executives, they are subject to the outbursts of a bullying Minister for Energy. The same tactics are employed against those energy corporations that defy the Minister's directives that they must not shutter power stations that are reaching the end of their economic life. The Minister for Resources, Water and Northern Australia has fallen into line with what the conservatives have decided is anti-fossil fuel rhetoric, vetoing government funding for a wind farm project whose application for funding assistance had been successfully scrutinised and found worthy of support by the independent Clean Energy Finance Corporation.

While the Coalition has shifted its focus to advocating gas as the transition fuel, and more recently hydrogen, it has still not fashioned a policy out of the *Roadmap* The future development of the energy system has by default been, or at least was, placed in the hands of the AEMO and the ESB. Both were bound by their legislated conviction in advocating the competitive NEM which, as we have noted, precluded any need for addressing emissions. Not that this mattered much because policy was going to be determined by the Minister even if the actual Coalition policy was never formally legislated but acted on through a furtive series of discretionary funding allocations. The Monash group[5] formed within the Coalition party room to exercise it policy preferences presented another impasse. It formed around a 'business as usual' strategy, save for the fact that it demanded government support to finance the construction of new coal-fired power stations. Subjected to these constraints, the *ISP* and the

pumped hydro capacity is questionable and the cost of the expansion is exaggerated by the plans to invest in transmission links to the renewable energy zones being developed in western Victoria and south-western New South Wales (Boot 2018; Foote 2020).

[4] One analysis of the proposed Kurri Kurri gas plant estimates it would run at a loss of $150 million a year (Foley 2021).

[5] The group references Sir John Monash, a civil engineer who commanded Australia's military involvement in World War One was instrumental in the extraction of brown coal and the electricity hub in the La Trobe Valley in eastern Victoria. The brown coal belt was represented as the potential Ruhr of the Australian political economy.

post-2025 Market were contained to trying to capture what is happening in the industry and eschewing a more engaging and imaginative vision of the future. This is certainly reflected in the first public presentation by the new chair of the AEMO, Daniel Westerman, who has acknowledged that renewable energy technologies are presently capable of meeting 70 per cent of energy needs at particular periods of the day and that this can be expected to climb to 100 per cent by 2025 given current investment patterns (Toscano and Foley 2021).

Restoring a Role for the State in the Liberalised Energy Market?

As the proactive agencies in the NEM governance architecture, AEMO's and ESB's oeuvre is quite narrowly conceived. They focused on the effective operation of the free market, even though as argued in Chaps. 3 and 11 this was a contrived construct, and they accept that the energy sector has already set upon a particular trajectory even though it was generally agreed that this was problematic. The AEMO-ESB scenario paints, or at least anticipates, a more radical shift in the organisation and restructuring of the energy economy though this is one in which some form of the liberal market and state coordination would still prevail, and, incidentally, one that sits uncomfortably on the liberal mantle. From 'a centralised system of large fossil fuel (coal and gas) generation [the energy market is moving] towards an array of smaller scale, widely dispersed wind and solar generators, grid batteries and demand response' according to Clare Savage, the chair of the AER and member of the ESB.[6] This transformation necessitates enhanced coordination to ensure compliance with the competitive spirit, and this has become apparent with the growing magnitude of fines imposed on operators for not complying with market rules (AER 2021). In this liberalised market, the state has become more interventionist, and is it any wonder that these agencies' criticisms of the Coalition's extra-market interventions are so muted.

Since the halcyon days of a vibrant climate politics of over a decade ago (Rosewarne et al. 2014), there has been little critical reflection in Australia

[6] Although as will be discussed in concluding this is being seriously contradicted with at least four massive hybrid solar and wind farms being proposed in Western Australia and the Northern Territory, each successive development announcement claims to be delivering what will be the largest by far hybrid renewable energy project in the world.

on how best to manage the transition from coal. Certainly, the Coalition government's loyalty to the fossil fuel industrial complex has been unrelenting, and the forces that give voice to the complex have done their damnedest to check any challenge to this. The same forces have moved to forestall any resumption of the climate activism that featured in the first decade of the new millennium by imposing restrictions on the right to call climate rallies let alone the right to protest, the COVID-19 pandemic having provided a convenient cover for this assault on freedom of expression. The endeavours to contain the expression of energy democracy continue to be an unscrupulous business. However, behind this attack on democratic rights there has been a quiet revolution afoot. While the resources of the state focused on defending the status quo, there has been a failure to query the transformation that has been unfolding as energy users have developed a range of strategies to take control of the generation of energy. From the installation of rooftop solar voltaic panels to the community-based energy storage schemes and establishment of energy cooperatives to state institutions and corporate enterprises negotiating Power Purchase Agreements that provided the security for energy generators to invest in renewable energy technologies, the relations of exchange that were the basis of interactions in the National Electricity Market have been increasingly superseded by a distinctly social form of interaction. The dominant, the transactional, forms of relations have increasingly been being displaced by energy relations being socialised, a process that has engaged renewable energy technologies and which bypasses the NEM.

The preoccupation of the energy governance agencies has been with how the physical infrastructure that underpins the NEM could accommodate this quiet revolution. The agencies' preoccupation has only incidentally sought to understand what is driving the retreat from the NEM and defecting from the grid connection. They have put the impetus for the NEM exits down to the price inflation that the liberalisation project had promised to subvert. They have also sought refuge in some untested propositions about reliability and security of supply which were always the concern of the big energy users. The agencies, with the backing of the federal and State governments, have sought solace in concentrating new investment in generating capacity in renewable energy zones and in supporting the construction of new interconnectors. Their solution to affordability, reliability and security concerns has been to advocate increasing the capacity to generate power and the capacity to transmit energy across the NEM from one State to another and improve distribution within the States, which is a model contingent on concentrating and recentralising energy generation and distribution, and this reverses the developing emphasis that energy users having a voice in how energy should be generated, distributed and shared or what can be encapsulate in the ideal of energy democracy.

This reframing of the NEM network easily accommodates the State governments, which have been so critical of the Coalition. State governments have been driving the increase in renewable energy generating capacity and the reframing of the NEM goes some way towards appeasing the States' ambitions to meet their power needs while becoming net exporters of power and building a revitalised regime of accumulation. South Australia is all but there; Tasmania has secured government endorsement to become the 'battery of the nation'; Queensland has declared its intention to become a 'renewable energy superpower' as has Victoria (Hydro Tasmania; Sky News 2020; Moore 2020; ARENA 2018a, b; Smythe 2019).

A further rationale for planning the development of the NEM has been the focus on generating investment opportunities for regional communities, a policy emphasis that serves to give some credibility to the junior party in the Coalition (Regional NSW n.d.; ARENA 2020; Environment Land Water and Planning, Victoria Government 2021). The expanded investment in generating and transmission is promoted as a means of revitalising local economies by injecting a new dynamism into those regions that are being designated as renewable energy zone hosts. The establishment of new (regional) zones of renewable energy is being represented as a win-win for regional economies as well as the State economies, especially when this enhances the force of the circular economy.

There is degree of pragmatism being played out in the designation of some regional locations as sites for generating electricity. Selection partly depends on the meteorological and biophysical conditions as a basic requisite, but suitable areas in which to establish a solar farm, wind farm or pumped hydro project are being influenced by partisan politics as well as the idiosyncrasies of the landholders with whom developers must negotiate to access land. The AEMO has assumed a more proactive stance in signalling suitable areas and endorsing these with State government support. Nevertheless, although the AEMO is drawn into a somewhat pragmatic decision-making process, it does so within some crucial parameters. Most importantly, accepting the development of the physical foundations of the energy market entails decentralisation of capacity, and yet effective governance of the renewable energy systems weighs heavily on reinstituting some measure of the centralised, industrial-scale model that has defined the organisation and structure of the NEM.

The Anarchic and Inequitable Bases of Planning

Orderly development requires planning, and this appears to have been eclipsed by the development and planning system which was not fully prepared for concentrating new generating capacity in regional renewable energy zones. The haste with which the renewable energy juggernaut has taken off contrasts with procedures required prior to proceeding with the investment. Orderly development does not come easily nor in line with some set timeframe and as discussed in Chap. 11, the *laissez faire*-ism that has eased capital's entry into the regional renewable zones has not unfolded in an orderly manner. The motivations for investing in regionally based energy generation range from the speculative, buying into the industry with the objective of making a capital gain when on-selling; buying in at a particular stage of the energy value chain to bring to the project a competitive advantage that draws on the investor's labour force skill profile, to one that is committed to a long-haul engagement with renewable energy. The ease of entry can exacerbate the anarchic character of the industry and restoring some sense of stability and order calls for more social forms of industry governance, and this is evident in the increasingly interventionist role of the agencies that are the governance backbone of the NEM: the AEMC's proactive role in updating rule making; the AER's enthusiasm for enforcing rules; and the oversight and coordinating role of the AEMO.

The drawn-out development and approval process has been exacerbated by the substantial increase in solar and wind farms which has resulted in generation capacity getting ahead of the established network feed-in links and transmission capacity. This has led to AEMO delaying connections and stirred mounting criticism of network governance. The Clean Energy Investor Group, for instance, has taken AEMO to task for placing restrictions on access to the grid which the Investor Group alleges is action being taken in the interest of the coal-fired generators (Garnaut 2019). The AEMO has sought to get around the problem by regularising development processes and restricting the development of new farms to designated regional renewable energy zones (REZ), but it cannot hide from the fact that the designation of some sites as suitable, and others as not fit for purpose, is contestable, especially if development has already got underway. Moreover, State governments have to be party to REZ site selection, and this can introduce political uncertainties into the decision-making especially if a project is linked to the realisation of some promised regional development agenda. This amplifies the distributional implications that are rooted in the (dis)organisation of the *laissez faire* energy market.

There is a similar tension around the adoption of Demand Management as a strategy to minimise the call for additional investment in generating capacity. The management of demand, by compensating energy users suspending their contractual rights to draw on electricity, is presented as a means of reducing pressures on the network during those relatively limited periods when demand peaks to test the limits of generating capacity. While the AEMC and AEMO looked to all energy users participating in such a scheme, it was the larger energy users that were first approached. This reflected the likelihood that transaction costs in negotiating the variation in supply costs for a large volume user would be substantially less than for a comparable volume delivered by a numerically larger group of smaller energy users. This, needless to say, raises concerns about the potential inequities that will be produced with the demand management proposal. The large energy-intensive enterprises, such as aluminium smelters, would have to be handsomely paid to make it worth their while to agreeing to cut their energy intakes, and the compensation would certainly be more than what they pay for their power. While such arrangements are commercial-in-confidence, it is reasonable to assume that the rate of compensation would be in a range that was commensurate with the prevailing wholesale market price of electricity. These equity considerations are compounded by such enterprises already benefitting from contracting with generators and governments for the supply of power at discounted prices. Equity matters because, by way of contrast, under new rules AEMO or retailers have the authority to increase the call on households to feed greater amounts of power during periods of peak demand or blocking feed-ins when there is a surfeit of supply, and unlike the adoption of demand management for large energy users the new rules that apply to households do not include provisions for compensation, although the matter of compensation is being reviewed by the AEMC.[7]

[7]The AEMC is considering the household sector as a critically important source of electricity that could be drawn on to meet peak demand periods, although the AEMC proposal would be managed by aggregators rather than households.

The plans to ensure a greater degree of stability in the *laissez faire* energy market by restoring some semblance of centralised energy governance raises contradictory pressures that have inequitable consequences. The recommendations to invest in additional interconnectors that are designed to cohere the NEM and provide greater capacity that enables power to be transmitted across State borders to meet peak load demand is not necessarily all that efficient. Energy is lost to heat when transported over transmission lines (AER 2021). Pumped hydro is energy-intensive. In the case of Snowy Hydro 2.0, as much as 40 per cent of the power to be generated by the Snowy 2.0 pumped hydro project will be expended in pumping water through 27 kilometres of tunnels between the two reservoirs to enable pumped hydro. More energy will be lost in transmitting power to the distant key demand centres of Melbourne and Sydney. Nor does such investment in interconnectors to cohere the NEM obviate the need for more capital investment in distribution networks. New interconnectors will often be linked to transmission lines, such as is the case with the tie-in of Snowy Hydro with a new interconnector plan for South Australia and the renewable energy zones being developed in western Victoria and south-western New South Wales.

The reinstitution of energy market order is a primary responsibility of the AEMO and the ESB. One of their principal objectives is the determination to weave the REZs into an industrial-scale energy model that requires a greater degree of coordination than might otherwise have been the case. This is being progressed at some cost because it is delimiting the options available to market participants and, in some instances, happening in complete opposition to the signals that are at play in the energy market. AEMO's coordination of the market can compromise the voice of those energy users who are proactive. AEMO interventions, backed by changes to market rules, tend to introduce new obstacles to the freedom to choose how one engages with the NEM. The objective of strengthening reliability of supply affects the options available to households that have installed rooftop solar panels and battery systems and how they elect to utilise or dispose of energy. However, system reliability requirements have resulted in AEMO gaining the power to determine the timing and the magnitude of the energy flows between households and the market with these exchanges mediated by retailers. The formation of municipal council consortia to negotiate as a single entity, with the objectives of reliability and affordability as well as the intention of striking agreements for the power to be drawn from renewable sources, can be frustrated by AEMO's decision to hold off approving the new installation's connection to the grid. Community energy projects are in a similar position, requiring not only endorsement by the State-based regulators but the approval of AEMO as well.

Many of these anomalies are the consequence of the policy shortcomings that arise in the context of having an electricity market whose founding principles are well-defined but silent on the issue of emissions. The energy governance agencies must occasionally seek to respond to the tensions created by the issue of emissions affecting the dynamics of the market and, in the face of the policy void, this can complicate how the agencies reconcile extra-market phenomenon with the competition dictates while continuing to defer to the requirements of the ostensibly liberal market. Delays in the time required to get a development approval and begin generating electricity does not inhere much confidence in the NEM, and this can encourage investors to explore alternative means for getting around the delays. Power Purchase Agreements, based on direct negotiations between an investor/operator and the energy user, are not organised through but bypasses the market. This extra-market route to engaging in the developing national market has resulted in some 40 per cent of electricity supply contracts in recent times being resolved as PPAs according to the Clean Energy Council (2021).

The NEM appears to lack the market coverage that establishes its principles and practices with an institutional force and traction that was the promise of the liberalisation project. The NEM seems to be becoming an increasingly spent force and certainly not an arrangement that can begin to accommodate the climate change challenge which is after all the cause of energy users abandoning the market and contributing to its impoverishment. The efforts to recover some measure of the force of *laissez faire* points to the bankruptcy of the Coalition's conviction in the liberalised market ideal and its defence of coal as well as the means for reconciling the place of coal in the future shape of the energy economy. This energy market construction places coal either at the centre of the market or demands its excision, and it is the latter that is prevailing. Faced with this dark future, the coal-industrial complex's hegemony has been immeasurably weakened. This reflects both the force of alternative democratic ways of doing energy as well as an opening up of opportunities for other alternatives. The immediate focus turns to the realisation that the coal-industrial complex is in its death throes and the energy market is being radically transformed.

Normalising the Reckoning of Greenhouse Gas Emissions as Political Project and Risk Management

There have been several alternative market scenarios canvassed, though the extent to which these have had traction rests more on the material manifestations rather than to explicitly articulated shifts in market policy paradigms. There were obvious limits to the adoption of these alternatives. For a start, such alternatives could potentially be considered in violation of the spirit if not the letter of the electricity market law. As we have observed, the ACCC could hardly be expected to rule otherwise than determine that the formation of the energy user consortia was on balance not a process of cartelisation designed to distort market outcomes. The consortium, formed to try to negotiate price reductions charged to users. Incidentally, this also drew attention to the illegitimacy of price settings that were not consistent with conventional environmental economic theory, and this presented a radical challenge to the principles for the competitive NEM. Yet, it was highly unlikely that trying to change the rules governing the energy market to ensure some reckoning of the cost of emissions would, or could, be incorporated into financial reports to have much traction. As reasonable as a move to change the market rules may have seemed, as if it could reinstate some legitimacy on price formation, it was unlikely to have much appeal. It would entail a radical recasting of the philosophy that had entrenched the spirit of competition for over a quarter of a century, and which had informed the privatisation of state utilities, and in omitting any reckoning of the economic cost of emissions, was a crucial factor in the making of international competitive advantage.

A different tack to addressing this anomaly drew on an activist tradition of shifting the concerns around enterprise-generated emissions to raising questions about the enterprise's reputation by exposing how they were contributing to climate change. Over the course of the second decade of the new millennium, climate activists launched campaigns against individual enterprises that highlighted their contribution to the global fossil fuel supply chain. Companies were lobbied to divest from supporting the extraction and combustion of coal. In a shift in their strategic focus, activists turned to exposing the critical role that financial institutions were playing in fostering the development of coal mining ventures and coal-fired generators. Financial support and the provision of insurance accounted for a significant proportion of these institutions' business. They were initially resistant to the calls for them to divest their financial support in the industry, but the divestment campaigners persevered and began to have traction. Enterprises' annual general meetings provided occasion to protest and draw the broader community's attention to the cause. Led by various activist groups, with the Friends of the Earth offshoot Market Forces assuming a lead role, a series of determined campaigns, have been waged (Market Forces 2019, 2020a, b, 2021). The campaigns also connected with similar movements in other global climate action sites, so they have become international in their organisation. Market Forces, in particular, has brought considerable pressure to bear on the big four banks. In 2019, it 'filed resolutions with National Australia Bank, ANZ and Westpac calling [upon them] to align their lending and exposure to fossil fuel projects with the goals of the Paris accord' (Toscano 2019). Its campaigns to discourage banks, especially international banks, providing finance for new coal mines and coal-fired generators have been quite successful securing public commitment from several of the financial institutions to divest from coal-related ventures (Goh 2019; Clayton 2020; Morton 2020; Jamasmie 2021; Buckley 2018; Market Forces 2019, 2020a, b). One by one, the big Australian banks, and many global banks, have announced intentions to reduce or halt lending to coal mining ventures and coal-fired generators (Bloomberg NEF News 2021; Potter 2020; Verrender 2020). The Australian Bankers Association submits that the Paris Accord has forced a rethink on lending because of the increased risk associated with fossil fuels (Hyland 2021).[8] Insurance giants have been the focus of similar lobbying and leading global companies have agreed to no longer insure enterprises engaged in thermal coal extraction and coal-fired power (DW 2018; Insure Our Future 2018; Pinson 2018; Wilby 2018; AAP 2020). The increased significance of institutional investors, and especially superannuation funds, provided another source of leverage through divestment campaigns (Grieve 2020, 2021; Friedlander and Egerton-Warburton 2017; Gilbert and Tobin 2018).

[8] In fact, so successful has this focused campaign been that coal mining companies and coal-fired generators complain that banks and insurance companies are discriminating against them by refusing finance. They are appealing to expand the remit of key government agencies to provide financial assistance along similar lines to that provided for renewable energy projects by the CEFC (Ludlow and Macdonald-Smith 2016; Hyland 2021). The National Party is pressing the Coalition to provide government funds to underwrite the construction of a coal-fired power plant in northern Queensland to 'back-up' renewable energy capacity and to step to provide loans that now cannot be sourced through financial institutions (Coorey 2021; Murphy 2021).

The divestment movements soon extended their oeuvre with a strategic shift that brought climate activists closer to the board at annual general meetings through shareholder activism. Attention turned dissecting the financial statements and a corporation's provisions for risk management. Invoking the principle of corporate social responsibility invited the possibility of revisiting the issue of emissions and climate change as a corporate governance issue. This has driven a veritable sea change in attitudes with respect to accounting for emissions. Campaigning has gradually refocused the lobbying from outside the entrance of venues where annual general meetings were being held, to inside on the floor of the corporation meetings after activists bought small share parcels or obtained the proxy for other shareholders. This forced boards to respond and speak to the issue. This climate change activism begot shareholder activism and strengthened the pressure climate activists could muster by providing them with the opportunity to ask questions of the board and tease out more details of the enterprises' contributions to the forces driving climate change (Allens 2014; Leibler 2017). This began to turn the tide in the debate on the burning of fossil fuels. The risks associated with climate change have now been placed at the forefront of understanding the potential costs of an economy structured around the coal-industrial complex.[9] Environmental, societal and governance (ESG) issues are being taken up by shareholder activists, and climate activists have been at the forefront of this. According to the 2018 Gilbert and Tobin report on shareholder climate action, climate change has become one of the key focuses of ESG campaigns.

The focus on risk and disclosure has been reinforced with the formation of a number of business associations with the explicit purpose of pressuring corporate leadership on climate. The Investor Group on Climate Change, which represents institutional investors who have total funds under management of more than $2 trillion, has driven the rethinking in corporate Australia to report on the financial risks associated with their carbon footprint. The Australian Council of Superannuation Investors has set clear guidelines on climate risk disclosure obligations. Crucially, this is exposing the material and financial risks that climate change presents for companies and has been reinforcing action to mitigate this risk. The conventional wisdom is now accepting of the need to protect the long-term

[9] Market Forces is now turning its attention to lobbying gas companies to report on how they are planning to meet the Paris Accord emissions reduction target (Macdonald-Smith 2021).

value of the company by undertaking financial risk assessments and to reassess the potential costs of having capital locked in coal assets (Toscano 2019). The Gilbert and Tobin report notes that six of the top ASX 50 companies were pressured into addressing a climate change orientated resolution at their annual general meeting held for the 2018 financial year.[10]

These developments injected more momentum into the increasing attention given to climate change concerns across the corporate world. This has been particularly the case with institutional investment funds that have been awakened to the potential challenges and costs that could be associated with climate change. The increased momentum for changes in ESG reporting has been underscored by the efforts of umbrella groups projecting their activities into the global arena for improvements in prudential governance. In Australia, these and other international corporate organisations, such as the Carbon Disclosure Project, and more recently, Climate Action 100+, formed in 2017 are engaging with the world's largest corporate greenhouse gas emitters to curb emissions, strengthen climate-related financial disclosures and improve climate change governance. Companies and financial institutions have signed on to deliberations on standards being developed by the Task Force on Climate-related Financial Disclosures, and then there is the ASX's *Corporate Governance Council Corporate Governance Principles and Recommendations*, which sets out some benchmarks for risk disclosure reporting, including consistent climate-related financial risk disclosures, to inform investors, lenders, insurers and other stakeholders. The Australian Energy Transition Initiative has brought together 16 of the largest greenhouse gas emitters, including BHP, Woodside, BlueScope Steel, BP Australia, Fortescue and Orica, which account for one-fifth of Australia's industrial emissions, to become proactive in efforts to reduce emissions (O'Reilly 2021). The Australian chapter of the Climate Leaders Coalition is a larger grouping bringing together 33 corporations and organisations committed to meeting the Paris Accord's emissions reduction target set upon keeping the global temperature from rising above 2 °C.[11] Others have emerged out of

[10] A Full Federal Court ruled in the matter of Australian Centre for Corporate Responsibility v Commonwealth Bank of Australia [2016] FCAFC 80, that there were some issues that should remain the remit of the board of management and not shareholders unless otherwise stipulated in the constitution of the board.

[11] The Australian chapter of the Coalition includes 33 of the largest companies in Australia, which account for 22 per cent of the nation's emissions.

shareholder initiatives, grassroots activism 'from below', such as the Australasian Centre for Corporate Responsibility. Notwithstanding these declarations, most focus on reining in emissions to meet a 2050 net-zero target, the decidedly unambitious target, while prevaricating on committing to cut emissions by the earlier target of 2030 by which time runaway climate change may well be on course. This obviously raises the question as to whether these corporate initiatives are little more than a 'green washing'.

A similar refocusing has occurred within the state system, where there has been a dramatic change in rethinking the potential import of climate change on business and reporting. Notwithstanding the Coalition government's persistent resistance and procrastination in formulating clearly defined policies to mitigate risk, a panorama of government departments and agencies have begun to address the potential deleterious implications of climate change. Quite early in the piece, the Defence Department raised the possible strategic implications (Department of Defence 2016; Ferguson 2015; Reynolds 2020). As early as the *2010 Intergenerational Report*, Treasury was taking up the concerns that climate change would present (Treasury 2010). In 2019, the Reserve Bank of Australia's Governor called for energy companies to 'come clean' on carbon, having drawn attention in 2019 to climate change 'exposing financial institutions and the financial system more broadly to risks that will rise over time' (Debrelle 2019; Wade 2019; Williams and Rogers 2021). These interventions have roused concerns that companies have not been investing enough in climate risk assessments (Kadhem 2019).

This shift in the boundaries being set to redefine climate governance is also evident in the recommendations of a wide range of federal regulatory standards authorities which display a remarkable degree of convergence with industry associations on the need for companies to adopt reporting requirements regarding disclosures on emerging risks of climate change. The Australian Prudential Regulation Authority has advised all APRA-regulated institutions of its outline of plans to develop a prudential practice guide focused on climate-related financial risks, as well as a 'climate change' vulnerability assessment, and has now released for consultation its draft guidance to banks, insurers and superannuation trustees on managing the financial risks of climate change (APRA 2021). In 2018, the Australian Securities and Investments Commission turned its attention to governance and disclosure and financial risks and stability associated with climate change (ASIC 2019). The Australian Accounting Standards Board

(2018) and the Australian Auditing and Assurances Board (2019) are seeking more robust assessments of climate change financial risks. These new standards should drive action that takes into account the increasing non-viability of coal as an energy source. These developments in planning benchmarks reflect a consolidation of the overwhelming public sentiment in Australia that has been calling for action to deal with the threat of escalating climate change and the deleterious consequences of the reliance on coal as a principal energy source.

This cultural shift in how the state engages with the climate change challenge has brought an equally significant shift in emissions governance and is now being reflected in the assessment of development applications for emissions-intensive projects. Among the first noteworthy development was the Western Australia Environment Protection Authority issuing a revised Environmental Factor Guidance on greenhouse gas emissions in 2019 (Environment Protection Authority, Western Australia 2019). Ostensibly a technical manual on assessing Greenhouse Gas Emissions and appropriate measures for their abatement, the Guidance was explicitly designed to do what the national policy does not, and that is to provide greater certainty and 'a more stringent approach' by fossil fuel extractive enterprises to reducing and offsetting emissions with a view to meeting 'Australia's international obligations on reducing emissions' The Guidance sets clear thresholds on emissions standards and what mitigation measures will likely be required. While withdrawn by the Premier following industry protests, the EPA has indicated that, with government backing, it will proceed with industry consultations and re-introduce the guidelines (Environment Protection Authority, Western Australia 2021).

There was a similar watershed moment in New South Wales in May 2019, when the New South Wales Land and Environment Court rejected an appeal by Gloucester Resources challenging a decision by the planning department to not approve their Rocky Hill coal mine development application. The grounds for the decision were that the mine would have significant planning, amenity and social impacts, and, in a first, the ruling also declared that what is urgently needed 'is a rapid and deep decrease in greenhouse gas emissions … in order to meet generally agreed climate targets'. This was 'because the greenhouse gas emissions of the coal mine and its coal product will increase total global concentrations of greenhouse gas emissions at a time when … [the] dire consequences should be avoided' (McGowan 2019). In a subsequent case in the Federal Court brought by several children in an endeavour to block an appeal against the

New South Wales Independent Planning Commission's approving the expansion of Whitehaven's Vickery coal mine, the judge 'found that carbon emissions released from mining and burning fossil fuels will exacerbate climate change causing personal injury and death to Australian children' but the Court did not block the expansion, although the judge invited further submissions (AAP Medianet 2021; Readfearn 2021). In responding to the submissions, Judge Bromberg declared:

> that the Minister [for the Environment] has a duty to take reasonable care to avoid causing personal injury or death to Australian children when exercising her powers to approve the Vickery extension project under national environment law. (AAP Medianet 2021; Readfearn 2021)

The Minister intends to appeal the judgement (Australian Government News 2021).

These determinations, and especially the references to Australia's international obligations and future generations, have resonated within the resources and energy sector. The unanimity that had hitherto defined how the state and the mining and resources corporate world should approach the question of coal and climate change has fractured. The war of attrition between the coal stalwarts and climate change activists has heated up and the tide has turned against the coal-industrial complex. Climate activists have concentrated their strategic focus on the industry associations, the Australian Coal Association (ACA) and the Minerals Council of Australia (MCA). Facing sustained criticism for its hard nosed, 'take no prisoners' approach to lobbying on behalf of the black coal industry in Queensland and New South Wales, and its leadership in blocking the Gillard Labor government's proposed resources rent tax, activists question the organisations' social legitimacy in making a case for 'clean coal'. In 2013 the ACA was merged into the MCA, but this only served to narrow the Council's lobbying for coal and provided a bigger target for the climate campaign against, exposing the MCA to criticism in the public domain. The Australasian Centre for Corporate Responsibility ramped up its efforts in impressing upon institutional investment funds the urgent need to pressure companies to end their membership of the Minerals Council of Australia. BHP was the first and most recognisable resources corporation to respond to the calls that it should demonstrate its commitment to emissions mitigation' by ending its membership of the Minerals Council of Australia and other coal lobbyist organisations. Keen to present itself as a mining company with a sustainable development resolve, BHP resigned from the World Coal Association and the ensuing public debate about the future of the MCA forced a change in its leadership (Thomson 2018; Readfearn 2019). Supporting the development of the resources industry more generally rather than coal in particular, the Council has sought to sterilise criticisms of its advocacy of coal by publicly backing the *Roadmap* (Minerals Council of Australia 2019).

This subtle shift in the pro-coal lobby is not without its difficulties. The Coalition is increasingly at odds with industry's declining confidence in their being a future for coal. The call for state backing of coal-fired power plants sits uncomfortably with preeminent energy corporations in Australia having begun to abandon their interest in coal. Rio Tinto has sold its investments in thermal coal and has committed to disengage from the fossil fuel resource sector (BHP 2020; Ker 2019; Ker and Stevens 2017; Rio Tinto 2018). BHP also began a medium-term restructuring of its business focus indicating its intention to withdraw from mining and exporting thermal coal. It recently sold its interest in the Cerrejón thermal coal project in Colombia to Glencore and is predicted to dispose of its coking coal interests (Zhou 2021; Ker 2019, 2020; Toscano 2021; McCarthy 2021). Both mining behemoths recognise they have contributed to the increasing concentration of greenhouse gases in the global atmosphere and that they have resolved to meet the Paris target and will work with downstream supply chain partners to cut 'scope 3 emissions' (Rio Tinto 2020). In line with its pledge to become a net-zero-emissions enterprise, BHP was among those that felt the need to signal its displeasure with the Mineral Council's robust pro-coal profile and forced a reorientation of the Council's industry development agenda (Lewis and Jessop 2019).[12] The decisions to divest follow a concerted campaign by climate activists waged throughout much of the 2010s to lobby resource corporations to end coal mining and to pressure financial institutions, the banks, investment and funds management companies and shareholder funds to not invest in any new coal-fired power stations.

Another iconic symbol of the retreat from coal has been AGL's decision to split the company into two separate legal entities (Macdonald-Smith 2020). As the single biggest emitter in Australia, AGL will hand the

[12] The Minerals Council subsequently released its Climate Action Plan in 2020, a report on meeting the Paris Accord (Foley and Toscano 2020; Minerals Council of Australia 2018, 2020).

mantle for a carbon neutral enterprise managing its retail and renewable generation facilities and gas generation businesses to AGL Australia, while the coal-fired power plants will be run by new company, Accel Energy.

The Coalition government has lost any credibility in its capacity to lead the nation on energy and climate change policy. In the lead up to the COP26 November meeting in Glasgow, peak business associations have voiced their strong concerns that the Coalition government has to submit a concrete plan and drop the hubris, that Australia will meet its commitment 'in a canter'. Many of Australia's largest corporations have outlined robust emission mitigation plans in line with COP26. In October, the Business Council of Australia released a Blueprint *Achieving a net zero economy* which mapped a transition path for key sectors (Business Council of Australia 2021). The Australian Industry Group called for a plan that would move beyond the target of reducing emissions by between 26 and 28 per cent by 2030 from to be on track to achieve net-zero emissions by 2050 (Willox 2021).

State governments have continued to press ahead in setting ambitious emissions reduction targets and in promoting renewable energy generation. The ACT, Victorian, South Australian, New South Wales and Tasmanian governments have each dissociated their ambitions from the federal government in setting net-zero emission targets by 2050 or earlier (Williams 2021a; Department for Environment and Water, Government of South Australia 2021; Department of Planning, Industry and the Environment 2020; Point Advisory 2021).

The Material Manifestations of Coal's Demise: The Ascendancy of Wind and Sun and the Makings of a Renewable Energy Future

Over the course of the preceding chapters, we have identified various responses to the liberalisation of the Australian energy market that do not fit with the contention that the 'market framework remains fit for purpose'. Historically, the reliance on coal has been justified by a group of hard-core resource industry champions and econocrats arguing the case for exploiting Australia's international comparative advantage based on the rich endowment of resources, including coal and now gas. Figure 2.1 charts how, with state support and the refusal to put a price on the emissions generated in the process of producing electricity, has translated into

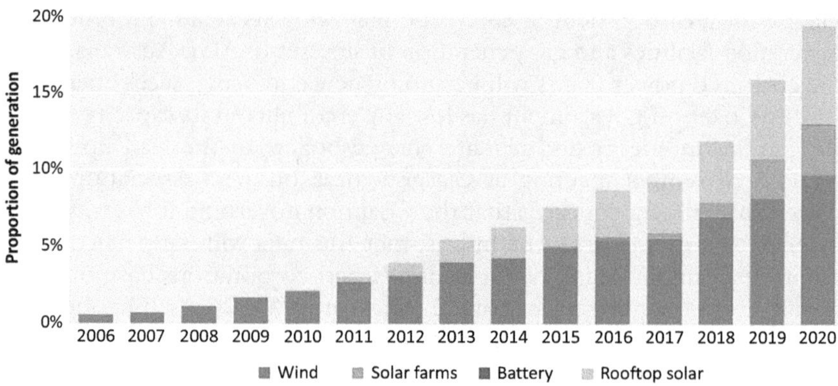

Fig. 13.1 Renewable electricity generation in the National Electricity Market. (Source: Australian Energy Regulator State of the Market 2021)

an international competitive advantage, evident in coal produced for domestic consumption and in the sustained increase of coal exports feeding the development of East Asian economies. Figure 2.1 is testimony to the argument developed that the privileged position of coal in the Australian political economy started to unravel over the course of the 2010s.

A clear impression of the declining significance of coal as a source of electricity can be gleaned by reference to Fig. 2.1, which maps the increased significance of renewable energy generation, and this is underscored by the fact that there has not been any investment in new coal-fired generating capacity since 2012. Figure 13.1 identifies the different sources of renewable energy that are challenging the material foundations of the hegemony of the coal-fired *laissez faire* market.

Household investment in rooftop solar voltaic panels continues to grow at an exponential rate and shows no signs of abating (Clean Energy Council 2021). Investments in wind farms dominated the increase in utility-scale renewable energy generating capacity while utility-scale solar farm investments have taken off as the cost of this technology has decreased. This resulted in a fourfold increase in the relative significance of renewable energy over the last decade, accounting for a little less than 5 per cent of all energy generated in 2010 to almost 20 per cent in 2020 and as a proportion of generating capacity, this is predicted to climb to as much as three-quarters in the next few years according to then chief executive officer of AEMO. There was, however, a fall-off of investment in wind and solar farms over 2020 as signs of excess generating capacity become apparent and with the decrease in power prices, which at times have been negative. The lack of policy clarity has been a contributing factor to this (Clean Energy Council 2021). Battery storage is only just starting to figure as an important source of energy, and governments are placing more interest in supporting the uptake of batteries by households, in communities and as utility-size storage back-up capacity for industrial-scale renewable energy (Ludlow 2021d; Williams 2021b). Batteries are proving to be an important stabiliser in providing frequency and ancillary services, especially with utility-size batteries, and one of the more lucrative aspects of the transitions (Clean Energy Ministerial et al. 2021).

The failure to include some reckoning of the environmental externalities in the design of and the rules governing the market was justified in terms of an impost that would undermine Australia's international competitive advantage or, at least, Australia's energy-intensive industries. This shortcoming is now being addressed, albeit taken up in a different language as a risk management concern. This is proving a gamechanger because securing finance for coal mining and coal-fired power plants has become more difficult as financial institutions have adopted more circumspect-lending practices (Hyland 2021; Macdonald-Smith 2021f).

Notwithstanding the significance of this linguistic shift, the conventional wisdom has been contested by the formulation of doing energy differently, in ways that are more inclusive, more democratic, and which subscribe to more equitable interactions with energy to invoke climate justice as a corollary of climate democracy.

The NEM's Emergent Antagonism: Beyond the Market and the Socialisation of Market Relations

A striking feature of the diminishing confidence in the NEM is that it does not seem to have registered with the AEMO and the ESB that there have been two critical developments that are at odds with the workings of the conventional economic construct, the *laissez faire* market in which the caricature *homo economicus* navigates his way through the options that present. This model provides next-to-no space for considering that there might be alternatives to navigating within the realm of the market. Yet, a stand-out feature of the turn to renewable sources of electricity is the extent to which participants are exploring or reaching outside and beyond the market to meet their power needs. The evolution of the prosumer captures quite neatly this metamorphous of exchange relations. Household investment in the installation of rooftop solar panels, and now the increasing interest in battery storage as a supplement, is unique to Australia's place as the most significant per capita household rooftop generator in the world. The relative significance of household power generation compared with utility-size wind and solar farms is captured quite neatly in Fig. 13.1. The pace of rooftop panel installation has not levelled off, but rather picked up. This reflects a radical cultural shift in residential Australia, transforming the residents from occupying a household as a passive site of consumption into prosumers and establishing the household as a site of enterprise. The investment provides the means of energy generating capacity that enables the residents to exercise greater control over fulfilling their energy needs as well as the option of feeding power into the grid as a supplementary source of income, although this option is being challenged by the reduction in feed-in tariffs and the threat of households being charged for the ability to fed power into the grid. The household can also decide to exercise independence, to defect, from the grid and the prospect of having to pay to be connected will serve to encourage defection. And all the while, the household is reducing its carbon footprint.

The new-found capacity for autonomy from the grid underscores two remarkable developments. Place in the energy system has been refashioned into more social forms of interacting around the generation and distribution of electricity, including the possibility of being able to disengage from the NEM. The ability of households to generate electricity frees the residents from being dependent on the market to meet their energy needs. This is not only with respect to the purchase of power at an affordable price but also the opportunity to restrict their electricity use to power from renewable sources, and this can provide opportunities for deepening the social character of endeavours to decarbonise. Moreover, because the exercise of climate agency residence can manifest in enlivening citizenship, establishing connections across the value supply chain can be combined with households doubling up on pursuing their aspirations in the public sphere of the political economy. Agency can be expressed across a range of planes. The historically distinctive feature of the energy market is that what were essentially contracts of supply resting on the anonymity of market participants based on individualised contracts are being displaced by household engagement with energy as a decidedly more social process.

The energy system now provides a range of options for the solar household on how to engage with others in decarbonising and demonstrating ecological sensibility, or some combination of these (Clean Energy Council 2021). Housing project developers are incorporating the principle of shared, socialised energy generation and storage capabilities into the design of multi-occupancy residential developments and estates (Johnson 2019; The Urban Developer 2019) Several State governments are incorporating energy generation and storage capacities as aspects of energy self-sufficiency into community housing projects (Mazengarb 2019; Synergy 2019).

State governments have also been providing support to help the implementation of community energy initiatives as regional communities have become more outspoken on the need for government to show some leadership in underwriting community projects. The state has been fostering the development of climate agency in communities by extending the reach of social projects that bring households together to develop collective responses to the climate change challenge. One model outcome has involved funding the establishment of community renewable energy hubs, information centres, that develop the capacity to identify and support and coordinate the development of local grassroots networks ideas for energy decarbonising, self-sufficiency and autonomy through to project stage. The Victorian Labor government has financed the setting up of several energy hubs to mobilise resources among local activists and support local environment groups to develop community-owned and controlled renewable energy projects in regional Victoria (Sustainability Victoria 2021a, b, c). The scheme has helped enhance the ability of groups of households forming into communities to work together to provide more sustainable sources of electricity and more resilient communities. The energy hub social enterprises are filling a gap in the provision of reliable power supplies by elevating efforts to support transitioning from the reliance on ageing distribution infrastructure and the limited service reach of the NEM in regional and remote areas to local renewably sourced energy. To date, the hubs seem to have had limited traction in fostering and stimulating energy autonomy and in promoting greater equity in accessing reliable power supplies, although one region highlights the potential of the energy hubs idea. The unparalleled enthusiasm for community renewable energy programmes in and around Yackandandah in north-east Victoria is one success story. Community confidence in the future of zero-emissions energy generation is evident in a host of neighbouring communities exploring options for investing in greater energy self-sufficiency.

The phenomenon of more social engagements with transitioning is also being replicated in state initiatives to promote the potential for small communities to consolidate a collective interest in solar panel installation and wind power with the installation of community batteries. The proposals are being trialled with the expressed purpose of utilising the storage to help stabilise the market. The cooperation of renewable energy companies in the Mooroolbark Community Mini-Grid trial is an example of the development of a decarbonised energy self-reliant community that is not reliant on being connected to the grid (Greensync 2016). The Western Australian government is more advanced in exploring community-based options in remote areas. It has been trialling the installation of solar and wind power units that are linked to storage capacity to permit these communities to operate off-the-grid generation systems, be electricity self-reliant and do away with the need for diesel-powered generators, and this is providing some lessons for the east coast (Western Power 2020; Parkinson 2020a). Other States have introduced similar schemes, although some are more ambitious and larger and industrial in scale than others (Parkinson 2020b, c).

The installation of community-scale battery storage systems is proving to be another appealing variation on the theme of socialised responses to transitioning from emissions-intensive sources of power. The installation of battery storage facilities in urban areas aims to provide the means of promoting more reliability and stability in the market. Households can pool and store some of their generating capacity, which can be drawn on to assist in insulating households from some of the vagaries of the market as well as levelling out recourse to the grid (Department of Environment, Land, Water and Planning 2021). Energy corporations see advantage in diversifying the services they provide—and, no doubt, regard such diversification as crucial to maintaining a place in the market. EnergyAustralia is experimenting with a trial involving connecting rooftop solar generating capacity with community batteries in Sydney (Kurmelovs 2021). The Labor Party has flagged a potential policy to finance the installation of community batteries. If elected to government, $200 million would be allocated to finance some 400 community battery systems, as a component of its 'Power to the People' programme (Mazengarb 2021a).

Such projects reflect the more active interest of residents as climate citizens in local government politics. This projection of the climate citizenship voice has fermented debate within local government on formulating decarbonisation initiatives that has seen the formation of consortia based on regional climate alliances. These have been morphing in rhizomatic fashion, steeling this level of government to give form to climate change concerns and to build upon these shared sentiments to mount concrete challenges to the coal-industrial complex. The Victorian model demonstrates what is possible in collective action, and the Local Government Association is contributing to support efforts by councils to place climate change as the preeminent item to be actioned on their calendar. Through the efforts to reduce the municipal councils' and communities' carbon footprint, municipal councillors have demonstrated the logic in developing more social forms of approaching the climate change challenge. The socialisation of the calls for action on climate change is manifest in council initiated PPAs demonstrating the variety of ways in which the transition to renewable energy is being progressed by engaging multiple partners, taking climate politics into new domains. This has proved strategically significant in revealing the merits in focusing on more than just the price of the commodity. The Queensland government's plan to hasten that State's transition to a low-emissions trajectory, based on fast-tracking the construction of renewable energy generating capacity and amortising the cost of doing so and doing so by means of a PPA was a straight-forward means of accelerating the decarbonising of the State energy system (Power Technology 2020a, b; Vorrath 2020b).

These more social forms of the contractual relations driving the turn to renewable energy can illuminate structural inequities in energy market relations and demonstrate greater candour in deliberations over what is to be transacted. The NEM invites neither of these possibilities because it is organised based on meeting the competitive impulse criterion, and the presumption that the 'invisible hand' will expunge anti-competitive practices, as the ESB suggests. It also conceals the substance of what is being traded, what the then Treasurer, now Prime Minister, holding up a lump of coal in the parliament declared as something not to be afraid of: 'This is coal, don't be afraid' (9 February 2017). The NEM provides very few vantage points from which to look beyond this thing that has 'ensured for over 100 years delivered an energy competitive advantage that has delivered prosperity for Australian businesses'.

The more social forms of contractual relations that tend to be the bases of PPAs have implicitly looked beyond the appearance of the coal commodity, because the fetishisation of this 'thing' that we should not be afraid of, concealed an ugly truth. When combusted in the process of generating electricity, coal is Australia's most significant source of greenhouse gases as it was for the global economy, and it is the combustion of coal that is the premier force that is driving climate change. This was not new news. It had been the object of ongoing efforts to expose and undertake some accounting of emissions well before Kyoto. There was always a possibility for an emissions trading scheme being adopted in Australia, especially when UNFCCC governance of emissions was captive of an economic orthodoxy that demanded governance be subsumed within the free market. But this was feint hope given the constitution of the National Electricity Market.

Relief from this economistic sentiment that saw only the bases of competitive advantage and wealth creation being delivered from the extraction, combustion and export of this commodity could not be sought through the NEM. The NEM was closed to this possibility, and this is why engagements beyond the market are so significant. They challenge the narrow, instrumental and anti-environmentalism that is the foundation of the coal-industrial complex. State governments have been more cognizant of this as have local municipal councils which have both intervened in market dynamics to inspire investment in rebuilding an energy system that draws on the bounty of the ecology, the wind and sun in particular.

It is somewhat of a paradox in this story of transitioning that there has been more interest demonstrated in Australia's natural bounty among global capital than Australian resident companies. Not uncommonly, much of the investment in developing the renewable energy potential has been based on international corporation in partnerships—another manifestation of how the climate change challenge is fostering more social forms of engagements. Ranging from established energy corporations, corporations that had built up their reputation by investing in the development of renewable energy technologies, through to the gamut of international financial institutions and investment funds that have been investing in wind and solar farms in Australia, ventures that have an enduring future, international capital has crafted a paramount presence on the ground floor in the turn to renewable energy. In a revisiting of the *Lucky Country*, domestic capital has tended to take for granted the store of

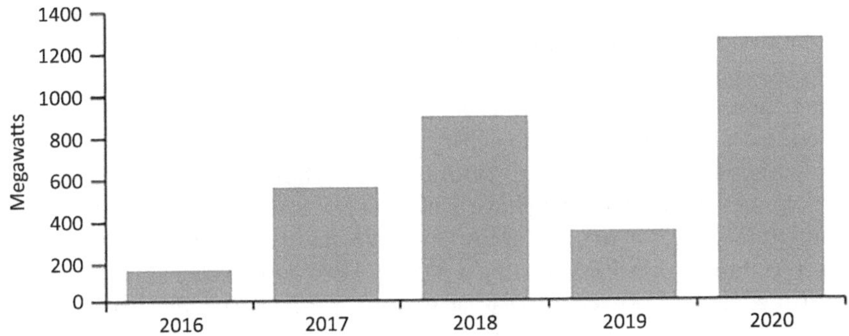

Fig. 13.2 Corporate Power Purchasing Agreements, Australia 2016–2020. (Source: Clean Energy Council (2021) *Clean Energy Australia Report 2021*)

wealth that could be extracted, and it has been international capital that has been spearheading the new bases of wealth creation.

One qualification to this has been the growing significance of Purchase Power Agreements as a means of funding investment in wind and solar farms. Figure 13.2 maps the import of PPAs as an instrument that enables more social forms of engagement between generators and users, and which provides the bases for concluding medium- to long-term power supply contracts. Contracts of such duration which contrast with the NEM wholesale spot market with clearing prices averaged over every five minutes. PPAs provide much needed certainty in the revenue flows that can be expected from an investment in a wind or solar farm and thus underwrite the necessary confidence that an investment to expand generation capacity will deliver a reasonable return. But they also underwrite the commitment to transitioning, and this has been a paramount motive of municipal councils which have shown significant leadership in striking Purchasing Power Agreements that have accelerated the progress of the transition from coal according to the clean energy (Clean Energy Council 2021).

The Socialisation of the Climate Change Challenge—Constraining the Reach of the Market

The formation of consortia highlights the limited efficacy of the *laissez faire* market in dealing with the inequities that are reflected in the asymmetry in the power relations within the NEM and the uncertainties that flow from this. In partnering with other energy users, Power Purchasing Agreements have strengthened the bargaining hand of electricity users. Councils have exercised their organisational capacity to give voice to their constituents' predilection for a low, if not, net-zero-emissions neighbourhood. Local and State and Territory governments have each backed these citizen endeavours to cut the community's carbon footprint. And it is not just municipal councils that have benefitted from such contractual opportunities. Private for-profit enterprises have also found to their advantage and reputation that PPAs can be a ploy to fast-tracking the shift from relying on coal-fired power to solar or wind power. There were 26 corporate renewable PPAs announced in 2020, directly contracting 1.3 GW and supporting more than 4.5 GW of renewable energy generation, a slight increase on previous years but notable because of the resilience of this arrangement. Corporate PPAs are accounting for some 40 per cent of the market volume (CEC 2021).

The positive consequences of this are mapped in Fig. 13.3, which highlights the extraordinary increase in renewable energy generating capacity. Compare this with Fig. 2.2, which indicates the extent to which investment in coal-fired generating capacity had been abandoned. All the indications are that PPAs will continue to grow given the escalation of pressure to meet risk management standards, and this is important because PPAs provide a level of confidence that emission reduction targets can be met more quickly by negotiating explicit timeframes for the supply of energy from renewable sources.

Moreover, the determination to meet such targets seem to outweigh other considerations, as the Clean Energy Council has noted: Those

> [b]uyers [organising PPAs] are less motivated by immediate electricity price pressures and more by ambitious sustainability targets and/or a longer-term perspective on the value of the PPA as a hedge against future electricity price volatility. (CEC 2021: 47)

There is thus a greater likelihood that renewable energy will continue to attract capital.

A remarkable feature of the turn to renewable energy is the increasing proportion of power that is being generated but not being fed through the wholesale market. Clean energy proponents, householders, citizens and for-profit business enterprises view the NEM as an institutional barrier to progressing the transition to a zero-emissions future (Clean Energy

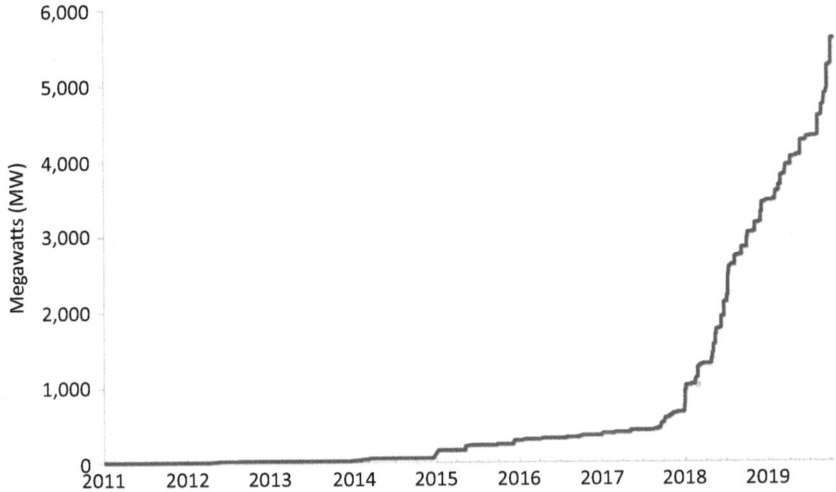

Fig. 13.3 Cumulative capacity of Clean Energy Regulator accredited large-scale solar power stations. (Source: Clean Energy Regulator (2020), Historical Large-scale Renewable Energy Target supply data. Figure 3.7 Department of Industry, Science, Energy and Resources (2020))

Council Report 2019: 40). Navigating energy generation and distribution as a social process, rather than transactional in the NEM, appears to be involving more diverse parties who seek to bypass the formally constituted market.

Market relations are being re-engineered with extra-market considerations engendering the recasting of the technological bases of energy generation with the objective of bringing an end to electricity generation as the preeminent source of emissions and climate change. Different parties that hitherto had no commercial reasons for joining forces are now partnering with others to both strengthen their bargaining power in negotiating the supply of electricity and, in aggregating the volume of power they want supplied as a class of users, determining the mode of electricity generation. Energy transactions have become framed by socially defined ambitions. As has been the case with households installing rooftop solar panels and, increasingly, battery storage systems, the decisions to invest are motivated by a more considered and conscientious endeavour to take control of the bill and doing so in a way that is decarbonising. As is evident from

Fig. 13.1, this social transformation is becoming more comprehensive in that more households are enlisting in the reconstitution of energy social relations to give expression to a common cause and collective will.

This is also setting an example that others are taking up, forcing the pace of the energy transition involving energy consumers becoming electricity generators and exercising control over energy. The initiatives to establish virtual power plants are not simply moves to strengthen the coordination of power flows and orchestrate stability in the NEM. The push to establish virtual power plants is as much a process of socialising market interactions designed to re-assert the pivot of corporate power in the market. The same holds for the scenario that Schott painted of intermediaries doing what the household cannot readily do and that is to transact efficiently and optimally. This notion of the efficient market at work in calling forth these intermediaries overlooks that the insertion of this third dimension, as it is for the originator of the virtual power station, a grasping of an opportunity to capitalise on the benefits of the household generating energy electricity.

Curiously, it does not seem to have registered with the AEMO and the ESB that an increasing range of energy users, and this includes government and private enterprises, are choosing to look beyond, even exiting, the market to meet their power needs. From households to communities to municipal councils to for-profit enterprises, there have been a growing number of energy users that are themselves electing to engage in the generation of their power needs or linking directly and contracting with energy generators to supply power but doing so outside the bounds of the NEM.

These include the record number of households investing in rooftop solar and battery systems and forming virtual power plants, the State and Territory government and municipal governments involved in extra-market interventions and contracting directly with generators, and the formation of energy cooperatives defecting from the grid. The distinguishing feature of these arrangements is that they are predicated on a spirit of cooperation. The cooperation is reflective of what O'Connor represented as the more social forms of interaction that can emerge in moments of crisis. The socialisation of the market can be posited as response to the disorder engendered by the generation of emissions and to the contradictions spawned by the liberalised energy market.

The transformation from the fossil fuel energy system to reliance on renewable energy is based on a technological transformation but it is one that is being driven by the social responses to the threat of climate change. The transformation is being led as much from 'below' as it is by the corporate sector, and as has been sketched this entails a range of ways of doing energy, more transparently, less transactionally and more socially, and democratic and empowering. Looked at from the perspective of the diverse social forms and the range of technologies these engage in generating energy, a different, though contested, future is unveiled. It is one that invites the possibilities for foregrounding the democratisation of energy and reigniting the commitment to energy justice.

References

AGL (2021) 28 April.

Allens (2014) 'Shareholder activism in Australia' 24 July.

AAP [Australian Associated Press] (2020) 'First State to divest thermal coal assets' 9 July.

Australian Accounting Standards Board (2018) 'Climate-related and other emerging risks disclosures: assessing financial statement materiality using AASB Practice Statement 2' 2 December.

AAP MediaNet [Australian Associated Press] (2021) 'Climate Change Duty of Care Enshrined into Law' 8 July.

Australian Auditing and Assurances Board (2019) 'Climate-related and other emerging risks disclosures: assessing financial statement materiality using AASB/IASB Practice Statement' 2 April.

Australian Energy Regulator (2021) *State of the Energy Market 2021.*

AER [Australian Energy Regulator] (2021) *Draft distributed energy resources (DER) integration expenditure guidance note*, July.

APRA [Australian Prudential Regulation Authority] (2021) 'APRA releases guidance on managing the financial risks of climate change' 12 March.

ARENA [Australian Renewable Energy Agency] (2020) 'Regional NSW's renewable energy sector is booming, with local demand for wind and solar energy trebling over the last five years' 7 February.

ARENA [Australian Renewable Energy Agency] (2018b) 'Tasmania interconnector project sets sail' 18 September.

ARENA [Australian Renewable Energy Agency] (2018a) 'Hydrogen to become Australia's next big export?' 20 August.

ARENA [Australian Renewable Energy Agency' (2017) 'Can we export renewable energy' 23 July.

ASIC [Australian Securities and Investment Commission] (2019) 'ASIC updates guidance on climate change related disclosure' 12 August.

Bakhtar News Agency (2021) 'South Australia to install free Tesla batteries in homes without rooftop solar' 2 June.
Baxter, T. (2019) 'The emissions reduction fund was deeply flawed—and no rebranding will change this' *The Guardian* 25 February.
BHP (2020) 'BHP Climate Change Briefing' 10 September.
Bloch, M. (2020) 'Queensland's First Community-scale Battery Project Nears Prime Time' *Solar quotes blog* 8 January.
Bloomberg NEF (2021) *Climate Policy Factbook.*
Boot, S. (2018) 'Inside the economics of Snowy 2.0' *The Saturday Paper* 15–21 December.
Buckley, T. (2018) *Over 100 Global Financial Institutions Are Exiting Coal, With More to Come* Institute for Energy Economics and Financial Analysis.
Business Council of Australia (2021) Achieving net zero economy October.
Clayton, R. (2020) 'ANZ's climate policy reveals steps away from coal to support net zero emission by 2050' *ABC News* 29 October.
Clean Energy Council (2021) *Clean Energy Council Australia 2020.*
Clean Energy Ministerial, Bloomberg Philanthropies and Bloomberg NEF (2021) *Energy Transition Factbook*, Prepared for the 12th Clean Energy Ministerial 31 May–6 June 2021.
Coorey, P. (2021) 'Government should underwrite coal-fired power: Nats' *The Australian Financial Review* 30 June.
Debrelle, G. (2019) 'Climate Change and the Economy' *Reserve Bank of Australia* 12 March.
Department of Defence (2016) *Environmental Strategy 2016–2036.*
Department for Environment and Water, Government of South Australia (2021) *South Australian Government Climate Action Plan 2021–2031.*
Department of Environment, Land, Water and Planning, Victoria (2021) Victorian Neighbourhood Battery Initiative.
Department of Environment, Land, Water and Planning, Victoria (2020) Traditional Owner Renewable Energy Program 30 November.
Department of Industry, Science, Energy and Resources (2020) *Growing Australia's Hydrogen Industry.*
Department of Planning, Industry and the Environment, New South Wales Government (2020) *Net Zero Plan Stage 1: 2020–2030.*
DW (2018) 'Allianz stops insuring coal companies' 4 May.
Eddie, R. (2021) "I didn't expect it to be possible': The Melbourne apartments using solar' *The Age* 27 June.
Energetic Communities (2021) 'Sewing a common thread'.
Environment Land Water and Planning, Victoria Government (2021) *Victorian Renewable Energy Zones Plan: A direction paper* February.
Environment Protection Authority, Government of Western Australia (2021) Review underway of greenhouse gas guidance 30 June.

Environment Protection Authority, Government of Western Australia (2019) *Environmental Factor Guideline: Greenhouse gas emissions* 7 March.

Ferguson, P. (2015) Submission to the 2015 Defence White Paper Consultation Process, Centre for Policy Development.

Foley, M. (2021) 'Snowy Hydro rejects modelling showing Kurri Kurri gas plant would lose $3b' *The Sydney Morning Herald* 14 July.

Foley, M., and N. Toscano (2021) 'Falling cost of renewables creates coal test for federal government' *The Age* 2 July.

Foley, M., and N. Toscano (2020) 'Miners' late no towards climate risks not enough for investors' *Thee Sydney Morning Herald* 22 June.

Foote, C. (2020) 'Snow job: why the cover-up over Snowy 2.0?' *Michael West Media* 11 August.

Friedlander, D., and M. Egerton-Warburton (2017) 'Shareholder activism' Australian Institute of Company Directors 1 May.

Garnaut, R. (2019) *Superpower: Australia's low-carbon opportunity* La Trobe University Press.

Gilbert and Tobin (2018) 'Australian Shareholder Activism Report 2018'.

Goh, Jassmyn (2019) 'Commbank to exit thermal coal by 2024' *Money Management* 9 August.

Greensync (2016) 'Greensync partner in community microgrid trial' 1 March.

Grieve, C. (2021) 'Funds back push for firms to air climate change risks' *The Sydney Morning Herald* 29 June: 30.

Grieve, C. (2020) 'How the global divestment push is testing Australia's resolve' *The Sydney Morning Herald* 7 March.

Hannam, P. (2021) 'Rooftop solar still booming but big renewable projects dropping' *The Sydney Morning Herald* 8 July.

Hicks, J., N. Ison, P. Gilding, and F. Meys (2014) 'A short guide to establishing a community renewable energy project' NSW Office of Environment and Heritage.

Hyland, A. (2021) "Savage blow': Coal companies hit back over banks' flight from fossil fuels' *The Sydney Morning Herald* 15 May.

Insure Our Future (2018) 'Leading insurers undermine climate goals by supporting the tar sands industry' 30 May.

Ison, N., and Langham (2015) 'Communities are taking renewable energy into their own hands' *The Conversation* 18 June.

Jamasmie, C. (2021) 'Australia's Macquarie to exit coal by 2024' *MiningDOTCom* 7 May.

Johnson, P. (2019) 'Sustainability served seven ways in new Nightingale project' *The Fifth Estate* 1 March.

Jotzo, F. (2020) 'Morrison government dangles new carrots for industry but fails to fix bigger climate policy problem' *The Conversation* 20 May.

Kadhem, N. (2019) 'Australia's biggest companies failing to plan for climate change risks: report' *Renew Economy* 18 February
The Kenny Report (2021) 'ANZ should play to their 'quiet majority' shareholders not their 'activists' Sky News Australia 7 April.
Keogh, J. (2021) 'Big investors warn Australia on climate change' *The Australian Financial Review* 5 July.
Ker, P. (2020) 'BHP confirms coal exit as profits underwhelm' *The Australian Financial Review* 18 August.
Ker, P. (2019) 'BHP says cutting emissions good for' *The Australian Financial Review*.
Ker, P., and M. Stevens (2017) 'Rio Tinto sells thermal coal assets to Yancoal' *The Australian Financial Review* 24 January.
Knaus, C., and L. Cox (2021) 'Coalition granted $21m to Liberal party donor to frack Beetaloo Basin' *The Guardian* 8 July.
Kurmelovs, R. (2021) 'Community batteries: What are they and could they help Australian energy consumers?' *The Guardian* 5 April.
Leibler, J., and A. Bloch Leibler (2017) 'Shareholder activism in Australia: A review of trends in activist lending'.
Lewis, B., and Jessop (2019) 'Top BHP investor piles on climate pressure ahead of shareholder meeting' *The Sydney Morning Herald* 10 October.
Ludlow, M., and A. Macdonald-Smith (2016) 'AGL Energy boss says new rules needed for wind, solar' *The Australian Financial Review* 27 July.
Macdonald-Smith, A. (2021f) 'Alinta calls for federal fund as banks desert coal power' *The Australian Financial Review* 16 June.
Macdonald-Smith, A. (2020) 'AGL under fresh fire on coal as BP raises the bar' *The Australian Financial Review* 5 August.
Market Forces (2021) *Is your bank investing in dirty fossil fuels?*
Market Forces (2020a) 'Macquarie fails to commit to coal and gas phase outs' 30 July.
Market Forces (2020b) 'Media Release: ANZ, NAB face climate change shareholder' 15 December.
Market Forces (2019) 'Analysis: Singapore banks signal an end to new coal financing' 18 April.
Mazengarb, M. (2021a) 'Federal Labor promises to slash taxes for electric vehicles, build community batteries' *The Guardian* 30 March.
Mazengarb, M. (2019) 'Solar and storage to power NDIS in first of its kind pilot' *Renew Economy* 8 May.
McCarthy, J. (2021) 'Analysts tips BHP about to sell its prized BMA coal assets' *In Queensland* 22 March.
McGowan, M. (2019) 'Court rules out Hunter Valley coalmine on climate change grounds' *The Guardian* 8 February.
Minerals Council of Australia (2020) *Climate Action Plan*.

Minerals Council of Australia (2019) 'Submission: Updating the Climate Change Authority's on meeting Australia's Paris Agreement commitments' 9 September.

Minerals Council of Australia (2018) *Energy and Climate Policy.*

Moore, T. (2020) 'How Queensland could become an energy superpower' *The Sydney Morning Herald* 19 August.

Morrison, S. (2021) 'Remarks: Minerals Council Parliamentary Dinner' Transcript 3 June.

Morton, A. (2020) 'Westpac to withdraw from thermal coalmining after 2024' *The Guardian* 4 May.

Murphy, K. (2021) 'Liberal MPs scorn National's $250bn plan for taxpayers to underwrite fossil fuels' *The Guardian* 7 October.

Narayan, M. (2019) 'The Demise of Coal' Energy Networks Australia 18 July.

O'Reilly, G. (2021) 'Mitigate the risks and reap the rewards of the energy transition' *The Guardian* 22 September.

Parkinson, G. (2020c) 'Biggest solar farm outside main grids nearly complete, miner counts huge savings' *Renew Economy* 10 November.

Parkinson, G. (2020b) 'Australia's biggest renewable energy hub to deliver low-cost power for ACT' *Renew Economy* 8 September.

Parkinson, G. (2020a) 'West Australia puts community batteries at top of new energy roadmap' *Renew Economy* 5 April.

Pinson, L. (2018) 'Allianz and AXA: An embarrassing affair with RWE' *Allianz* 26 October.

Point Advisory (2021) *Net zero emissions pathway options for Tasmania: Background Paper* 19 March.

Potter, B. (2020) 'Banks are putting the squeeze on fossil fuels' *Livewire* 16 December.

Power Technology (2020a) 'Stanwell signs power purchase agreement for wind farm in Queensland' 26 August.

Power Technology (2020b) 'Neoen power purchasing agreement with CleanCo Queensland' 6 October.

Readfearn, G. (2021) 'One more mine does make a difference': Australian children argue for the environment—and the law agrees' *The Guardian* 10 July.

Readfearn, G. (2019) 'BHP shareholder vote raises pressure to quit Minerals Council' *The Guardian* 18 October.

Regional NSW. Government (n.d.) 'Regional NSW's renewable energy sector is booming, with local demand for wind and solar energy trebling over the last five years'.

Reynolds, L. (2020) 'Q&A, Defence Strategic Update' Australian Strategic Policy Institute 3 July.

Rio Tinto (2020) *Climate Change Report: Our Approach to Climate Change.*

Rio Tinto (2018) 'Rio Tinto completes sale of remaining coal assets' 1 August.

Rosewarne, S., J. Goodman and R. Pearse (2014) Climate Action Upsurge: The Ethnography of Climate Movement Politics London: Routledge.
Smyth, J. (2019) 'Rural America could power a renewable economy – but first we need to solve coal debt' Clean Cooperative 2 July.
Sky News (2020) 'Tasmania to become the 'battery' of the nation' 15 December.
Sustainability Victoria (2021b) 'New Regional Community Hubs announced' 8 July.
Sustainability Victoria (2021a) 'New funding for community organisations to reduce costs and greenhouse gas emissions, 8 July.
Sustainability Victoria (2021c) Community Power Hubs Program.
Synergy (2019) 'Synergy teams up for affordable housing project' 26 September.
Tevfik, H. (2017) 'Rise of Shareholder Activism' *Squark Box* CNBC.
The Urban Developer (2019) 'Australia's Largest Smart City Battery Development Underway' 15 October.
Thomson, J. (2018) 'BHP formally leave the World Coal Association over climate policy' *The Australian Financial Review* 5 April.
Thornhill, J. (2021) 'An Australian oil giant will be paid to bury its own emissions' *Bloomberg* 2 November.
Thornhill, J., and M. Burgess (2020) 'Banks don't want to lend to Australian coal miners any more' *Bloomberg News* 28 October.
Toscano, N. (2021) 'BHP begins thermal coal exit with sale to Glencore' *The Sydney Morning Herald* 28 June.
Toscano, N. (2019) 'From barricades to boardrooms: The climate activists targeting corporate Australia' *The Sydney Morning Herald* 12 October.
Toscano, N., and M. Foley (2021) 'Grid aims to be 100% clean in four years' *The Sydney Morning Herald* 14 July: 3.
Treasury (2010) *Australia to 2050: Future challenges 2010 Intergenerational report.*
Verrender, I. (2020) 'The future of coal has already been decided in boardrooms around the globe' *ABC News* 20 January.
Vorrath, S. (2020c) 'Fortescue plans to run Pilbara mine on solar only during day, but battery the key' *Renew Economy* 6 August.
Vorrath, S. (2020b) 'Massive Clarke Creek wind farm inks deal with Queensland coal generator' *Renew Economy* 5 August.
Vorrath, S. (2020a) 'Massive Pilbara wind and solar export hub gets environmental green light' *Renew Economy* 5 May.
Wade, M. (2019) 'The year the Reserve Bank sounded the climate change alarm' *The Sydney Morning Herald* 30 December.
Westerman, D., (2021) 'A view from the control room' AMEO-CEO-Danial Westerman's CEDA keynote address, 14 July.
Western Power (2020) 'Community batteries delivering big benefits'.
Wilby, G. (2018) 'Is coal mining a bad risk?' *Australian Mining* 8 October.
Williams, P. (2021e) 'Boost for batteries under power system reforms' *The Australian* 15 July: 16.

Williams, P. (2021d) 'Pressure builds for industry to act now on climate change' *The Australian* 15 June.
Williams, P. (2021c) 'No more ad hoc power plans: Taylor' *The Australian* 25 May: 20.
Williams, P. (2021b) 'Boost for batteries under power system reforms' *The Australian* 15 July: 16.
Williams, P. (2021a) 'Victoria goes it alone on climate' *The Australian* 2 May.
Williams, P., and D. Rogers (2021) 'Companies need to come clean on carbon, Reserve Bank warns' *The Australian* 18 June: 13.
Willox, I. (2021) 'Both bad climate policy and no policy will see Australia lose jobs and investment overseas' *The Guardian* 16 October.
Zhou, V. (2021) 'BHP seeks scope 3 emissions cut from iron ore' *Australian mining* 13 January.

CHAPTER 14

Capturing the Renewable Energy Surge: Navigating a Future for Renewable Energy

Australia's energy market has undergone a radical transformation over the last 25 years. This was initially driven by a conservative agenda for freeing the sector from the control of the state through a process of corporatising and then privatising state-owned utilities to form an ostensibly perfectly competitive free market, pave the way for private investment to modernise energy generation and creating a new source of accumulation. Constituted in the image of the conventional economic construct, the liberalisation of the energy market coincided with the international debate on climate change. The debate, globally and within nation states, increasingly pivoted around formulating governance measures that were informed by deference to the same conventional economic construct. Energy market liberalisation was conceived as a means of consolidating the international competitive advantage of Australia's resources sector and energy-intensive industries. A crucial aspect of this advantage, and which underpinned this goal, was that competitive advantage should be based on the price set as a reflection of the actual monetary cost of extracting the resource or of firing the power plants. This precluded taking into account any reckoning of the greenhouse gas emissions that were generated as fugitive emissions released in the course of extracting coal and gas and in the combustion of these fossil fuels to generate electricity. A core element of conventional economic theorising, namely the adoption of a proxy for calculating the economic costs of the environment being degraded as the direct consequence of economic activity, was abandoned. How to manage emissions as the driving force of

S. Rosewarne, *Contested Energy Futures*,
https://doi.org/10.1007/978-981-19-0224-6_14

climate change became the prime focus of the debate in the UNFCCC, and, ironically, the debate took its cue from conventional economic theory, centring on the merits, or otherwise, of capping emissions in conjunction with establishing an emissions trading scheme (ETS). The failure to reconcile the conflicting approaches to economic theory has created an enduring and fundamental tension that has frustrated any progress in reconciling energy market governance and mitigating emissions.

The opposition to any meaningful measures to cut emissions has held steadfast. None of the commitments to introducing a form of an ETS have either not been honoured, nor come to fruition, or been quickly abolished following the election of successive Coalition federal governments. Yet, there has been a dramatic increase in the generation of renewable energy in Australia and the pace of this transition shows no sign of abating. Some quite innovative and novel strategies have transformed the geography of energy Australia, and these form the key focus of this study. The corollary has been the effective resistance of conservative forces to support any consideration of substantive emissions management in Australia and the continuing assertion of the political might of those forces coalescing in the coal-industrial complex. This battle is not uniquely national in its oeuvre, but global, and the preoccupation of climate change discourse has varied over the last couple of decades, although it is the concrete developments that we need to look towards to garner a more critical impression of the progress that is being made, or not, in meeting the climate change challenge. Those associated with the coal-industrial complex have been reorientating their oeuvre to embrace gas, and the success of their endeavours, to remove restrictions on surveying prospective reserves, on the issue of exploration licences and development of gas extraction, especially by fracking, is testimony to the resilience of this complex. Still, there have been some remarkable developments that have not been hamstrung by this juggernaut and which could portray Australia as one of the world's surprising success stories.

The dethroning of coal as king confronts this unfolding political saga as an increasing proportion of Australia's energy needs are being generated from renewable sources. The Australian Energy Market Operator's new chief executive, Daniel Westerman, has acknowledged this, observing that 'Australia is leading the world in many aspects of the energy transition' (Westerman, 2021). At the forefront of this transition is the extraordinary uptake of energy drawn from the power of the sun and the force of the wind with:

> a one thousand per cent increase in the number of large-scale solar farms in the last three years, and the number of windfarms has doubled in that same period. … Solar systems are now fitted to over 2.5 million Aussie homes, the fastest growing part of the generation mix. It is a stunning democratisation of power. (Westerman, 2021)

The question of the adequacy of this progress is a separate issue. It is, however, obvious that Australia's transitioning could almost be one of the world's surprising success stories, but let us not get ahead of ourselves. This has been a contest in which renewable energy has all but triumphed in spite of the determination of successive Coalition governments, backed by the resources industry, to retain coal as king. In concluding the story, it would be remiss if we did not acknowledge the extraordinary transformation that is being launched with the massively scaling up in the size of solar and wind farms as the basis for manufacturing hydrogen.

There is, however, a dark side to this story that has held back the transition from the combustion of coal, the details of which we have outlined to highlight the contested nature of the transition. Australia, or at least the federal Coalition government, is just not committed to meeting the emissions targets set out in the Paris Accord. Still, there are some significant developments that distinguish Australia, and in concluding we reflect on the potential for these various endeavours to continue breaking through the policy impasses to construct a more emissions-benign energy future.

Australian Renewable Energy Leadership

More than any other country, residential Australia has led the world in the installation of rooftop solar voltaic panels and has the highest per capita take-up and is the basis of Westerman's accolade. The pace of the installation of rooftop panels shows no sign of abating. While investment in utility-size solar and wind power farms had dropped off a little over 2020 the number of projects in process is equivalent to current aggregate renewable energy capacity. Numerous energy analysts are predicting that renewable energy's relative share of the market will continue to increase and that this will force coal's early exit as the chief source of electricity, a scenario that AEMO is preparing for. Nonetheless, the Coalition government stands firm in its refusal to concede ground on its commitment to fossil fuels, although it is coming under immense pressure to concede some ground. The Queen's quiet word that: 'It's very interesting when they talk but they don't do' could well have been a direct reference to Australia's Prime Minister who, at the time of completing this study, was still deliberating over whether he should attend the Glasgow event and still did not have as plan.

In the course of mapping the new geography of the National Energy Market, a picture of uneven development and a fracturing of the energy system have become apparent. The market has become more anarchic. In the enthusiasm to join the transition, wind and solar farms have been sprouting up in regional areas without much consideration given to the capacity of the local distribution network's ability to accommodate them. State governments have encouraged such investments as evidence of their commitments to a decarbonised energy system. The rapid increase in residential rooftop solar was equally problematic, those investing in solar panels doing so to lighten the economic burden of increasing power costs but also reflecting a new awareness of their place in the biophysical world. These developments have had distributional implications, with some landholders benefitting from getting in on the first tranches of land leased for the development of wind and solar farms and others missing out. Households benefitted from installing subsidised solar systems and being rewarded with lower power bills and feed-in tariffs whereas other households have been paying a premium that was imposed to cover the cost of the 'environment' subsidies, and these residents face the prospect of 'energy poverty' and, worse, of being disconnected from the grid. The liberalisation of energy markets coincides with the escalating increase in power prices matched by a degree of profiteering by the energy corporations that had come to dominate the market. The eventual outcome of the investment surge has, however, resulted in electricity prices becoming among the cheapest in the world, benefiting low-income households.

The liberalisation project was premised on a doctrinaire rendering of conventional economic theory that held much install for the spirit of competition, the promise of delivering affordable power and doing so efficiently. However, the project's integrity was compromised by its failure to synthesise into the rules governing the National Electricity Market an appreciation of the workings of the biophysical and atmospheric world in which the newly formed market was embedded. How to account for the cost of emissions that were an intrinsic feature, a coproduct of the generation of electricity, remained a point of bitter contention and continues to do so. The benefits of the liberalisation of the energy market were supercharged by rules lauding the competitive impulse while precluding any reckoning of environmental externalities. The international competitive advantage that privileged the resource sector was an accounting fiction, and yet this fiction prefigured the structure and dynamics of the Australian economy. It continues to do so with the coal mining corporations enjoying the benefits of record prices and paying nominal royalties for the right to extract nature's 'free gift'.

More recently, the capture of some measure of emissions under the terms of the more anodyne notion of risk management has broken an impasse that had proved intractable. Carbon disclosure reporting, alongside the European Union's decision to introduce a tax on imports that have not been subject to a carbon impost connected to an emissions trading scheme, the 'carbon border adjustment mechanism', could be game changers.

The Building Blocks of Renewable Energy: Democratisation and Social Grounding

Notwithstanding the Coalition's obstinacy, the liberalisation of the energy market has made possible some quite extraordinary developments. This has clearly been on display in the residential sector as households invested in the ever less costly rooftop solar voltaic panels, and now battery storage systems, to lead the world in per capita household electricity generation. This has given a voice to the household, in what Westerman describes as the 'democratisation of power'. State governments are now more cognizant of the benefits that can be generated from these more social forms of engaging with electricity and putting more resources into supporting the uptake, which continues to increase. They are also more forthright in addressing some of the anomalies that have arisen with home ownership being the prerequisite for being able to access State subsidies and other support measures. Equitable access has become a foregrounded priority with a range of programmes on offer in most States and the ACT. These include the provision of rooftop solar panels as part of the package in the provision of public housing, subsidising the installation of rooftop panels and storage batteries for low-income households or government rebates paid to landlords who install solar panels on rental properties, although landlords have demonstrated comparatively little interest in the schemes (Smethurst 2021). However, the measures that are being adopted to promote reliability of supply are not without their contradictions.

The resilience of the solar household in being able to maintain their standing as the doyens of energy democracy is going to be affected by the success or otherwise of them becoming 'partners' in virtual power stations and the extent to which the energy retailers have been establishing the virtual power stations as a means of regaining control over the residential energy market, to capture the benefits of residential renewable energy generating capacity. AEMC's planned rule changes that propose charging households for the right to feed power into the grid and for AEMO to sequester power from households to alleviate power shortages during periods of peak load demand shift the risk of generating electricity onto households with negligible compensation (Under the Retailer Reliability Obligation the responsibility for requisitioning power now falls to retailers). This would be to the disadvantage of solar households and could discourage further investment in residential generating capacity or prompt households to defect from the grid and abandoning their participation in the NEM. Energy retailers that established virtual power stations are not averse to lobbying governments and the AEMC for rule changes that would enhance their management of virtual power stations and their place in the market. New rules would refashion the principles that informed the design of such schemes, make them more instrumentally focused and enable corporate energy retailers to shift responsibility for network stability onto the residential sector (AGL 2021). Yet, this may not unfold so neatly because if residential Australia was to pull back from feeding power into the National Electricity Market retailers would be forced into resuming investing in generating capacity in order to meet AEMC reliability requirements and this would surely have to include some margin to avert risks in supply.

As we have argued here, the transformation has not been restricted to the private sphere of the household. It has been heard loud and clear in other domains in unique ways. The proactive roles of municipal councils in forming consortia to strengthen their collective voice and unite in common cause have accelerated the extent to which communities have opted for cleaner sources of electricity. When this takes the form of PPAs, the uncertainty associated with the vagaries of the market, and the management of risk through hedging against price fluctuations and other unanticipated developments are swept aside by the knowledge that investment in renewable energy technologies will yield a continuing and predictable flow of revenue. The estimate that some 40 per cent of energy trades are taking the form of PPAs is testimony to the benefits that can flow from extra-market dealings. PPA arrangements bypass the wholesale market, personalise direct relations between councils and energy generators and establish a more social *raison d'être*. Moreover, as is the case with residential Australia, the councils are engaging with electricity on terms that are not so instrumental, that acknowledge the importance of bringing an end to the destructive force of emissions and reducing their carbon footprint. The commercial appeal of PPAs is that they are based on the parties coming together to organise long-term power supply contracts—the wind and solar farmer establishing an outlet that assures a future revenue stream, the energy user getting in on the ground floor to secure the future flow of renewable energy. The commercial relations are more social than transactional and endure, yet they can prove to be stabilising because the very conditions that encourage the PPAs as more social forms of exchange diminishes the magnitude of power being traded in the wholesale market. While the wholesale trade continues to be a significant source that retailers purchase their accounts for as much as 60 per cent of power being traded, the relative bargaining power of the large wholesalers is enhanced because the number of retailers bidding for power in the market is reduced. The small retailers in particular confront three behemoths that are not averse to oligopolistic pricing. Similarly, the dominance of the large wholesalers in the future markets and their ability to influence future prices narrows the range of market-based options that can be explored to stabilise returns. The wholesalers are able to assert greater influence over pricing and hedging outcomes. The small energy retailers are more likely to feel the impact of this, and this highlights the illusion that grips the ACCC in its responses to concerns about the exercise of oligopolistic power. The ACCC dismissed these claims arguing the increasing numbers of small energy retailers had increased the intensity of competition and this would frustrate the exercise of market power by the wholesale electricity corporations.

This initiation of new forms of exchange relations through the formation of consortia that broker electricity supply contracts on the basis of power purchase agreements provides another vantage point from which to observe the instances of the radically reframed social forms of interactions that have taken shape in the liberalised market with the ending of state monopoly. Given the objective of PPAs is linked to decarbonising electricity generation, we should not underestimate the importance in recognising the negotiation of partnerships as a dialogical process. This is a process that calls forth the voices of those electing to join a consortium before commencing discussions with a prospective PPA partner. It invites discussion on the challenge of climate change. The multilayered partnering breaks with the tradition of transactional relations and inheres the democratisation of climate and energy as key to the agreement. The interest of energy users setting up a dialogue with a renewable energy generator can also foster energy generators and users to look beyond the conventional economic oeuvre as they elect to make carbon-free power an integral aspect of their negotiations. Certainly, this is in marked contrast to the performance of the Prime Minister in the Australian parliament when presenting something that appears as little more than an object, a lump represented in its fetishised form, valued only in terms of the dollars it returns as a commodity to be exported.

Consortia generally do not have much say in the location of wind or solar farms, although in canvassing options, they do set up a dialogue that reflects the socialisation of energy relations. With the location of wind and solar farms concentrated in mostly regional areas, the turn to renewable energy prefigures a significant redistribution of the material benefits that can flow from the construction and operation of regionally based renewable energy generators, and this will be at the expense of the ageing coal-fired power plants and the communities that service the power stations. But both spectrums of this transition, of renewal and of closure, invite conversation, although discussion on likely closures has been sorely absent.

The possible downside to the PPAs, at least for market economists, is that the negotiation of lengthy supply agreements can compromise market contestability, but this possibility is easily overcome by incorporating into the contract provisions that enable dialogue to be a continuing feature of the supply arrangement. Contestability does not necessarily stand in the way of PPAs locking in more ambitious pathways for reducing emissions.

The more social forms of engaging with energy are obviously manifest in the founding of energy cooperatives. Cooperatives are mushrooming across Australia, many in the wake of wildfires or other extreme weather events that have been experienced by communities in rural regions, especially those that have been hit hardest by the conflagrations. The cooperatives inject more potency into transitioning as a democratisation process. Indeed, climate democracy is the founding principle of cooperatives. Formed on a platform of member participation, the general consensus among members is that any surpluses beyond agreed revenue distributions should be invested back into the community in keeping with the objective of strengthening democratic forms of organising and engaging more of the local community in accessing the power of the sun or the force of the wind and, crucially, enhancing the ecological wellbeing of the community (Daley 2020). These principles and practices are serving as an inspiration impelling other communities to join a cooperative or form their own as has been the observed with Enova and its lead in supporting several energy cooperatives being formed in the Northern Rivers area of New South Wales, and similarly with TRY, Totally Renewable Yackandandah, in its communications with neighbouring communities in north-east Victoria. TRY's success has underpinned the efforts of the local parliamentary member in the electorate of Indi, Helen Haines, to progress state support for assisting the establishment of energy cooperatives. A bill drafted by Haines, the *Local Power Agency bill*, is presently before the Australian Parliament captured in the cooperatives' constitution.

Energy cooperatives have not gained the broad support of communities as is the case in Denmark, Germany or the United States, but there is obviously renewed and expanding interest. The instances of specifically energy cooperatives being established are explicit and strategic responses to the climate change challenge. Haines' bill recognises that a reasonable measure of State government financial backing and technical support provided by grid operators has helped facilitate the formation of energy cooperatives (Environmental Justice Australia 2021; Daley 2020). The bill and a re-energised peak body, the Community For Renewable Energy, ramping up its efforts to press governments to set targets for renewable energy and for community renewable energy systems are reflective of a new grassroots' urgency to address the climate challenge (Luckman 2021).

Renewable Energy and Decentralising Energy-Based Accumulation

The transition from a coal-fired economy to an economy that is increasingly reliant on renewable energy sources has also been a process of decentralising energy. In some respects, this has been an integral feature of the democratisation process. The overwhelming number of cooperatives is located in regional and rural Australia and is being touted as a means of rejuvenating the local economies. Enova has been quite upfront in arguing that one of the enterprise's ambitions was to retain within the local community a greater proportion of the revenue generated by cooperatives in place of city-based energy corporations capturing the revenue. This argument in favour of constructing a circular economy as the complement to the cooperative community reintroduces the extent to which the social permeates the different facets of community life and how this has been being woven into the responses to the challenge of climate change. State governments are demonstrating renewed interest in supporting community-based renewable energy projects in response to growing demand. This is evident in the New South Wales government backing the Orange Community Renewable Energy Park, which will provide energy independence and deliver power from local renewable sources for over 2000 households in the central west of the State (Hambrett 2020; Dalton 2020; Queensland Council of Social Service 2018). For the some 80 per cent of the over 100 cooperatives operating in Australia formed in rural communities, and these cooperatives are strengthening the vibrancy of communities and cementing a sense of purpose and solidity (Daley 2020).

A similar ambition has informed support for the establishment of renewable energy farms in regional and rural areas (Dalton 2020). The farms are seen as injecting new opportunities and reinvigorating life in these communities. Truth be told, the stimulus to local employment opportunities is mostly restricted to the construction phase as there is not much labour required once the farms are up and running. But there are increased efforts to support such development. Victorian government funding support for regional community energy hubs is a conscious initiative to resource communities to themselves research interest in and explore opportunities for developing community-controlled renewable energy enterprises and emissions mitigation initiatives.

The allocation of $1.1 million under the terms of the Traditional Owner Renewable Energy Program provides funding for Indigenous groups to determine how they want to be part of Victoria's renewable energy boom (Peacock 2021a). The programme speaks to the issue of how Indigenous self-determination might be affected by decentralisation. It formally places the pursuit of climate justice on the political agenda. This raises the more general issue affecting residential and small business operators who are experiencing or facing the prospect of energy poverty and the possibility of being disconnected from the grid. Historically, requiring energy suppliers to meet consumer service obligations was one means of protecting households from losing access to what is after all an essential and necessary good in electricity. Decentralisation of generation and the privatisation of grid and network distribution operators and retailers have eroded the force of community service obligations. This is an issue that has attracted the concern of the Australian Energy Regulator

(AER 2021), and in prompting the Regulator to engage with the principle of 'climate justice' engendered a shift in its priorities and in its oversight of the market in a similar way to that of AEMO's new chair introducing the notion of energy democracy.

This recovery of the clarion call for *climate justice* demonstrates a conviction in and the worthiness of the social as a strategic response to the climate change challenge. *Climate justice* was such an important motif of the climate change movements in the political excitement that was captured in the lead-up to the Copenhagen UNFCCC. The movements coalesced around a radical shift in the focus of debate that took up the translation of the climate change challenge into an issue of uneven development associated with climate change, and the responsibilities and consequences across the global community. In addressing the inequities, the political pendulum has swung back around to the campaign that sought to expose the climate imperialism of the OECD countries, and *climate justice* has once again become a goal in the climate movements' lexicon. The Traditional Owner Renewable Energy Program speaks to this, as do the State governments' initiatives to involve low-income households and renters, through their landlords, to join the rooftop generation revolution.

More than the Local: International Investment in Australia's Energy Transition

The establishment of utility-size solar and wind farms accounts for a substantial proportion of the renewable energy generating capacity that has come on-line in the past decade. This is captured quite neatly in Fig. 13.1. A distinguishing feature of this is that a substantial proportion of the investment in utility-scale renewable energy is the overwhelming dominance of international capital (Fowler 2020). Only five of the bigger 20 projects are in Australian corporate hands. This should not come as a surprise given federal government policy is to encourage international investment in the resources and energy sectors. The Coalition government's *National Resources Statement* released in February 2019 outlined government policy, emphasising that the development of the resource and energy sectors as never being so important to the nation's economic wealth and prosperity. Attracting international investment is seen as crucial for growing the sector. This reliance has both positive and negative implications for the transition.

International investment has proved a positive force insofar as international capital has not been constrained in their investment options to the same degree that domestic sources of capital have. Historically, national financial institutions have been critical sources of development capital, and this has tended to lock financial capital into supporting particular industry sectors, with the resource sector an historically leading sector drawing disproportionately on local financial institutions. One reflection of this has been the banking sector's responses to divestment campaigns in their reluctance to respond to such demands. The big four banks have a long and established history of being the main financiers of the mining and energy sectors and that has continued to be the case until very recently. In some respects, banks have been so locked into underwriting the resource sector, and so deeply enmeshed in the political economy and culture of the industrial and energy-intensive national economy, that to withdraw could severely devalue their existing commitments. The structural and cultural affinity that the banking system has had with the resources and energy sectors is based on banks more or less assured returns on the banks' financial backing, and the banks have clearly resisted the calls to divest their holdings in coal (BankTrack 2021a, b; Market Forces 2020).

International flows of capital have helped to break this structural and cultural deference to fossil fuels (Edis 2020). In a number of instances, local investors have only been able to proceed with the development of a renewable energy project by striking a partnership with an international counterpart as the means of raising the necessary capital to get the project off the ground. Indeed, the injection of capital has proved to be one dimension of the advantages to be had from forming such partnerships with established energy corporations. The partnerships have been based on the package of benefits that are brought to the project, including the technological developmental knowhow and of the technical skills and expertise of personnel that have been imported as components of the investment package. This is evident in the case of several of the world's premier energy corporations specialising in wind turbines, solar farm enterprises and battery manufacture in effect leading the development of the sector, with local investors effectively being silent partners that simply provide the entrée into the Australian economy. Some State governments have played a similar role, inviting international energy companies to tender for projects, contracting with the companies to establish an enduring connection and then leveraging off support by negotiating the setting up of local assembly operations or manufacture of components as a condition of State financial backing.

The commitment of international capital to the transformation of Australia's energy industry has also been important in successfully establishing another pressure point on government with global corporations lobbying government to break domestic resistance to experimenting with the deployment of novel technologies that would give the 'investing venture a first-leader advantage'. This was the case with the installation of the utility-scale Lexus battery at the Hornsdale wind farm, which helped to confirm the South Australian State's decision to break with the historical commitment to underwriting reliance on fossil fuels. International capital has successfully leveraged governments to make good the Coalition's 2019 Natural Resources Statement that stressed the need to attract international capital by establishing a 'globally attractive destination for resource projects'. Corporate leveraging has not always been based on equanimity. Some international investors have threatened to withdraw their capital should government not accede to accommodate capital's investment preferences (Kehoe 2021). Similar scenarios have been outlined by Australian investment funds that have raised the prospect of Australian capital moving offshore should governments foster an environment that made investment in the renewable energy sector a more challenging proposition (Collett 2021).

International capital has been the vanguard in opening up regional and rural Australia as key components in the nation's energy sector's development. There are, however, some downsides to this reliance on international capital in fostering investment in the development of renewable energy. For one, while international energy corporations have been prominent in breaking new ground, a large proportion of the offshore capital being invested in renewable energy ventures is being led by institutional investors. Investment institutions and pension funds have been a critically significant financiers of wind and solar farms. Their interest, however, is primarily one of maximising returns in the short to medium term and not so much in committing to the long term, and this can introduce a degree of volatility in investment and industry development plans and with this industry-wide uncertainty.

Australian governments offer a range of incentives designed to attract capital and support investment in the development of the sector, and one important aspect of these is that they do not discriminate on the basis of origin, and this has proved to be a critical factor in the scale of foreign investment in renewable energy generating and storage capacity. This is especially the case with the upfront financial assistance provided to some projects by the CEFC and ARENA. There has not been any serious reflection on the merits in having an open tender that restricts or imposes conditions on funding international capital ventures to circumvent the possibility that government support does not simply encourage opportunist investments that capitalise on the financial rewards that are being explored, a support and which would otherwise lay the foundations for a strategic investment that is the first stage in a planned stream of investments in the sector. Similarly, the fast-track depreciation write-off policies that make investment in a greenfield site venture an attractive proposition are equally as attractive for minimising taxation obligations and risking the 'lock-in' of profiteering as has been made abundantly clear with the investments in the North-West shelf gas fields (Clean Energy Ministerial et al. 2021). In some instances, international investment has been opportunistic, trialling an experimental technology in Australia because of the subsidies and other financial support on offer or because the allowable fast-track tax write-offs enable ventures to reduce their capital investment commitments that turn greenfield ventures into non-taxable credits.

It must be acknowledged, however, that the momentum of transitioning in the development of utility-size solar and wind farms in Australia has been possible only because of the international financing of these developments. This has been despite the Coalition government's intransigence in refusing to rise to the occasion and join other nations in agreeing to a 2030 emissions reduction target that would have a material effect in slowing down the increase in global temperatures. Just as the Coalition cannot insulate Australia from global climate, in being wedded to an open-door policy with respect to international trade and investment, so this has facilitated Australia being drawn an international regime of accumulation based on a constructing a renewable energy future.

This paradox is also evident in the demands by conservatives within the Coalition that agriculturalists should be rewarded for sequestration of carbon dioxide. This has become an incentive for trading in carbon emissions offsets which is proving attractive to international investors seeking carbon credits. A more recent phenomenon that has acted as an incentive to attract international investment has been the link between investing in Australia and being awarded carbon credits for investments in emissions reducing ventures that can be used as offsets to meet emission targets in their home-headquarter countries (Greber and Macdonald-Smith 2021; Macdonald-Smith 2021d). This has become an opportunity for leveraging with several major American and Canadian investment and pension funds being quite transparent in declaring their interest in carbon credits and publicly advising of the possibility of the preparedness to divest from Australia if the government does not give ground on its opposition to endorsing the Paris Accord (Kehoe 2021).

Australia policy parochialism is not beyond reproach when it comes to climate policy. The precarious state of the energy investment horizon given the Coalition's resistance to endorsing the Paris Accord is probably a factor in the fall-off of institutional investment in the second half of 2020.[1] The warnings by American and Canadian investment and pension funds have been reiterated by the Australian fund manager, Australian Ethical, predicting Australian investors in funds management could also be prompted to shift investments overseas if government policy continues to support fossil fuels to the detriment of renewable energy (Collett 2021). The energy governance architecture, the AEMC and the AEMO, are acutely aware of this being a delicate moment in the history of the liberalised energy market. The AEMC has responded to calls to remove obstacles that stand in the way of boosting renewable energy-based investment. This was the case with the rule changes following the installation of the Hornsdale Tesla battery, which will make investment in batteries an even more attractive proposition (Ludlow 2021). Australia is facing mounting international criticism for its refusal to set an emissions reduction target that can be aligned with the COP26 ambitions. Even the Coalition's former Finance Minister and now Secretary of the OECD has swung behind the International Energy Agency's appeal. But it is the refusal to resolve the tension that defines the turbulent state of the energy industry and which will force it to act, and this is coming to a head with the development proposals around clean energy, green hydrogen and ammonia that will make the recent history look like a teddy bears picnic.

Resolving Emergent Tensions in a Diverse Energy System

AEMO's Westerman has argued that Australia is on the cusp of another energy revolution, with four massive clean energy projects under development, and AEMO's most recent modelling includes an accelerated transition scenario in response to these developments (Murphy 2021; Williams

[1] Investments by international energy corporations held up as actual construction work continued apace and will continue into the immediate future.

2021b). The energy landscape is on the cusp of a potential radical transformation with the development of 'green' hydrogen, hydrogen produced from the energy generated by solar and wind farms or pumped hydro. Green hydrogen was placed on the energy development agenda in 2019 when COAG commissioned the government chief scientist to report on the potential of hydrogen as an alternative source of clean energy, especially as natural gas could not be considered an enduring solution.[2] Alan Finkel, who had been researching techniques for manufacturing hydrogen more cheaply, argued that, if it acted quickly, Australia could become a 'major global player by 2030' (*National Hydrogen Strategy* 2019). At the same time, Ross Garnaut, ever the entrepreneur and architect of the Rudd Labor government's *Carbon Pollution Reduction Scheme* and advocate of *Zero-missions*, published *Superpower: Australia's low-carbon opportunity*. *Superpower* which reiterated Finkel's vision of Australia exploiting its natural endowments—of land mass, sun and wind—to develop 'green' hydrogen. Once the cost of the technology had been reduced, hydrogen could be developed to provide an inexhaustible source of zero-emissions transportable power. *Superpower* envisioned a development model along the lines of previous resource booms, but one based on moving up the value-added ladder. Garnaut contended that, given the spatial location of mineral resource deposits in Australia, wind and solar farms could be established within reasonable proximity to these resources to enable the first stage of resource-processing being undertaken more or less *in situ*. The processed commodity would enjoy the standing of international competitive advantage as an export in place of the raw material. Processing would be clean, the cost of the commodity reduced and there would be global savings on substituting the cost of shipping the bulk unrefined resource, compared with cost of shipping the raw material. The Australian economy would benefit by moving up a rung of the global commodity supply chain, transport costs and the emissions linked to shipping reduced.

Under Finkel's stewardship, the government has encouraged greater coordination of State and Territory research and development of hydrogen and supported the establishment of hydrogen energy hubs and industry-led clusters to both strengthen the research effort and foster demand for hydrogen, although the extent to which this can be considered a systematic development plan is some way off. Some projects are

[2] The limits of gas as an enduring source of energy are self-evident. It is finite and the extraction of gas, and particularly by means of fracking, is highly contested.

already underway and reasonably advanced, such as Jemena's Western Sydney Green Gas project which is developing to the point that it will be economically competitive to use solar and wind power to produce hydrogen via electrolysis. Fortescue Metals has committed to developing the technology, taking it to a different level by doing what Finkel and Garnaut have advocated, spearheading the production of green steel (and green aluminium) using energy generated in massive hybrid solar and wind farms. Fortescue Metal's immediate interest is in reducing its energy costs by substituting its current reliance on diesel and gas as its chief power sources in its Pilbara iron ore mining venture and, in the longer term, developing the ability to process iron ore *in situ* to capture more value from its iron ore mine interests in the Pilbara region of north-west Western Australia.

The promise of a hydrogen-powered future is driving what could be the most radical transformation in the energy landscape since the post-World War Two development of open-cut mining and the industrial-scale coal-fired power electricity generation system. All parties involved in the large-scale hydrogen ventures, the federal government, State and Territory governments and private investors, envisage Australia becoming the world's preeminent clean energy producer and exporter by 2030.[3] Four massive hybrid wind and solar farm projects are currently being planned or developed with the earliest up and running by the mid-2020s.

In October 2019, Fortescue Metals released plans for developing in partnership with Alinta Energy an energy transmission hub in the Pilbara area of Western Australia that would be linked to the 60 MW solar farm being constructed as the pivot for the Chichester Solar Gas Hybrid Project (Fernyhough 2021). This was to be coupled with building a further 150 MW solar plant and 150 MW gas plant, as components in the US $700 million investment in the Pilbara Energy Connect programme. With 42 MW big battery back-up, the Project would deliver 25–30 per cent of Fortescue's stationary energy requirements from solar power, and two new iron ore projects will also be linked by the Pilbara Energy Connect transmission line programme. (ARENA contributed $24.2 million in funding towards the Solar Gas Hybrid Project to demonstrate the potential of large mining projects' energy needs being based on renewable sources (ARENA 2021).)

[3] There are several nations making similar claims that are further down the development track.

While this project was being digested details of plans were released for an even more ambitious project to build a solar panel farm across 6500 square kilometres in the East Pilbara region of Western Australia. Initially proposed as a source of power that would be transmitted by undersea cable to Singapore and Indonesia, the Asian Renewable Energy Hub will now concentrate on the manufacture of hydrogen and ammonia for export and domestic use (Mazengarb 2020). The Western Australian government's Environmental Protection authority approved the development application and it was classified by the Coalition government as having Major Project Status in September 2020 (Environmental Protection Authority, Western Australia 2020). However, the federal Minister for the Environment has declined to approve the project on the grounds that it could have a negative impact on a Ramsar Convention declared roosting area for migratory birds, and the development application is now being redrafted (Vorrath 2021b). The $50 million-plus project involves a consortium that includes InterContinental Energy, CWP Energy Asia, Vestas and Macquarie. It would be built over ten years and would generate 26 GW of wind and solar power, or 98 TWh of cheap, clean, renewable energy per year with most being used in the production of clean hydrogen and ammonia, for local and export markets. It would employ 20,000 workers during its construction phase and 3000 workers when completed.

A third massive renewable energy project being led by Singapore's Sun Cable, has been launched to be constructed in a location near Tennant Creek in the Northern Territory. The Australia-ASEAN Power Link (AAPowerLink) is being led by Singapore's Sun Cable, partnering with Fortescue Metals and Atlassian co-founder Mike Cannon-Brookes. The PowerLink 10,000 hectares solar array constructed on a cattle station would generate 10 GW of power with 30 GWh battery back-up (James 2021). When first announced, it envisaged exporting power by constructing the longest undersea cable (3700 km) to deliver clean energy to Singapore and other south-east Asia customers, but its immediate focus is on transmitting power to Singapore by undersea cable (Parkinson 2020; Vorrath 2021a; Murtaugh 2020). It will be capable of supplying up to 20 per cent of Singapore's 24/7 power needs with batteries installed in Darwin and Singapore to ensure load-balancing for continuous daily dispatch. The Northern Territory government declared AAPowerLink major project status in July 2019, and this was endorsed by the Coalition government in July 2020.

A more subdued announcement that boasted an even grander solar and wind farm ambition was the $100 billion Western Green Energy Hub, involving the consortium CWP Global and InterContinental Energy (Hong Kong) with plans for a 50 GW solar and wind farm spread over 15,000 square kilometres to produce hydrogen and the more transportable ammonia in the Western Goldfields-Esperance region in south-west Western Australia (Peacock 2021b). Mirning Green Energy Limited, a wholly owned subsidiary of the Mirning Traditional Lands Aboriginal Corporation, will have a 'meaningful carried equity stake' in the project and a permanent seat on the project's consortium board (InterContinental 2021). These unprecedentedly scaled-up clean energy projects will dramatically change the geography of the Australian energy system. Although none of the projects are to be established within the bounds of the National Energy Market (NEM), they will have significant consequences for the NEM.

Towards a New and Clean Energy Nirvana: The Remaking of *The Lucky Country*?

As the NEM has evolved, so it has coincided with the making of a more fractured or disarticulated energy system than that that prevailed under state control. The NEM is now constituted by relatively distinctive and discrete components each operating according to different *raison d'être*. The residential sector has become increasingly self-sufficient with the potential capacity to defect from the national energy market. State governments have responded to the popularity of this more social form of household engagement with energy by supporting investment in a range of community energy projects. Investment in consolidating the storage capacity of rooftop solar panels in neighbourhood battery systems is one example of the endeavour to which energy systems, while not established as cooperatives, can ferment a sense of community energy contributing to reducing the carbon footprint. Such initiatives can also begin to question the alienating character of power supplied as a commodity on the basis of transactional relations with a distant energy retailer. The ascendancy of climate consciousness is also affecting a change in the supply of electricity to households. Stand-alone systems are being trialled in a number of locations and especially those on the periphery of the grid with the installation of stand-alone household solar panel and wind generating systems replacing investment in expanding the poles and wire infrastructure. These might not be formally founded on the principles of co-operative ventures, but they do replicate the abandonment of the transactional relations that characterise the system of commodity exchange typifying the *laissez faire* market. The mushrooming of energy cooperatives in regional and rural Australia has become the means of moving beyond the episodic and fragmented exchange relations that define the commodified world to inject the spirit of sociality and a common cause in living within the constraints and richness of the biophysical world.

Municipal councils have responded to the pressures that were created with the ascendancy of the vertically integrated energy corporations by backing the establishment of solar and wind farms to secure their own sources of clean energy at more economical prices. Built on the basis of partnerships pursuing shared goals of reducing their carbon footprints, these initiatives reinforce the import of more social forms of engaging with energy to confront the challenge of climate change. And, as we have observed, the appeal in partnering to negotiate Power Purchasing Agreements designed to mitigate emissions and enhance climate profile has not been the sole province of the not-for-profit sector. PPAs have proved an appealing mechanism for advancing the for-profit enterprise's desire to project the profile of a good corporate citizen.

Not all large polluters are so wedded to the idea of socially-grounded initiatives, such as PPAs, as the motivation for reducing their carbon footprint especially when they have established a rapport with the state. This is notably the case for the 215 large greenhouse gas emitters that have agreed to emissions targets under the terms of the Emissions Reduction Safety Scheme. Whereas they were required to purchase carbon credits when they exceeded their agreed baseline target, the Coalition relaxed the requirement and permitted the companies to increase production and baseline emissions. This was justified on the basis that there was an increase in turnover which could be expected to produce more emissions while simultaneously becoming more efficient with respect to emissions. In large measure, the rules and regulations governing emissions in the NEM have been organised in such a way as to meet these enterprises' particular approaches to generating energy, so there is little reason for them to engage in more social forms of generating energy. The existing modes of organising production have in effect been sanctioned by the Coalition's inaction. This reinforces the argument developed here that energy governance has underlined the disarticulated character of the NEM, and this has been exaggerated by the Coalition's decisions to have the (federal) state re-enter the business of generating, transmitting and distributing electricity as is occurring in the instance of Snowy Hydro. The ham-fisted attempts by the Minister for Energy and Emissions Reduction to bully the energy corporations into postponing shuttering power plants that are fast becoming economically unviable highlight the shambolic state that the energy order has been reduced to. The ESB has recommended a more measured approach to establish some sense of order to guarantee minimum generating capacity and have the government follow the lead of Germany and lock in a payment to the power-plant corporations to maintain the generating capacity of coal and gas plants that could be called upon to power up when needed (Macdonald-Smith 2021a, 2021c; Williams 2021a). The ACCC and the Department of Environment recommend the government enter into low fixed-price energy off-take agreements for the later years (say 6–15) of appropriate new generation power projects. This could prove a relatively economic proposition if the generator was able to negotiate the supply of a reasonably large volume of power with a large and established user because a supply contract would secure future revenue flows and this would provide the necessary collateral to obtain debt finance (ACCC 2019; Department of Environment and Energy 2018). Viewed as a stop-gap temporary measure to ensure energy security, the proposal clearly signals a lack of faith in the ability of the market to deliver supply reliability. Like other fossil industry support schemes—financial backing for Snowy Hydro, support for Carbon Capture and Storage, the commissioning of an inquiry into government support and ownership of a north Queensland coal-fired power station, and the list goes on—that reasonably be considered to be throwing good money after bad, the proposed off-take insurance policy might be considered a more measured approach to minimising supply uncertainty but it is still reflective of the anarchic state of the energy market. An even more costly proposal designed to shore up reliability of the NEM has been to get federal and State governments to invest in the construction of additional interconnectors to increase the transmission capacity across the east coast grid. Another solution, which appeals to remedy the anarchistic expansion of renewable energy generating capacity, has been to invest in constructing more interconnectors but this cannot easily overcome the different *raison d'être* defining the operations of the various cohorts constituting the NEM. Should the massive mega hydrogen projects be developed as planned, they are likely to reinforce the disarticulated character of the Australian energy system and the political economy more generally.

This critique highlights the limits of the transition. As much as it is a contested process, the contest plays out on an uneven field organisationally and structurally. Fossil fuel capital has been able to define the energy order by drawing on its financial resources and through leveraging this economic power through the state, especially through the medium of the coal-industrial complex. Well-resourced renewable energy corporations have ready access to state officials as we have seen with the positive reception given to Elon Musk who had promised a solution to the variability in supply generated by wind and solar arms. Others have access to well-worn paths to the offices of Prime Minister's and the Minister for Energy and Emissions Reduction. And most operators in the energy market have the opportunity to sit at the same table as AEMO to discuss their concerns as paid-up members of AEMO, and AEMO as the agency responsible for securing the orderly development of energy generation, transmission and distribution is required to hear and respond to these concerns. And it is important to note who in this energy governance system is not party to this organisational link. Municipal councils are not members of AEMO, although prior to the reconstitution of COAG the Australian Local Government Association had a seat at the COAG table which likely permitted municipalities to give voice to their concerns through COAG to the Energy Ministers Council. The one municipal generator in Australia, the Sunshine Coast Council, which operates the only-state-owned solar farm, is not listed as an AEMO member. There are a couple of energy cooperatives listed as members. Households are not represented despite it is likely that they will be charged for the privilege of feeding power into the grid and should be reclassified as operators. Households do have the opportunity of exercising a voice through Energy Consumers Australia. Formerly funded by COAG, though probably out of AEMO's budget, Energy Consumers Australia was granted the right to be heard by and make submissions to the COAG Energy Council. However, it shares the same epistemological standpoint as AEMO, the conventional neoclassical model, and could thus be relied upon to endorse the status quo and the positive merits of an unfettered market. The AEMO has also recommended strengthening the transmission and distribution networks, including the construction of two new interconnectors, developments, and a third, the Marinus link which will enhance power transmission capacity from Tasmania to the mainland, is being considered.

To make matters worse, some fundamental inequities have surfaced in the ostensible attempts to restore some measure of order and stability in the face of the transition from the coal-fuelled energy gaining ground and it is invariably households that are being expected to pay the price. AEMO's decisions to assert the right to draw energy from households to meet peak load system shortfalls, to block households drawing on energy from the grid when 'the wind isn't blowing and the sun isn't shining', and to impose a charge on households for being connected to the grid highlights the fundamental inequities in this centralised energy system. This has been determined at the same time that the AEMC has introduced the rule that rewards—compensates—the big industrial energy users who forego power that they have contracted to be supplied in anticipation that this will avoid blackouts when peak load demand is predicted to exceed supply. And to complement this policy mix, the ESB and ACCC are proposing that coal or gas-fired generators be paid to maintain some capacity in reserve to be called upon to meet any supply shortfall. These merely highlight the continuing hold that fossil fuels have on visions of the energy future.

References

ACCC [Australian Competition and Consumer Commission] (2019) *Retail Electricity Pricing Inquiry: Final Report.*

AGL (2021) 'AGL responds to Victorian Neighbourhood Battery Initiative Consultation' 23 April.

ARENA [Australian Renewable Energy Agency] (2021) 'Alinta Fortescue Solar Gas Hybrid Project' 20 January.

Australian Energy Regulator (2021) *Annual Report 2020-2021*, October.

Bairich, A. (2021) 'Deloitte points to 'sea change' in mining sector ESG reporting' SNL Metals & Mining Daily: West Edition 9 February.

BankTrack (2021a) 'Banks and Coal' 3 June.

BankTrack (2021b) 'Banks and coal mining' 19 July.

Clean Energy Ministerial, Bloomberg Philanthropies and Bloomberg NEF (2021) *Energy Transition Factbook* Prepared for the 12th Clean Energy Ministerial 31 May–6 June 2021.

COAG [Council of Australian Governments] Energy Council (2019) *Australia's National Hydrogen Energy Strategy.*

Collett, J. (2021) 'Australian Ethical calls for more climate action' *The Sydney Morning Herald* 21 July.

Daley, M. (2020) 'The amazing world of our thriving co-operatives, especially in the regions' *The Fifth Estate* 28 September.

Dalton, T. (2020) 'Local social licence cooperatives are taking over pubs, stores, solar farms to save country towns' *ABC News* 19 December.

Department of Environment and Energy (2018) 'Underwriting new generation investments: Public consultation paper' October.

Doleschai-Ridnell, J. (2021) 'Collaboration drives electrification, automation' *Business News* 15 October.

Edis, T. (2020) 'Australia had a real future as an energy superpower' *The Canberra Times* 30 January.

Energy Security Board (2021) *Post 2025 Market Design Options—A paper for consultation Part A* 30 April.

Environmental Justice Australia (2021) 'Co-operative energy models—Locally owned and controlled energy'.

Environmental Protection Authority, Western Australia (2020) 'Statement that a proposal may be implemented: Asian Renewable Energy Hub' 15 October.

Fernyhough, J. (2021) 'Fortescue to produce green hydrogen, and targets steel by 2023' *Renew Economy* 15 March.

Fowler, E. (2020) 'Foreign firms flock to Australian renewable energy' *The Australian Financial Review* 6 July.

Greber, J. and A. Macdonald-Smith (2021) 'Taylor fast-tracks use of green carbon credits' *The Australian Financial Review* 25 June.

Hambrett, M. (2020) 'Renewable energy park for Central West NSW community part of 'energy democracy' movement' *ABC News Central West.*

Intercontinental Energy (2021) 'Mirning Traditional Lands Aboriginal Corporation'.

James, F. (2021) 'Sun Cable signs development agreement with NT Government to build giant solar farm, but intended customer Singapore yet to commit' *ABC News* 28 January.

Kehoe, J. (2021) 'Funds turn up heat on emissions' *The Australian Financial Review* 6 July: 3.

Luckman, D. (2021) 'C4CE's Submission to the Inquiry into the Australian Local Power Agency Bill' 9 July.

Ludlow, M. (2021) 'Rule change lets batteries earn more' power into the grid' *The Australian Financial Review* 15 July: 2.

Macdonald-Smith, A. (2021d) 'Carbon offset prices reach record level as demand grows' *The Australian Financial Review* 6 July: 16.

Macdonald-Smith, A. (2021c) 'Schott stands firm on support coal power to ease transition' *The Australian Financial Review* 8 May: 20.e

Macdonald-Smith, A. (2021b) 'Users, Taylor clash on energy reliability' *The Australian Financial Review* 1 May: 21.

Macdonald-Smith, A. (2021a) 'Mild summer, rooftop solar hit Origin sales' 1 May: 25.

Market Forces (2020) 'The giant stirs. First State Super to divest thermal coal mining' 9 July.

Mazengarb, M. (2020) 'Massive Asian Renewable Hub grows to 26MW wind and solar' *Renew Economy* 16 October.

Murphy, K. (2021) 'Australia's energy market operator plans for net zero emissions by 2050 as Morrison stalls' *The Guardian* 30 July.

Murtaugh (2020) 'Australia fast-tracks plan to send solar power to PNG' *The Australian Financial Review* 20 July.

Parkinson, G. (2020) 'Iron ore giant Fortescue sets zero emissions target by 2040, eyes for more renewables' *Renew Economy* 16 June.

Peacock, B. (2021b) "Historic' 50 GW renewable hydrogen hub proposed for WA' *PV Magazine* 13 July.

Peacock, B. (2021a) 'Victorian government funds Indigenous groups $1.1 million to determine communities' renewable projects' *PV Magazine* 7 July.

Queensland Council of Social Service (2018) *Community Energy in Queensland* August.

Smethurst, A. (2021) 'Calls to overhaul solar panel scheme for renters amid slow take up' *The Age* 23 July.

Vorrath, S. (2021b) '"Clearly unacceptable": Morrison government blocks massive Pilbara renewable hydrogen hub' *Renew Economy* 21 June.

Vorrath, S. (2021a) 'Victoria to fund "neighbourhood-scale" battery projects up to 5MW' *OneStepOffthe Grid* 3 March.

Westerman, D. (2021) 'A view from the control room' AMEO-CEO-Daniel Westerman's CEDA keynote address, 14 July.

Williams, P. (2021b) 'Energy players battle for power' *The Australian* 30 July: 19.

Williams, P. (2021a) 'Power grid shake-up to boost supply' *The Australian* 30 April: 18.

CHAPTER 15

Challenging Fossil Fuel Capital: The Promise of the Renewable Energy Surge

The enthusiasm with which federal and State governments are embracing research and development of hydrogen as an alternate to coal and gas signals a new episode in the history of energy, one that foreshadows the end of carbon democracy. The presumption that hydrogen will mark a smooth and orderly transition is reminiscent of Donald Horne's *The Lucky Country*, published in 1964, at a time when a mineral and resources boom was underwriting the nation's economic prosperity. *The Lucky Country* contended that the preoccupation with lauding the boom and how this was being driven by the exploitation of the rich bounty of natural resources, the 'free gifts of nature', was a misreading of history. Horne criticised the complacency that arose because business and governments could rest on their laurels. In an introductory observation that is probably more prescient today than it was all those years ago, Horne noted that 'Australia is a lucky country run mainly by second rate people who share its luck'. The Coalition government's *Technology Investment Roadmap* promises to deliver a technological revolution that will scale up clean energy projects and deliver a low-emissions future. This is a vision that has increasingly become contingent on the cost of producing green hydrogen declining if it is to become the pivot of the net-zero carbon economy. Yet this does not stop the Prime Minister and the Minister for Energy and Emissions Reduction arguing that Australia is on track to 'meet and beat' its unambitious 26–28 per cent 2030 emissions reduction target (Slezak 2021). The *Roadmap* also looks to relying on some complementary measures to meet

S. Rosewarne, *Contested Energy Futures*,
https://doi.org/10.1007/978-981-19-0224-6_15

this target, such as carbon capture and storage and geosequestration for which the government is providing additional financial support, adding millions in state subsidies that have virtually nothing to show for the past two decades of state assistance. Notwithstanding this 'meet and beat' hubris, the future is predicated on fossil fuels continuing to form the nucleus of the economy, coal maintaining its value as an export because of its international competitive advantage and gas becoming a more important source of energy and holding, if not strengthening, its competitive position in Asian markets as a 'transition' fuel.

The truth, of course, is that the international competitive advantage is, in part at least, a fiction, the result of the rejection of attempts to provide a reckoning of the cost of greenhouse gas emissions emitted in the process of generating electricity. The establishment of the National Energy Market that excluded any explicit reference to emissions as a corollary of the coal-fired power generation has been the lie in the construction of the model *laissez faire* energy market. This has been key to the coal-industrial complex's hegemony. It has set in play a politics defined in terms of the defence of carbon that has deceptively relied on the exercise of unfettered choice in the market while pulling the strings in behind-closed-doors politicking. This leverage has recently engaged a new carbon cadre drawn from the offices of the Mineral Council and the more assertive Australian Petroleum Production and Exploration Association, as well as the Australian Pipeline Industry Association and the Australian Gas and Pipeline Association. They have placed gas on the political agenda as the fuel of choice, a resource that is being misrepresented as a clean and dependable resource for driving the transition to a low-emissions economy.

The ideology of the freedom of choice being the mantle on which the preferred source/s of energy has been founded places carbon as the key driver in the exercise of a democratic will. Climate politics has reconstituted the momentum of the democratic process. In opposition to the complacent reliance on the extraction of coal and gas, climate activists have prosecuted an alternative oeuvre that confronts this extractive carbon vision with the call to 'keep it in the ground', to pressuring financial institutions to withdraw their capital from the fossil fuel industry and to refashion corporate governance practices to embrace risk management as an essential reporting requirement. This moves beyond the scenario of the free market in which choice is exercised by individual transactional exchanges to more social, organised and planned measures that intervene in the *laissez faire* market, in what O'Connor has encapsulated in the notion of the socialisation of responses to the environmental crisis

(O'Connor 1998b). This is not the predetermined destiny of a dialectic in play. The more social forms of responses to the emergent climate crisis are not preordained but are contingent upon the different forces and rationale that have prompted a range of collective initiatives that plan and organise measures to mitigate emissions. In the Australian context the deployment of a range of renewable energy technologies has proved a popular recourse. The metamorphoses of residential Australia into becoming a significant generator of electricity was not the product of some central planning agency but one of the few ways which households could assume some control over their energy destiny, especially when Coalition governments excluded the possibility of an emissions trading scheme that limited households' options to express their preference for clean energy.[1] The investments in community power storage systems are designed to aggregate generating capacity to introduce a more orderly form of energy distribution. Such systems are being advocated as a response to the anarchy that had come to define the National Electricity Market, but equally, because they do not neatly fit the centralised, industrial-scale generating and transmission system, they fuel the anarchy.

Coalition governments have been beholden to the coal-industrial complex. Historically, they have defended the unfettered operation of the *laissez faire* National Energy Market in the belief that inviting generators and other polluting enterprises to voluntarily engage in emissions mitigation would prove a more attractive means of reducing emissions and avoid the potential threat of an industry exodus should a reduction target be mandated. The Labor Party, in opposition, following the unsuccessful attempt to be re-elected to government in 2013, has been fearful that its working class support base in key electorates in coal belts would abandon the Party if it revisited measures to cut emissions.

Emissions trading schemes have never not been considered part of the debate on global warming and climate change. Emissions trading schemes were being flagged in the Ecological Sustainable Development Working Group, raised as a serious possibility in the negotiations over the Kyoto

[1] Red Energy, the retail arm of Snowy Hydro, in a recent advertorial, has recently promoted the Federal government owned generator as the leading source of clean energy, but admits that: 'When you flick on a switch in your home, there's no way of knowing exactly where that power comes from'. The clean electrons cannot be distinguished from those generated by the combustion of coal. Red Energy appeals to prospective customers to trust that it will match the promise by generating an equivalent amount of clean energy as that that customers contract to buy (Red energy 2021).

Protocol, considered in reports commissioned by the Australian Greenhouse Office, recommended by Prime Minister Howard's ETS Working Party, advocated by Ross Garnaut in his reports prepared for the Labor Party, their introduction planned as key elements in Rudd's Carbon Pollution Reduction Scheme and finally introduced in Gillard's Clean Energy Package, only to be jettisoned by Abbot's conservative government. Securing the international competitive advantage of Australia's emissions-intensive industries has justified the suppression of initiatives to introduce an emissions trading scheme while hiding behind the masquerade that the National Energy Market had been established.

In focusing on the *laissez faire* market stalwarts, the intention here has not been to make a case for injecting some conceptual integrity into the theoretical framing of, or the rules and regulations governing the NEM by advocating the adoption of an emissions trading scheme. On the contrary,[2] the concern has been to highlight the hypocrisy of the arguments that have underscored the unfettered operation of the energy system. Emissions trading schemes have been rejected by fossil capital because ETS impose costs on capital. But the costs are invariably of relatively greater significance for lower-income users. Carbon imposts take the form of a regressive tax that disadvantages low-income users. Unless their imposition is linked to compensation packages, ETS compromise the pursuit of climate justice. Moreover, assuming that an ETS is eventually introduced and does provide a reckoning of the real cost of emissions-intensive commodities and that results in impost that incentivises consumers to reduce their consumption of these commodities, the conventional economic analysis takes too much for granted. It presumes low-income electricity users have the same capacity to accommodate inflated power prices as high-income users and that individual consumer behaviour is sufficiently flexible to respond to changes

[2] The critique of the rules and regulations governing the operation of the National Electricity Market being based on a selective drawing on conventional economic theory to exclude any mechanism that providing a reckoning of the economic costs of emissions should not be regarded as an argument in favour of an emissions trading scheme. Emissions trading schemes amount to a regressive tax that disadvantages low-income users, and thus are to the detriment of climate justice goals. Moreover, assuming that an ETS will provide a reckoning of the real cost of emissions-intensive commodities and that the resultant impost will incentivise consumers to reduce their consumption of these commodities takes too much for granted. It presumes that individual consumer behaviour is sufficiently flexible to respond to changes in relative prices, that there are substitution options available and that the aggregate effect of the price is sufficient to have a meaningful impact on emissions (Perry et al. 2013). A carbon price does little to check the fetishisation of electricity.

in relative prices to substitute low-carbon embodied goods in the place of carbon-intensive goods. It presumes that there are substitute goods available and within the financial reach of individuals when we know that the income effect acts as a constraint on the substitution effect. And we know neither of these to be the case (Longden et al. 2021). Furthermore, the constraints on equitable outcomes can be exaggerated when the exercise of choice is limited by a market that is dominated by oligopolistic enterprises (Perry et al. 2013). Then there is the more vexed issue as to whether the aggregate effect of individuals transitioning to lower-carbon commodities will have sufficient impact to have a meaningful impact on emissions. Just as importantly, the 'market solution' of putting a price on carbon does little to check the fetishisation of electricity. Introducing a carbon impost in the free market does not invite consumers to think beyond the price of the commodity.

The omission of emissions reflects the historical authority of the coal-industrial complex and how it has been able to set the terms of the debate on emissions governance. The confidence that the *Technology Roadmap* can ensure the continued extraction of both coal and gas demonstrates a hubris that goes well beyond an exaggerated complacency. *Carbon Democracy*, or in terms that capture the thrust of the analysis here, *energy* or *climate democracy*, provides an alternative framing for understanding the power politics in the transition from a coal-fired economy to one that draws more momentum from the sun and wind. This is observable in the unfolding focus of climate politics (Mitchell 2013). On the one hand, the climate campaigns that highlight the impact of emissions help to legitimise the rationale for the pitch 'keep it in the ground'. This in turn has forced a shift in the fossil fuel agenda with the turn from coal to gas as the fuel that can be exploited to facilitate the transition to a low-emissions economy. The Energy Minister has argued the shift from coal to gas will be determined in the market; the *laissez faire* National Energy Market provided that there are not any obstacles established that could block the exercise of consumer choice. If only it was so straightforward. The deference to the free market overlooks the proselytising of gas as a more emissions-benign fuel that has occurred with a shift in the personnel make-up of the fossil fuel industrial complex and the more intensive lobbying by industry associations and particularly the Australian Petroleum Production and Exploration Association and the Australian Pipeline Industry Association as well as the Australian Gas and Pipeline Association. The success of this invigorated lobby group is evident with the Coalition government approving which will enable

Empire Energy to proceed with exploration in the Beetaloo Basin gas field in the Northern Territory, in what possibly could be the largest land-based gas project in Australia. The approval has been backed up with a generous government subsidy even before a final environmental impact assessment has been finalised endorsing Empire Energy's extraordinary exploration licence prior to the completion of the environment impact assessment. The approval permits some exploration to commence and this has been backed up with a generous subsidy (Mercer and Borrello 2021; Morton 2021; Knaus 2021). The government applauded the announcement by BHP that, in partnering Woodside, it was pushing ahead with the $16 billion Scarborough to Pluto liquified natural gas project. Both projects can be listed as pathbreaking—the Scarborough LNG project the largest oil and gas project in Australian history and the Beetaloo project one of, if not, the largest land-based gas projects on record. The likely level of emissions from these two projects will completely compromise Australia's Paris emissions reduction commitments.

Carving out a Reinvigorated Energy Democracy

There is another side to this reframing of the ground on which energy is being organised. The constitution of the NEM in terms of relatively distinctive and discrete components each operating according to different *raison d'être* reflects the traction of alternative measures for generating energy and repositioning communities in the broader ecology. The terrain on which energy democracy has been enlivened has introduced more social forms of engaging with energy as the range of energy users contest the coal-industrial complex. We observe this in the residential sector, which has become increasingly self-sufficient and has responded to the popularity of the household becoming directly involved in the generation of energy and entertaining the idea of community energy projects. These might not be formally founded on the principles of cooperative ventures, but they do replicate the abandonment of the transactional relations that characterise the system of commodity exchange typifying the *laissez faire* market. The mushrooming of energy cooperatives in regional and rural Australia has become the means of moving beyond the episodic and fragmented exchange relations that define the commodified world to inject the spirit of sociality and a common cause in living within the constraints and richness of the biophysical world.

Municipal councils have responded to lobbying by climate citizens and the pressures that were created with the ascendancy of the vertically integrated energy corporations by backing the establishment of solar and wind farms to secure their own sources of clean energy at more economical rates. As we have observed, the appeal in partnering to negotiate Power Purchasing Agreements that enhance the profile of the climate and through investment in wind and solar farms designed to mitigate emissions are a reminder that emissions abatement has not been the sole province of the not-for-profit sector. PPAs have proved an appealing mechanism for advancing the for-profit enterprise's desire to project a profile of a responsible corporate citizen and reducing its carbon footprint. The more social forms of engaging with the energy system, involving various levels of planning marking out a contrasting vision of a future, is one that seeks to contain the force of anarchy through cooperative endeavour in contrast with the conventional motif of let the competitive impulse rule (O'Connor, 1998a, b).

There is a discernable shift towards more social forms of farming as the scale of the solar and wind farms seeking development approval has been increasing. Proponents of the larger farms carry greater risk and the developers have been seeking to form partnerships with energy users by negotiating long-term supply and pricing agreements with energy users. This is not to argue that this is a universal practice. On the contrary, the older established polluters have agreed to setting emissions targets under the terms of the Emission Reduction Safety Scheme quite substantially with the result that there is an impetus for partnerships to be negotiated to lock in energy supply and delivery arrangements between enterprises establishing the farms and energy users. Still, there are many industrial enterprises that are not so wedded to meeting their individual benchmark emissions reduction targets and interpret this as an acceptable 'business as usual' scenario. NEM has been organised in such a way as to meet these enterprises' particular energy needs. These are corporations that have invested in generating capacity and power transmission and distribution infrastructure specifically to accommodate them and negotiating long-term discounted energy supply contracts to subsidise their presence. This underscores the disarticulated character of the NEM, made more profound by the Coalition government's decisions to have the (federal) state enter the business of generating, transmitting and distributing electricity as is occurring in the instance of Snowy Hydro. One solution to remedying the anarchistic expansion of renewable energy generating capacity has been to invest in constructing more interconnectors but this cannot easily overcome the different *raison d'être* defining the operations of the various cohorts constituting the NEM.

The sheer size of these projects establishing huge wind and solar farms to produce green hydrogen will underwrite quite substantial structural change in the Australian economy. The rush into green hydrogen is being underwritten by a level of confidence that is reminiscent of the complacency that coloured policy thinking that was the focus of *The Lucky Country*. Development draws resources, capital and labour and material inputs, at a cost to other sectors of the economy and especially to the detriment of the majority of the resident population on the eastern seaboard. We can expect uneven development across the nation, and in the absence of a national contingency plan this can be expected to have detrimental effects on the viability of other sectors. One possible positive outcome would be that this will have a negative impact on the future value of coal. Coal's days will be numbered as hydrogen displaces metallurgical and thermal coal as the key source of energy, although the pace of this retreat in Australia could perhaps be tempered by the continuing demand for high quality metallurgical coal. But, on the assumption that hydrogen lives up to expectations, that it does become less expensive to produce, and therefore a competitive source of energy, the more optimistic of scenarios contend that coal's death knell is already ringing. Subject to substantial technical advances that would bring down the costs of producing green hydrogen, although—we should not get ahead of ourselves because hydrogen produced at a competitive price is some way off and government support for developing green hydrogen is contingent on on regarding hydrogen as a complement to fossil fuels. While some communities that have been reliant on coal mining as the principal source of employment are developing transition plans in anticipation of coal closures, there is the very real prospect that in the absence of a national transition plan being put in place or resources provided to support local communities to develop their own transition programs; we can anticipate very uneven development into the future. We are simply putting most of our eggs in the hydrogen basket and placing our hopes in the rewards being generated 'trickling down' or spreading across the nation, and there is not much evidence of this occurring. In all likelihood, the prospect is a scenario of structural imbalances and uneven development not unlike those captured in *The Lucky Country*. Development will be mostly concentrated in the west and north of Australia on First Nations' country, and will prove extremely disruptive and destabilising for the great majority of the population. And yet, to compound this, the Coalition cannot see beyond a carbon-intensive future for despite all the hype and sudden confidence in the promise of hydrogen, the Prime

Minister could not help celebrating that he was 'jumping for joy' upon hearing the news that the new Woodside-BHP partnership was set to proceed with the massive Scarborough natural gas project on the North-West Shelf. This development will help to cement Australia's standing as a 'global energy superpower'. The celebratory jump is just another display of the naïve complacency coupled with a level of ignorance that blinds the Coalition government to the fact that the project is completely at odds with Australia's Paris commitments. The transition cannot be left to the market, to waiting for those invisible hands to again work their way through the energy system, nor to the hopes of the Coalition government that a technological fix will be orderly or a prospective Labor government that is fearful of being labelled as anti-fossil fuel.

Contradictions into the Future

There is an urgent need for action to be taking place now. It is already clear that coal-fired generators cannot compete with much lower priced energy produced from renewable sources and face early closure, as stranded assets, disrupting supply, potentially creating serious energy supply shortfalls (Macdonald-Smith 2021a, 2021b; Williams 2021). Gas is in a similar position. The Energy Security Board has recommended that with a view to guaranteeing minimum generating capacity, the government should follow the lead of Germany and lock in a payment to maintain the generating capacity of coal and gas plants that can be called upon when needed (Macdonald-Smith 2021a, 2021c; Williams and Rogers, 2021). The ACCC and the Department of Environment recommend the government enter into low fixed-price energy offtake agreements for the later years (say 6–15) of appropriate new generation power projects. This could prove a relatively economic proposition if the generator was able to negotiate the supply of a reasonably large volume of power with a large and established user because a supply contract would secure future revenue flows, and this would provide the necessary collateral to obtain debt finance (ACCC 2019; Department of Environment and Energy 2018). Viewed as a stopgap temporary measure to ensure energy security, the proposal clearly signals a lack of faith in the ability of the market to deliver supply reliability, and like other fossil industry support schemes—financial backing for Snowy Hydro, support for Carbon Capture and Storage, the commissioning of an inquiry into government support and ownership of a north Queensland coal-fired power station, and the list goes on—the proposal can reasonably be considered to be throwing good money after bad.

This reservation with respect to the force of the market returns us to the issue of the efficacy of the *laissez faire* market to engender fair and reasonable outcomes. The story outlined here is that we cannot rely on the *laissez faire* market to deliver such outcomes. Nor for that matter can we rely on the Coalition government's ostensible emphasis on market mechanisms to deliver reliability of supply on economical terms. More social forms of economic interactions involving energy, of planning, have been constructed to address the weaknesses and contradictions that surface in the unfettered market, but it is imperative that these be consolidated. More planning, more social and inclusive forms of planning that could both expedite and smooth the transition path are essential. The reliance on transactional interactions in the laissez faire 'free market' is a recipe for anarchy as seen in the Minister for Energy and Emissions Reduction's ham-fisted attempts to bully the energy corporations into postponing the shuttering of power plants that are fast becoming economically unviable. The more social forms can expose what is overlooked or set aside when engagement with energy is dealt with through the sterile medium of the transactional interaction. Nations' Green Nations 2021 climate climate

A not-so different agenda informs the cooperatives' platform in their claims to construct a circular economy, one which reinvests the value generated back into the community to enhance the wellbeing of all inhabitants and not simply extracted to serve the aggrandising interests of the coloniser. Garnaut argues in *Superpower* that the technology being developed to produce and cheapen the cost of hydrogen will provide more flexibility in the locational deployment of those enterprises, or communities, and, with this, the possibility of rebuilding those industrial cities that had once engaged in processing the extracted resource utilising low-cost power generated in coal-fired power stations. The promise of hydrogen, claims Garnaut, lies in it enabling more flexible use than historically has been the case with coal in terms of how and where it is to be used. It holds out the possibilities for redeveloping some of the former industrial resource-processing-based cities—the Newcastles, the Wollongongs, the Port Piries and the Whyallas—and building new ones closer to where the resource deposits are located. The potential problem with the massive green hydrogen industrial-fuelled centres is that, in the case of the four instances being developed, they will be developed as enclaves primarily exporting power in one form or another, more articulated into the global economy than the national economy, and likely to reproduce the anomalies and contradictions that have affected the aforementioned industrial processing centres. What is missing from this model for a green-energy-led development is planning as a social process that coheres the Australian political economy in ways that involve all market participants and communities within as well as beyond the energy market. This is not going to happen through the force of the invisible hand. As we are witnessing across the National Energy Market system, the rebuilding of a zero-emissions economy will most effectively be wrought through the construction of transition plans that give primacy to social forms of interacting in the biophysical world, to rebuilding the different energy economies on solid social foundations.

Many are global fossil fuel corporations that are hedging their bets by investing in renewable energy projects while retaining their investments in coal, oil and gas extraction. Others are beginning to make the transition. But they are also wedded to maintaining the industrial-scale energy generation model as we are witnessing with the massive wind and solar farms being developed in northern Australia. The objective of these enterprises is, first and foremost, to be able to report the return on their investment, to accumulate capital and this is a far cry from laying the foundations for an energy democracy that delivers *climate justice*. As we have argued, the corporate capital accumulation impulse is not universally shared across the spectrum of those participating in the energy market. Households investing in rooftop solar panels are motivated by the possibility of saving on their power bills although this is not necessarily in terms of the considered calculative logic that solar panel installation companies make out because the functioning of the household does not rest on whether there is a return on the investment. The desire to reduce the household's carbon footprint is a critically important influence in the investment decision. Similarly, the municipal councils' engagement in fostering the process of decarbonising and the turn to renewable energy may be justified in terms of this making economic sense, but the embrace of transitioning is driven by the challenge of arresting climate change and the social forces, the climate citizens, who placed this concern on councils' agendas. This motivation is more a defining feature in the formation of energy cooperatives although they are moved by plans to defect from a system in which energy is generated and sold as a commodity. In endeavouring to encompass more participants in their project, unlike the energy corporations they are not driven by the logic of capital accumulation nor the objective of capturing an ever-expanding share of the market but are rather motivated by an idealistic concern to reduce their carbon footprint and enhance equitable access to clean energy. Expanded accumulation does not define their *raison d'être*.

While each of these engagements with renewable energy signals that we are succeeding in charting transition paths to a cleaner future, they also highlight that the mapping of pathways is contested, to an extent blocked, and this as well as engenders contradictions. This is a classic case of different social forces competing to define the future shape of distinct energy regimes. On the one hand, the coal-industrial complex is holding onto the energy- and industrial-intensive model. This is being partly complemented by global renewable energy corporations that see the opportunity for investing in large-scale wind and solar farm projects as a model for the future based on a clean and energy-intensive regime of expanded capital accumulation. Counterposed to this have been the concerted efforts of households, municipal councils and energy cooperatives investing in generating capacity and energy use that is decentralised, more concerned with energy self-reliance, reducing carbon footprints and ensuring sustainability, and not with accumulation. New forms of grassroots engagement have emerged, such as solar gardens that invite the participation of climate citizens who might not otherwise have the opportunity for more direct engagement in clean energy generation as the overwhelming and defining objective. As we have argued, it is no-easy task for energy cooperatives to defect from the grid. Defection might well challenge the overarching energy capital-intensive model, but this does not necessarily translate into a counter-hegemonic challenge to energy capital's dominance. In some respects, it can be seen to be to capital's advantage by permitting energy distributors being relieved of having to maintain energy services for the more distantly located consumers and the necessary infrastructure in the market. The municipal initiatives, while they can compromise the market reach of energy corporations, do not really challenge the corporate model. Household installation of solar panels and battery storage establishes a degree of energy self-reliance that can present more of a challenge because, given their numbers, it strikes at the heart of energy corporations' market dominance. Nonetheless, as we have observed, energy retailers have been exploring and adopting some novel initiatives, such as inviting households to join virtual power stations, that return some control over energy flows while also enabling the corporations to shift the risk of investing in energy to residential Australia, and the AEMC has changed the rules that require retailers to ensure supply reliability. AEMO has been empowered to impose its will on household energy use and distribution, to assert the right to draw energy from households to meet peak load system shortfalls, to block households drawing on energy from the grid when 'the wind isn't blowing and the sun isn't shining'. Likewise, the AEMC decision to impose a charge on households for being connected to the grid is another instance of the household consumer being expected to meet more and more of the capital costs of upgrading energy infrastructure, more of the risks of maintaining generating capacity and highlights the fundamental inequities in this centralised energy system, which is more blatantly exaggerated by the plan to reward—compensate—the big industrial energy users to cut their energy use in the expectation that this will reduce the prospect of a shortfall in supply.

The contest over securing a dominant mode of organising our energy system is being played out under a governance architecture that presumes the consolidation of the energy-intensive model. The opportunity for alternative models to be debated, let alone explored, is quite limited. As significant as the household generators have become, they do not have the power and influence that the energy corporations have in debates on the dynamics of the NEM. As consumers they are granted a voice that can be projected into the COAG Energy Council, courtesy of the Energy Consumers Australia, but consumers do not have a voice within the AEMO. Energy cooperatives can be members of AEMO, and a couple are, but the membership is overwhelmingly representative of for-profit-focused energy operators (AEMO 2020). There is very little scope for dissenting or unorthodox perspectives to be aired in these forums. The rapidly declining value of feed-in tariffs, which in most instances are substantially less than rates retailers charge for supplying power is one indicator of the asymmetry in market power of households compared with retailers. Changes to market regulations by the AEMC illustrate further the lack of a voice of households compared with corporations. The recent changes in regulations regarding demand management, which are designed to avoid blackouts during periods of peak load supply shortages, that permit the sequestration of energy from households at the discretion of AEMO. This arrangement contrasts with the compensation packages negotiated with energy-intensive enterprises when they are asked to relinquish their call on power during high peak load periods of demand. Proposals to charge households for the right to connect to the grid are another indication of the asymmetry as to who is carrying the cost of maintaining the energy market.

Notwithstanding the limited scope for giving voice to the concerns about equitable and affordable access to clean energy, the corporate industrial-scale model is not without its contradictions. Corporations confront the dilemma that presents in drawing on the force of the wind and the radiated energy of the sun, a reliance on the 'free gift of nature'. Unlike the energy fossil fuel regime, the generation of renewable energy has become an extraordinarily inexpensive process and this is reflected in the effect this has on the realisation of the value embodied in the energy value supply chain. The investment in solar and wind generation has given rise to nature's 'free gifts' being reflected in market prices setting in train a questioning of the viability of this energy source for generating a return on capital. Negative prices have become a feature of the market as more wind and solar farms and household rooftop panels are installed and this can only be resolved by limiting or restricting access to nature's gift. This might be to the benefit of those enterprises investing in solar or wind power and looking for an investment return, but it would require something akin to privatising access to solar radiation or the force of the wind which will likely prove a much greater challenge than the established sources of energy in the form of fossil fuels. In fact, rationing access to the right to garner the radiant heat of the sun and the force of the wind is exactly what the AEMO is doing. Restricting the approval of solar or wind farms connecting to the grid is, ostensibly, designed to limit the prospect of congestion but it also becomes a means of regulating and limiting supply.

The contradictory features of the capitalist market run ever more deeply as is demonstrated in the investment surge in renewable energy which arises with the transition from an energy system faced with supply limitations to a situation where there is a surfeit of energy being generated. The investment in expanding renewable generating capacity marks a renewed impetus of energy as a catalyst in boosting capital accumulation, and this has been being celebrated by State governments. As enthusiastic proponents of transitioning by supporting investment in wind and solar farms and doing so on an ever-increasing scale, State governments have applauded the idea of an energy-led economic boom and a boom in which the State is envisaged becoming an 'energy superpower'. South Australia, Tasmania, New South Wales and Queensland have each invested in the notion that they can export surplus power to neighbouring States in the first instance and globally in the second (Morton 2020; Moore 2020). Capital accumulation remains the fundamental imperative, and governments of all political persuasions are wedded to supporting this imperative. This vision of an energy future is being replicated on an even grander scale in the massive solar farms that deliver electricity at negative prices. This is surely a manifestation of an imminent overproduction. This is a not an uncommon feature of capitalism and lies at the heart of capitalism's contradictory tendencies as theorised in classic Marxist theory. James O'Connor builds on this notion in two important respects in order to synthesise an appreciation of the environmental crisis. The contradictory tendencies play out with capitalism's dependence on extracting natural material resources, the essential inputs in production, as well as in the impositions on the natural environment as the receptacle for the wastes that are generated in the course of production and consumption. The overexploitation of natural resources results in an 'underproduction' of nature, manifest in shortages of natural resources to feed the system of production, while the impositions on the natural environment can also be conceived as underproduction because the accumulations of waste can collide with the diminution in the capacity of the environment to assimilate this waste. This is encapsulated in O'Connor's notion of the 'second contradiction', which captures very neatly the consequences of the increasing concentration of greenhouse gases in the global atmosphere that exceed ecological systems' ability to assimilate the emissions.

However, capital is constantly on the lookout for opportunities to escape the spatial limits of the market both with respect to what O'Connor has problematised as the inherent tendencies for capitalism's contradictory predilections for 'overproduction' (as an economic phenomenon) and of 'underproduction' (of nature). These phenomena can engender forces that endeavour to rein in the contradictory tendencies through planning processes or through more social modes of organising. We have mapped this impulse in the formation of more social forms of organising. This may manifest in the state assuming a more interventionist role, managing capital's extraction of natural resources or restricting the impositions on the environment. It can involve individual capitalist enterprises setting aside some measure of their competitive impulse to develop cooperative measures to temper the magnitude of drawings from or impositions on nature, such as has been demonstrated by the Business Council of Australia endorsing the objective of net-zero emissions by 2050. The Council acknowledges this will necessitate state leadership in planning. Similarly, the 'second contradiction' provides a constructive frame that assists in situating the grassroots climate campaigns defence of nature. In each instance, the state, capital or climate activists and citizens share a common outlook focused on restoring some sense of balance or sustainability in the economy-environment nexus although in different ways and to different degrees.

The idea of a 'second contradiction' provides a critical frame for interrogating capitalism's exploitative and destructive tendencies and the various strategic interventions that have been formed to counter these tendencies. The 'second contradiction' helps elucidate the climate change dynamic argued here, the contested nature of the energy system. In focusing on the merits of what is being represented as the solution to capitalism's assault on the environment, 'green hydrogen', there is the prospect of new contradictions emerging. Green hydrogen champions envisage its pre-eminence as a technology that will secure the future of humanity because it will displace the reliance on fossil fuels in the generation of electricity, of coal in the manufacture of steel and aluminium and rid oil as the base fuel in road, rail, sea and air transport, while also mitigating or even ridding the system of production of greenhouse gas emissions.[3] Australia is said to be in a unique position to take advantage of this new technology on an incomparable scale, with its rich bounty of landmass, sun and wind just waiting for the opportunity. The vista that this conjures is of a boundless sea of solar panel arrays punctuated by wind turbines. However, the construction of this vista is contingent on striking agreements with the First Nations peoples whose country will be transformed, a land that has provided the sustenance of a people for generations. As Paul Cleary has discussed in his study of Fortescue Metal's dealings with the custodians of the lands in the Pilbara region of north-west Western Australia, the Yindjibarndi, negotiations over the right to mine iron ore deposits were ruthless. They could be concluded as having been to Fortescue's advantage because of State legislation designed to facilitate development. Recent legislative changes, while shifting the balance in the right to negotiate the preservation of customary sites vis-à-vis mining still preserve capital's advantage (Cleary 2021; Toscano and Milne 2021). This is a classic illustration of an instance of accumulation by dispossession that is critical in the foundation of a new energy order (Harvey 2005). The construction of the massive solar and wind farms will entail an industrial transformation of country on a scale not seen since Australia's colonisation. It will be transformative, alienating First Nations' peoples from country, the establishment of solar and wind farms restricting traditional custodians' access to country. The extraordinary scale of the proposed farms in the Northern Territory and Western Australia amounts to a colonisation that has no contemporary counterpart, and it is a colonisation that is being aided and abetted by federal and State governments in their determination to find a solution to the climate crisis that does not compromise the pace of capital accumulation.

[3] The displacement of fossil fuels is also creating considerable between different fractions of capital as fossil capital lobbies for assurances that the State will not abandon support for the coal mining industry and the union representing coal mine workers pressures the State to establish transition programmes (Hines 2021).

This does not have to be a process of dispossession, appropriation and alienation. One planned solar and wind farm in south-west Western Australia, the Western Green Energy Hub, has been actively engaging with the traditional custodians, the Mirning people. The Western Green Energy Hub's website states that the Mirning people will have a "meaningful carried equity stake" in the project and a permanent seat on the project's board, although some care is warranted to not read too much into this commitment because this hardly goes to the question of being on country, on accessing and transforming the land that defines the Indigenous identity of who these people are. Other massive projects are following this lead having learned some lessons from the uproar that followed Rio Tinto's destruction of the 46,000 year Juukan Gorge sacred site in the Pilbara that the Puutu Kunti Kurrama and Pinikura people had been fighting to save. These proposed developments are offering similar or comparable representation and/or involvement, although when so much of the Indigenous land is usurped, heritage sites destroyed, in the process of installing solar panel arrays and banks of windmills and done so without due respect and consideration, this clearly falls far short of an example of energy democratisation (Cleary 2021). Any sense of climate justice is compromised, figuratively speaking thrown to the wind in the interests of capital accumulation.

The promise of new technological solutions to the climate crisis manifest in the twin crises of underproduction and overproduction is not unproblematic. The promise of hydrogen is seen to not be blocked by a deficit of nature's 'free gift'. The sun and wind know no limits. They are regarded as inexhaustible, although there has been little research into the impact of climate change on the future viability of these technologies. They will, of course, be vulnerable to extreme weather events as was demonstrated by the South Australia blackout. On the other hand, the tendency for overproduction could be seen as an obstacle to the sustainability of the hydrogen energy future save for one aspect that imbues this future with even more promise. The potential contradiction of overproduction is readily resolved and engenders still more benefits insofar as cheapening the cost of electricity will make the task of shifting reliance on fossil fuels as the source of power for a myriad of activities so much more within our grasp. The paradox of underproduction has the potential for resolution with the electrification of the economy, of ridding the world of the reliance on fossil fuels. Electrification could jump start the clean energy revolution with the substitution of fossil fuels in the most emissions-intensive sector in the economy.

The pervasive electrification of economies is held up as the solution. Progressing such a strategy is already being considered as a focus of transition planning (Doleschai-Ridnell 2021; Energy Insider 2016; Clarke 2020; Department of Environment, Land, Water and Planning 2019; Department of Environment, Planning and Infrastructure n.d.).

The transition to renewable energy is unfolding as a transformational moment in the history of industrial capitalism. The transition is not simply a matter of the substitution of one source of energy for another, of solar and wind power replacing coal and gas fired electricity generation as power plants are retired as they reach their use-by date or become economically unviable. The transition is increasingly being linked to displacing fossil fuel generated power across the spectrum of the economy, bringing an end to the reliance on coal, gas and oil as the principal energy sources. This is more than a matter of displacement and substitution. It is underwriting the electrification of economy, of consumption and production. Just as the turn to oil in the twentieth century transformed industrial capitalism and laid the foundations for a new era of capital accumulation, so the investment in cleaner sources of energy reduces the threat of climate change becoming a fetter on production and consumption, as a more economical source of energy electricity generated from the sun and the wind invites the electrification of a host of energy-fuelled consumption and production activities (de Lorenzo 2021). The electrification of the economy is already becoming the pivot of new patterns of accumulation. The rush to electric vehicles is one celebrated illustration of this process (Energy Insider 2016). This is a process that is propelling investment in renewable energy to ensure that there is sufficient generating capacity to meet the growing demand but which likely see the availability of inexpensive renewable energy inspiring the development of new energy-dependent technologies. This, of course, comes with its own contradictions because it can impel new levels of production and consumption which increases the draw down on the world's finite resources as well as the waste that will be produced. The advocates of electrification, especially as it is envisaged as an opportunity and as a solution to energy-overproduction, generally have very little to say about the ecological stresses that increased consumption and production will have on the biophysical world. Hydrogen is being fetishised in a way that coal has been, 'something not to be afraid of', and this simply amounts to another visiting of O'Connor's 'first contradiction' and 'second contradiction' (Van Pulvelde 2021; O'Connor, 1998a, b). This contradiction must be addressed and the lesson that can be drawn from reflecting on the energy transition in Australia is the constructive role played by those social forces that have challenged the hegemony of fossil fuels and sought to spearhead the transitioning to a clean energy future. The challenge that now confronts these social forces is whether they are sufficiently robust to assume the lead in carving a more sustainable and just future.

References

ACCC [Australian Competition and Consumer Commission] (2019) *Inquiry into the National Electricity Market* November.

AEMO [Australian Energy Market Operator] (2020) *Annual Report 2019–2020.*

Clarke, M. (2020) 'Australia to offer Pacific funds to PNG for ambitious plan to get power to the people' *ABC News* 17 February.

Cleary, P. (2021) *Title Fight: How the Yindjibarndi battled and defeated a mining giant* Collingwood: Black Inc.

Department of Environment and Energy (2018) 'Underwriting new generation investments: Public consultation paper' October.

Department of Environment, Land, Water and Planning, Victoria (2019) 'Electrification opportunities in Victoria's industrial sector' 22 October.

Department of Environment. Planning and Infrastructure, South Australia (n.d.) 'rail electrification'

Doleschai-Ridnell, J. (2021) 'Collaboration drives electrification, automation' *Business News* 15 October.

Energy Insider (2016) 'Decarbonisation policy pathways: electrification and fuel switching in the transport sector' *Energy Networks Australia* 14 April.

Harvey, D. (2005) *The New Imperialism* Oxford: Oxford University Press.

Hines, J. (2021) 'Coal country becomes renewables hotspot, but there are concerns communities may be left behind' *ABC Capricornia* 23 June.

Horne, D. (1964) *Australia The Lucky Country* Penguin.

Knaus, C. (2021) ''Grave mistake': climate scientists issue dire warning over Beetaloo Basin fracking plans' *The Guardian* 23 September.

Longden, T., S. Quilty, B. Riley, L.V. White, F. Klerck, V. Napaltjari Davis and N.F. Jupurrurla (2021) 'Energy insecurity during temperature extremes in remote Australia' *Nature Energy*

Macdonald-Smith, A. (2021c) 'Users, Taylor clash on energy reliability' *The Australian Financial Review* 1 May, 21.

Macdonald-Smith (2021b) 'Mild summer, rooftop solar hit Origin sales' *The Australian Financial Review* May: 25.

Macdonald-Smith, A. (2021a) 'Santos first target of wider fossil fuel exit campaign' *The Australian Financial Review* 4 February.

Mercer, D., and E. Borrello (2021) 'Woodside BHP Scarborough gas project off north-west WA gets final go-ahead' *ABC News* 22 November.

Mitchell, T. (2013) *Carbon Democracy: Political Power in the age of oil* London: Verso Books.
Moore, T. (2020) 'How Queensland could become an energy superpower' *The Sydney Morning Herald* 19 August.
Morton, A. (2021) 'Woodside's new Western Australian gas project 'a bet against' global 1.5C goal, report finds' *The Guardian* 3 December.
Morton, A. (2020) 'The massive projects that could make Australia a renewable energy superpower' *The Guardian* 14 November.
O'Connor, J. (1998a) *Natural causes: essays in ecological Marxism* New York: Guilford Press.
O'Connor, J. (1998b) 'Capitalism, Nature, Socialism: A Theoretical Introduction' *Captilism Nature Socialism* 1, 11–38.
Perry, N., S. Rosewarne and G. White (2013) 'Australia's Clean Energy Policy: taxing carbon and the illusion of the equity objective' *Ecological Economics* 90: 104–113.
Red Energy (2021) 'Renewable Energy Future' *The Sydney Morning Herald* 10 December.
Slezak, M. (2021) 'Australia looks like a climate laggard, as other countries ramp-up efforts' *ABC News* 3 November.
Toscano, N., and P. Milne (2021) 'Fortescue to face revolt over heritage' *The Sydney Morning Herald* 2 November: 22.
Van Pulvelde, D. (2021) 'What to do when electrification increases emissions' Energy Networks Australia 5 August.
Williams, P. (2021) 'Energy players battle for power' *The Australian* 30 July: 19.
Williams, P., and D. Rogers (2021) 'Companies need to come clean on carbon, Reserve Bank warns' *The Australian* 18 June: 13.

Index[1]

[1] Note: Page numbers followed by 'n' refer to notes.

S. Rosewarne, *Contested Energy Futures*,
https://doi.org/10.1007/978-981-19-0224-6

FSC
www.fsc.org
MIX
Papier | Fördert
gute Waldnutzung
FSC® C083411

Kolibri 360 GmbH
Ferdinand-Jühlke-Straße 7
99095 Erfurt, Deutschland
produktsicherheit@kolibri360.de